Fragment-Based Drug Discovery

RSC Drug Discovery Series

Titles in the Series:

1: Metabolism, Pharmacokinetics and Toxicity of Functional Groups
2: Emerging Drugs and Targets for Alzheimer's Disease; Volume 1
3: Emerging Drugs and Targets for Alzheimer's Disease; Volume 2
4: Accounts in Drug Discovery
5: New Frontiers in Chemical Biology
6: Animal Models for Neurodegenerative Disease
7: Neurodegeneration
8: G Protein-Coupled Receptors
9: Pharmaceutical Process Development
10: Extracellular and Intracellular Signaling
11: New Synthetic Technologies in Medicinal Chemistry
12: New Horizons in Predictive Toxicology
13: Drug Design Strategies: Quantitative Approaches
14: Neglected Diseases and Drug Discovery
15: Biomedical Imaging
16: Pharmaceutical Salts and Cocrystals
17: Polyamine Drug Discovery
18: Proteinases as Drug Targets
19: Kinase Drug Discovery
20: Drug Design Strategies: Computational Techniques and Applications
21: Designing Multi-Target Drugs
22: Nanostructured Biomaterials for Overcoming Biological Barriers
23: Physico-Chemical and Computational Approaches to Drug Discovery
24: Biomarkers for Traumatic Brain Injury
25: Drug Discovery from Natural Products
26: Anti-Inflammatory Drug Discovery
27: New Therapeutic Strategies for Type 2 Diabetes: Small Molecules
28: Drug Discovery for Psychiatric Disorders
29: Organic Chemistry of Drug Degradation
30: Computational Approaches to Nuclear Receptors
31: Traditional Chinese Medicine

32: Successful Strategies for the Discovery of Antiviral Drugs
33: Comprehensive Biomarker Discovery and Validation for Clinical Application
34: Emerging Drugs and Targets for Parkinson's Disease
35: Pain Therapeutics; Current and Future Treatment Paradigms
36: Biotherapeutics: Recent Developments using Chemical and Molecular Biology
37: Inhibitors of Molecular Chaperones as Therapeutic Agents
38: Orphan Drugs and Rare Diseases
39: Ion Channel Drug Discovery
40: Macrocycles in Drug Discovery
41: Human-based Systems for Translational Research
42: Venoms to Drugs: Venom as a Source for the Development of Human Therapeutics
43: Carbohydrates in Drug Design and Discovery
44: Drug Discovery for Schizophrenia
45: Cardiovascular and Metabolic Disease: Scientific Discoveries and New Therapies
46: Green Chemistry Strategies for Drug Discovery
47: Fragment-Based Drug Discovery

How to obtain future titles on publication:
A standing order plan is available for this series. A standing order will bring delivery of each new volume immediately on publication.

For further information please contact:
Book Sales Department, Royal Society of Chemistry, Thomas Graham House, Science Park, Milton Road, Cambridge, CB4 0WF, UK
Telephone: +44 (0)1223 420066, Fax: +44 (0)1223 420247,
Email: booksales@rsc.org
Visit our website at www.rsc.org/books

Fragment-Based Drug Discovery

Edited by

Steven Howard
Astex Pharmaceuticals, Cambridge, UK
Email: steven.howard@astx.com

Chris Abell
Department of Chemistry, University of Cambridge, UK
Email: ca26@cam.ac.uk

ROYAL SOCIETY OF CHEMISTRY

THE QUEEN'S AWARDS FOR ENTERPRISE: INTERNATIONAL TRADE 2013

RSC Drug Discovery Series No. 47

Print ISBN: 978-1-84973-908-5
PDF eISBN: 978-1-78262-093-8
ISSN: 2041-3203

A catalogue record for this book is available from the British Library

Published by The Royal Society of Chemistry,
Thomas Graham House, Science Park, Milton Road,
Cambridge CB4 0WF, UK

Registered Charity Number 207890

For further information see our web site at www.rsc.org

Printed and bound in Great Britain by CPI Group (UK) Ltd, Croydon, CR0 4YY

Foreword

In 1995 Glaxo Wellcome acquired Affymax, the pioneering Combinatorial Chemistry company based in California, for ~$500m in an attempt to increase the productivity of its research. The 1990s was a decade in which many industry executives believed that in order to increase the output of their R&D groups 'industrialisation' of the whole process was required. As such, significant investment was made in infrastructure for high-throughput screening and combinatorial synthesis in most major pharmaceutical companies. The belief was that the ability to rapidly screen libraries of several 100,000 compounds against key disease targets would yield many new drugs.

Combinatorial chemistry was an approach to generate those huge compound libraries. As one of the Glaxo Wellcome scientists working at that time in research, I was able to get an early insight into this brave new world of drug discovery. Despite the initial promise and the huge increase in apparent productivity, we soon came to realise that many lead compounds discovered using these approaches did not have optimal physico-chemical properties that would allow them to be developed into high-quality drug candidates. The subsequent years have shown that our initial concerns were valid as this approach of industrialising drug discovery has largely failed. It is clear that drug discovery is a personal endeavour, not a process.

One of the goals of screening huge combinatorial chemistry libraries was an attempt to increase the area of chemical space that was being sampled. Compounds in these early libraries were typically large (MW > 350 Da) as they contained multiple functional groups to increase the chance of finding interactions with the protein target. An alternative, and at the time contrarian, approach to crack this same nut was to consider screening compounds that were much smaller and which contained perhaps only one functional group. The potential advantage of this approach was that the functional group would not be sterically hindered; as in the case where it was part of

RSC Drug Discovery Series No. 47
Fragment-Based Drug Discovery
Edited by Steven Howard and Chris Abell

Published by the Royal Society of Chemistry, www.rsc.org

a larger molecule. However, the initial binding affinity against the target would be rather low, perhaps mM, which would pose significant challenges in detection. It turned out that there was an even bigger challenge; the cultural shift required in the minds of medicinal chemists to appreciate such a low-affinity starting point. There were many conversations in the early years during which this point was made; it was already difficult enough to optimise a μM hit from an HTS campaign, why would anyone consider starting with a mM fragment hit! Of course in those years many scientists continued to be seduced by ligand potency, not ligand efficiency.

As is now generally accepted, the fragment discovery approach is a deconvolution of the combinatorial methodology and should allow the same (or even more) chemical space to be sampled using significantly smaller numbers of compounds. Indeed, a library of 1000 fragments can be shown to represent a similar (or even greater) range of chemical space, when compared with a combinatorial library of 1 000 000 larger compounds. Therefore, the fragment approach is a more elegant and perhaps more intellectually satisfying approach which could explain why not only industrial groups but also academic groups have embraced it. A key challenge in fragment screening is the ability to detect the low affinity of the initial binding, which can be in the 5–10 mM range in the case of protein–protein systems, and then of course to develop those fragment hits into useful drug leads. However, many groups over the years have successfully overcome these challenges as highlighted by chapters in this book and also the many fragment-derived compounds that are now in clinical development. It is clear that the fragment approach is now established as one of the many technologies utilised by pharmaceutical companies to find new small-molecule drugs.

Dr Harren Jhoti
President and CEO
Astex Pharmaceuticals
Cambridge, UK

Preface

Since the seminal work by Fesik *et al.* 'SAR by NMR' in 1996, fragment-based drug discovery (FBDD) has become an established technique within both the pharmaceutical industry and academia. The concept is simple; that is small molecules, or fragments (MW < 300 Da), are more likely to form a specific complex with a given protein, than are larger molecules. A consequence of this is that chemical space can be sampled much more efficiently using fragments than using molecules of greater complexity. Fragment-based screening, and successive optimisation of hits towards lead molecules, has been applied to many different protein targets across a variety of therapeutic areas. The first fragment-derived drug, vemurafenib, an inhibitor of mutant B-RAF (for treatment of melanoma), is now approved and being used to treat patients. At least 10 other fragment-derived drugs are in various stages of clinical trials.

This book aims to 'take stock' of the latest advances in the field of FBDD. In the following chapters, leading practitioners in the field from both industry and academia describe the latest techniques and applications. The authors lay out case studies, critical review and opinions which will give the reader a thorough appreciation of both the principles and best practice within FBDD.

One consequence of screening smaller, simpler fragments is that their binding affinity is often relatively low (>1 mM). Developing assays of sufficient sensitivity, and fidelity, to identify fragments with low binding affinity has been a fundamental challenge of FBDD. As necessity is the mother of invention, this challenge has driven the development of highly sophisticated biophysical screening techniques based on X-ray protein crystallography, surface plasmon resonance (SPR) and nuclear magnetic resonance (NMR). SPR and NMR are covered in detail by Tony Giannetti and Isabelle Krimm in Chapters 2 and 3, respectively. X-ray protein crystallography features heavily in most chapters, both in a screening role and in support of structure-based fragment evolution. Comparing and contrasting the output from these

RSC Drug Discovery Series No. 47
Fragment-Based Drug Discovery
Edited by Steven Howard and Chris Abell

Published by the Royal Society of Chemistry, www.rsc.org

biophysical screening techniques is an area of active debate. This is discussed by Ian Wall *et al.* in Chapter 4.

FBDD continues to evolve rapidly and has recently seen new applications in areas such as epigenetics, G protein-coupled receptors (GPCRs), protein–protein interactions, antibacterials and the identification of novel allosteric binding pockets. In this book, experts from each of these fields discuss lessons learned based on both their own experience and key examples from the literature. These chapters describe aspects of fragment library design, screening techniques and hit validation. Examples include strategies for optimisation of fragments towards lead compounds, and, ultimately, drug candidates.

We would like to thank all the contributors to the chapters in this book for their outstanding effort and commitment to this project. We would also like to thank the staff at the RSC for their support in bringing this book to completion. We hope that this book will provide a useful resource for scientists who are looking to understand the practice of FBDD.

Steven Howard
Department of Medicinal Chemistry
Astex Pharmaceuticals
Cambridge, UK

Chris Abell
Department of Chemistry
University of Cambridge, UK

Contents

RSC Drug Discovery Series No. 47
Fragment-Based Drug Discovery
Edited by Steven Howard and Chris Abell

Published by the Royal Society of Chemistry, www.rsc.org

Personal Essay: Fragments in the Blogosphere

DANIEL A. ERLANSON*[a]

[a]Carmot Therapeutics Inc., 409 Illinois Street, San Francisco, CA 94158, USA
*E-mail: derlanson@carmot.us

1 Introduction

In July of 2008, Teddy Zartler, then at Merck, launched a blog called *Practical Fragments*. The mission statement was – and continues to be:

> *This blog is meant to allow Fragment-based Drug Design Practitioners to get together and discuss NON-CONFIDENTIAL issues regarding fragments.*

I had met Teddy just a few months before at Cambridge Healthtech Institute's annual Drug Discovery Chemistry conference in San Diego, and he invited me and a few other scientists to contribute posts to the new enterprise. Although I followed the literature closely, I hadn't spent much time reading blogs, so I was ambivalent. What purpose would a blog serve? In the spirit of experimentation, I decided to give it a try. Seven years and more than 450 posts later, *Practical Fragments* has left its small footprint on the web; the number of readers has grown steadily, and posts have even been cited in the primary literature. Still, the invitation to write for this book re-opened the original question of what purpose *Practical Fragments* serves. This chapter is an attempt to answer that question, and to touch on the broader question of what role social media can and should play in science.

RSC Drug Discovery Series No. 47
Fragment-Based Drug Discovery
Edited by Steven Howard and Chris Abell

Published by the Royal Society of Chemistry, www.rsc.org

2 A Living Review

Prior to *Practical Fragments* I had written or co-authored a few reviews on fragment-based lead discovery (FBLD), including one of two early reviews that attempted to distill most of the literature up to early 2004.[1,2] In 2006, I published a "chemical update", which picked up where the previous review left off with 32 new examples that had come out in the past 2 years.[3] With the increasing growth of fragment examples in the literature it looked like reviews would be increasingly out of date by the time they were published, and so blogging seemed like a good way of highlighting papers more or less in real time. Indeed, one of the earliest posts highlighted a paper describing the discovery of AT7519, a clinical-stage cyclin-dependent kinase (CDK) inhibitor from Astex Therapeutics.[4]

Personally I find these types of success stories useful and even inspirational, and they continue to be a mainstay of *Practical Fragments*. One of the nice features of a blog is that referencing earlier posts provides context to new discoveries. For example, in 2012, researchers from AstraZeneca published an elegant example of fragment linking to generate a nM inhibitor of the anticancer target lactate dehydrogenase A (LDHA), which was covered on the blog.[5] Less than a year later researchers from Ariad published their work on the same target, which led to molecules with interesting similarities and differences.[6] A parallel situation arose when scientists at Genentech published their discovery of a novel fragment-binding site on Ras, perhaps the holy grail of oncology targets.[7] Shortly thereafter, researchers at Vanderbilt University published their own independent discovery of this binding site and associated fragments.[8] Being able to click from one post to another can be useful to someone new to the target or the field. Indeed, a figure from the more recent Ras blog post comparing structures from both papers was reprinted by other researchers in a *Journal of Medicinal Chemistry* review on the topic.[9]

As entries have accreted over the years, several of the molecules highlighted have entered the clinic, which has led to periodic summaries of clinical compounds derived from fragments. In early 2015, at least 30 molecules derived from fragments had entered clinical development, of which at least 16 were still active.[10] One drug, vemurafenib, was approved for sale in 2011.[11] At the suggestion of one commenter on the blog, this list was given a permanent link on the side-bar for easy reference.

Practical Fragments is not just about drugs and chemistry: many of the biophysical techniques used to identify and characterize fragments have been discussed too, including mainstays such as surface plasmon resonance (SPR),[12] isothermal titration calorimetry (ITC),[13] nuclear magnetic resonance (NMR), differential scanning fluorimetry (DSF),[14] and X-ray crystallography. Some of these are among the most popular posts.

Practical Fragments also highlights new or emerging methods, such as mass spectrometry (MS),[15] computational screening,[16] weak affinity chromatography (WAC),[17] target-immobilized NMR screening (TINS),[18] microscale thermophoresis (MST),[19] capillary electrophoresis (CE),[20] inhibition in solution (ISA),[21] ultrafiltration,[22] and enthalpy arrays.[23] A nice feature of a blog is that when a new post refers to one of these techniques it is easy to hyperlink to the original post, which gives a newcomer to the field a full description, while not forcing an experienced researcher to read about SPR for the umpteenth time.

Finally, for reasons that remain obscure, fragment-based drug discovery attracts an inordinate number of reviews, special journal issues, and books. Many of these are covered as they are published, and the books (six counting this one!) are linked on the right side of the front page.

But *Practical Fragments* is about more than summarizing research findings. As described in the next section, a blog can be an excellent forum for discussing the limitations of experimental techniques. In particular, it can be a place where scientists offer each other tips, and warn against potential artifacts, anonymously if need be.

3 Warning Signs and Guideposts

What's worse than running a screen and coming up empty-handed? Ending up with false positives. To find low-affinity fragments researchers sometimes need to push techniques to their limits, which can lead to various types of artifacts. The problem is particularly acute given that fragment-based drug discovery is relatively new in many organizations, so people may not be aware of potential problems. If you're accustomed to screening compounds at 1 μM concentration, screening at 1 mM concentration could present unexpected challenges. Moreover, since fragment-based teams are often multidisciplinary, it can be easy for artifacts to creep through the cracks in expertise.

To arm the research community against such issues, *Practical Fragments* has highlighted a number of problems. Compound aggregation at high (and sometimes even low) μM concentration is a phenomenon that has led to numerous spurious reports in the literature and wasted efforts.[24] However, even gold-standard techniques like NMR and X-ray crystallography are subject to artifacts if one is not diligent.[25,26]

But it is not just a question of assays: some compounds are inherently reactive or likely to generate false positives, and if these compounds appear in a screening collection they will likely dominate any hits. Jonathan Baell has christened such molecules *pan-assay interference compounds*, or PAINS, and many of these are fragment-sized.[27,28] Some of the more insidious compounds can be reduced by common assay components such as dithiothreitol and then spontaneously re-oxidize in air to generate reactive hydrogen peroxide, which can confound many biochemical and cell-based assays. In response to a post on such molecules, one commenter wrote, "I learned something new today! Woot!".[29]

An enormous opportunity for blogs is to facilitate discussion among scientists more rapidly and less formally than in the traditional literature. A nice example of this was initiated by a post in 2013, entitled "Fragmenting natural products – sometimes PAINfully".[30] This covered a prominent paper in which the researchers 'fragmented' natural products into individual components and used these to generate a fragment screening library. Unfortunately, the authors seemed to be unaware of PAINS, leading to the inclusion of many dubious compounds. The post attracted comments from the first author as well as from the *Nature Chemistry* editor who had handled the paper. If nothing else, the exchange made more people aware of problem compounds.

Although the potential for discussion is inherent in the blog format, it is a rare post that receives more than one or two comments. This is in contrast to some other chemistry blogs such as Derek Lowe's excellent *In the Pipeline*,[31] which routinely receives dozens of comments for each post. Some of this is just a matter of scale: fragment-based drug discovery is a rather small niche, and *In the Pipeline* can receive more page views in a month than *Practical Fragments* has in its history. That said, Teddy did ask why people didn't comment more. Several of the responses were along the lines of, "I don't really have much to add. I mostly read the blog to keep current on what's going on in FBDD (something that your posts excel at), and it seems silly just to chime in to a post and say 'cool'." On a more charming note, one commenter wrote that "comments tend to come from controversy and I think the current FBDD community is quite a contented bunch".[32]

4 A Mirror for the Community

In addition to commenting, another useful feature of blogs is the ability to run polls to gather information of interest to the community. Of course, like polls everywhere, these can be subject to low turnout and self-selection among participants. There is another unfortunate (and unexpected) similarity to some political polls: it turns out that the *Blogger* platform on which *Practical Fragments* is housed has a nasty habit of 'losing' votes – a problem that accelerates over time – which means that poll results need to be captured and archived before degrading. All those caveats aside, *Practical Fragments* has been able to capture some interesting data, starting with the question of readership (Figure 1, left panel), which in 2010 was split roughly evenly between academia and industry and between practitioners and aficionados of fragment-based lead discovery.[33] A repeat poll in 2013 showed a similar distribution, though with a slight shift towards industry (Figure 1, right panel).[34]

Polls can also reveal which techniques researchers in the community use to find fragments. In 2011 *Practical Fragments* asked this question, resulting in the data in Figure 2 (blue bars). Respondents could select multiple techniques, and one striking finding was that the average respondent used between two and three different techniques, presumably to help weed out artifacts, confirm true positives, and obtain more detailed structural information.[35] A repeat of this poll in 2013 revealed similar findings, though with an increase in the extent to which crystallography was used as well as an increase in the number of individual techniques used (Figure 2, red bars).[34]

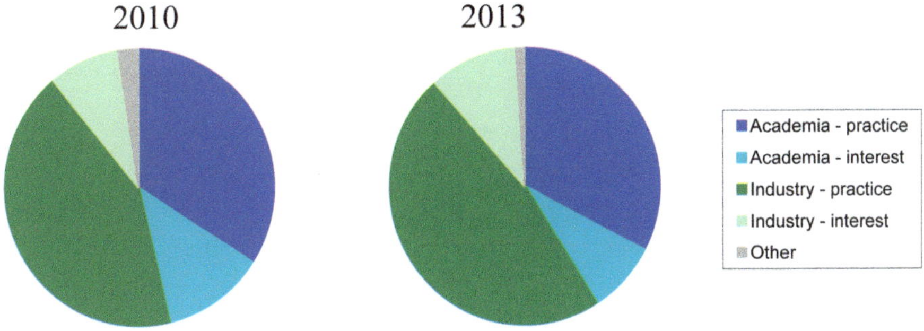

Figure 1 Demographics of readership. Left: Poll ran May–June 2010 and received 82 responses. Right: Poll ran December 2013 and received 95 responses.

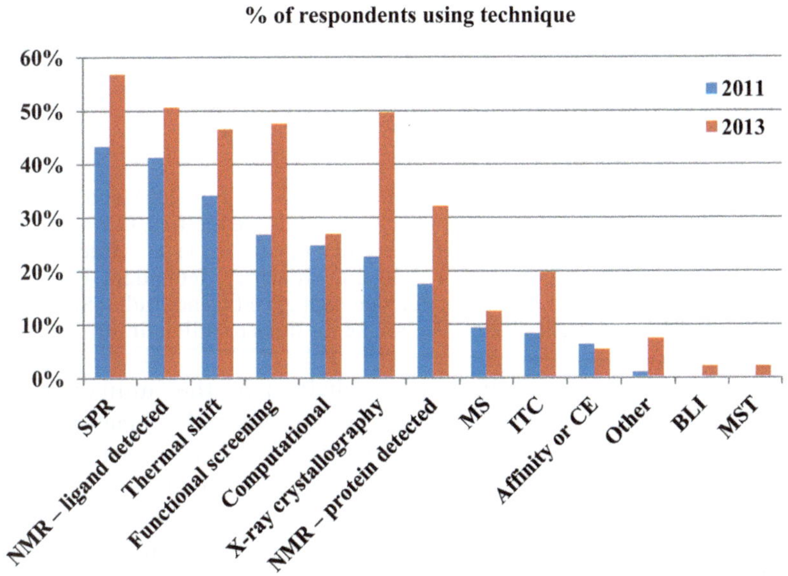

Figure 2 Fragment screening methods; respondents could choose multiple metrics. The 2011 poll ran September 2011 and received 97 responses (blue). The 2013 poll ran December 2013 and received about 96 responses (red). BLI = biolayer interferometry; other abbreviations as noted in text.

And just as there are many ways to find fragments, there are lots of ways to evaluate them too. Ligand efficiency is the simplest,[36] but a whole cottage industry has arisen to supply new metrics that incorporate more data.[37] Do people use these other metrics? A poll in 2011 found that ligand efficiency (LE) was the overwhelming favorite, trailed by ligand lipophilic efficiency (LLE, Figure 3).[38] Of course, these findings could change over time; the third most common metric, LLE_{AT}, had been published online only a month before our poll,[39] so perhaps we will revisit this question, particularly in light of some of the recent controversies around the appropriateness of LE.[40]

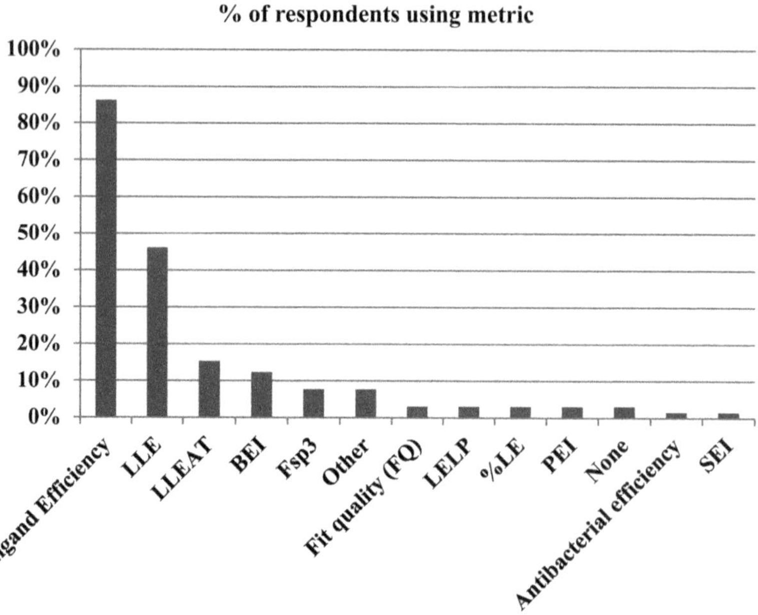

Figure 3 Metrics used to evaluate fragments. Poll ran July 2011 and received 65 responses; respondents could choose multiple metrics. LLE = ligand lipophilic efficiency; BEI = binding efficiency index; Fsp3 = fraction of sp3-hybridized carbons; LELP = ligand efficiency dependent lipophilicity; %LE = % ligand efficiency; PEI = percentage efficiency index; SEI = surface-binding efficiency index. See ref. 38 for links to full definitions.

LE is defined as the binding energy divided by the number of heavy atoms,[36] but some metrics use molecular weight instead of heavy atoms; the idea is that larger atoms such as bromine should suffer a penalty compared with smaller atoms such as fluorine. That said, in a 2013 poll, 27 out of 38 respondents used only heavy atoms, 1 used both heavy atoms and molecular weight, and only 2 used molecular weight alone (with the remaining ignoring size altogether).[41]

Several polls have explored fragment screening libraries. One of the selling points of fragment methods is that you can cover chemical space more effectively with a smaller library,[42] but what is the optimal library size? A poll in 2013 found a median of 1000–2000 fragments (Figure 4).[34] This is consistent with a 2013 analysis of published libraries, which found a median of 1300 fragments among the 22 libraries summarized.[43]

In terms of fragment library design, a frequent question is how large fragments can be. Astex's *Rule of 3* suggests an upper cut-off of 300 Da,[44] and a poll in 2012 found that most people put an upper limit of 20 atoms,[45] which translates to roughly 260 Da according to a Pfizer analysis that found the average non-hydrogen atom in their corporate collection had a mass of 13.286 Da (Figure 5).[36]

Perhaps just as interesting is the question of how small a fragment people would put in their library, the subject of a poll in 2013 (Figure 6).[46]

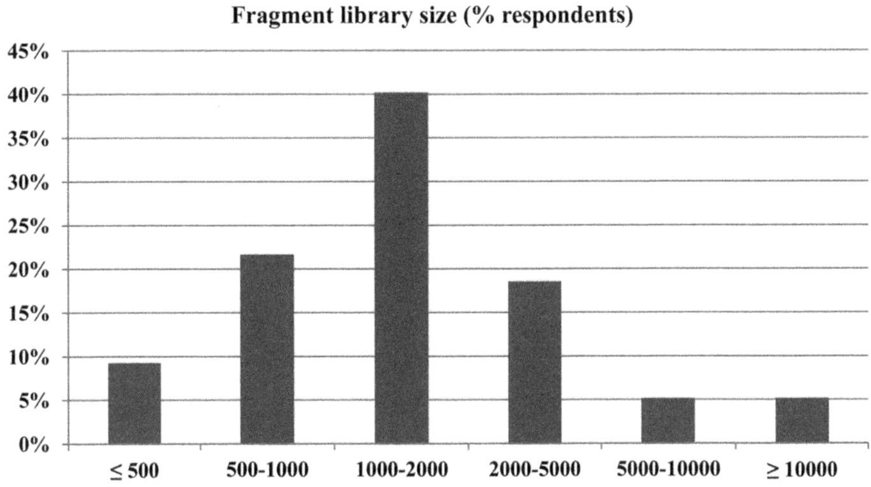

Figure 4 Fragment library size. Poll ran December 2013 and received 97 responses.

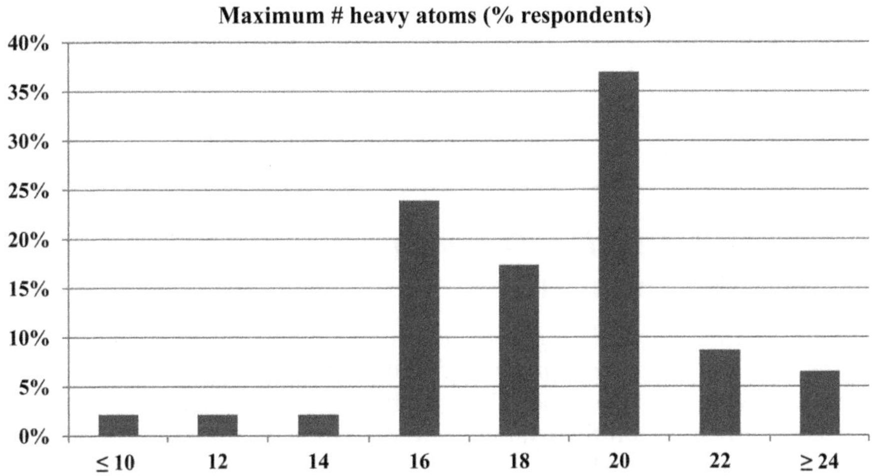

Figure 5 Maximum size of fragments allowed in a library. Poll ran May 2012 and received 46 responses.

Obviously you want your fragment to be as small as possible, but make it too small and you run the risk of having fragments that are so weak that you cannot detect them. As it happens, the median bottom bound is 7 or 8 heavy atoms, basically the size of 4-aminopyridine (marketed as fampridine) or 1,2-benzoquinone (if you didn't screen out your PAINS).

Finally, *Practical Fragments* is an ideal forum to publicize, summarize and discuss conferences and fragment-related events. One of the *Links of Utility* on the right side of the blog is to *Upcoming events*, which is updated frequently.

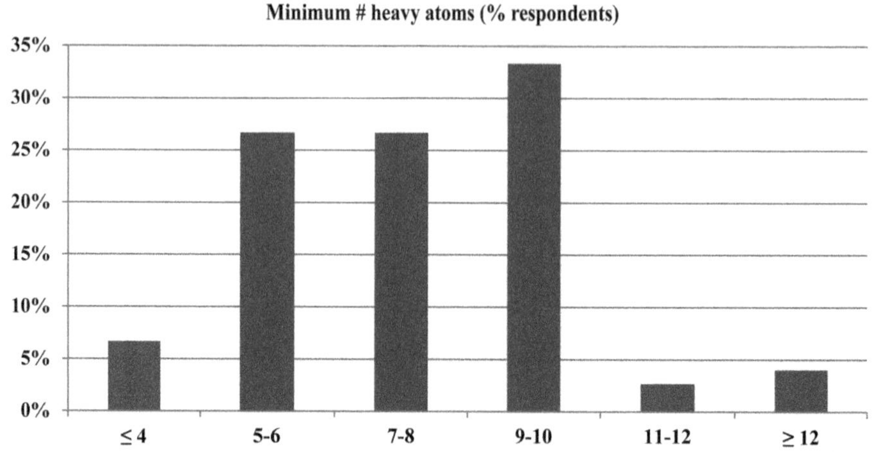

Figure 6 Minimum size of fragments allowed in a library. Poll ran May 2013 and received 75 responses.

Every FBLD-based event I've attended is summarized on the blog, sometimes co-written with others, and often with comments from attendees.[47,48] Although necessarily brief, these posts can give a flavor of the conference in this era of decreased travel budgets. As one commenter noted, "Thanks for the summary, much appreciated since I already hit my conference quota for the year".[49]

5 Beyond *Practical Fragments*: Fragments in the Broader Social Media

In addition to *Practical Fragments*, one other blog focuses heavily on the field: Pete Kenny's *Molecular Design*.[50] This was originally called *FBDD Literature*, and it retains a good bibliography for each post. The site tends to focus slightly more on theory, but there is significant overlap with *Practical Fragments*, and the two blogs often reference one another. Pete is also one of the most frequent commenters on *Practical Fragments*.

Around the time Teddy founded *Practical Fragments*, he also started a *LinkedIn* group devoted to fragment-based drug discovery; as of March 2015 it had 4115 members.[51] This group has had some spirited discussion, though without a powerful search feature these discussions tend to disappear more quickly than *Practical Fragments*, where posts from several years ago are still routinely accessed. Still, information from such non-traditional sources can be useful. Ben Davis and I recently wrote a review on things that can go wrong in fragment screening. We wanted to capture information that doesn't typically make it into the peer-reviewed literature, so we put out calls on both *Practical Fragments* and *LinkedIn* for personal stories and anecdotes, several of which made it into the published review.[52] Another powerful feature of *LinkedIn* is the ability to email all members of the group, which can be useful for alerting members to new conferences. Of course, such tools need to be used judiciously to avoid becoming obnoxious!

Three other sites also deserve special mention. In addition to *Practical Fragments*, Teddy also runs *Quantum Tessera Consulting*, which has some nice complementary posts.[53] Chris Swain's *Cambridge MedChem Consulting* covers a range of topics, some devoted to fragments.[54] One particularly useful feature is an annotated list of commercial suppliers of fragments, along with analyses of what these contain.[55,56] Finally, anyone interested in drug discovery would do well to check out Derek Lowe's *In the Pipeline*, which has been running since 2002 and covers just about everything even remotely related to the field, including fragments.[31] Derek prominently displays links to blogs and other online resources, which as of March 2015 numbered over 80 entries.

The scientific media are changing rapidly. Blogging is still a relatively new phenomenon, and, as Niels Bohr (and others) have observed, "it is exceedingly difficult to make predictions, particularly about the future".[57] That said, I do believe blogs will increasingly foster rapid and spirited discussion and facilitate the cross-referencing of topics. New forms of communication, such as the microblogging exemplified by *Twitter*, will increasingly point researchers to longer posts and papers. One indication of the importance of social media is the fact that an article entitled "An introduction to social media for scientists" was viewed nearly 50 000 times within 3 months of publication.[58]

Most of us have probably encountered social media as adults. It is exciting – and perhaps slightly daunting – to realize that the next generation of scientists will grow up viewing blogs as just another resource, alongside and perhaps even supplanting or blending into other traditional venues such as books and journals.

Acknowledgements

I would like to thank Teddy Zartler for establishing *Practical Fragments* and inviting me to join him, Monya Baker for helpful comments on this manuscript, and all of the readers of *Practical Fragments* over these past seven-plus years. Thanks especially to those of you who leave comments – keep 'em coming!

References

Note: as the focus of this chapter is social media, most of the references are links to solely electronic sources, all of which were active as of 17 March 2015. Where these summarize a journal article, links to the primary article can be found within the relevant blog posts.

1. D. A. Erlanson, R. S. McDowell and T. O'Brien, *J. Med. Chem.*, 2004, **47**, 3463.
2. D. C. Rees, M. Congreve, C. W. Murray and R. Carr, *Nat. Rev. Drug Discovery*, 2004, **3**, 660.
3. D. A. Erlanson, *Curr. Opin. Biotechnol.*, 2006, **17**, 643.
4. D. Erlanson, *Practical Fragments*, 2008, http://practicalfragments.blogspot.com/2008/08/clear-thorough-account-of-fragment.html.
5. D. Erlanson, *Practical Fragments*, 2012, http://practicalfragments.blogspot.com/2012/04/fragment-linking-leads-to-nanomolar.html.
6. D. Erlanson, *Practical Fragments*, 2013, http://practicalfragments.blogspot.com/2013/02/fragment-linking-for-ldha-ariads-turn.html.
7. D. Erlanson, *Practical Fragments*, 2012, http://practicalfragments.blogspot.com/2012/05/fragments-versus-ras.html.
8. D. Erlanson, *Practical Fragments*, 2012, http://practicalfragments.blogspot.com/2012/06/fragments-versus-ras-part-2.html.
9. Y. Wang, C. E. Kaiser, B. Frett and H. Y. Li, *J. Med. Chem.*, 2013, **56**, 5219.
10. D. Erlanson, *Practical Fragments*, 2015, http://practicalfragments.blogspot.com/2015/01/fragments-in-clinic-2015-edition.html.
11. D. Erlanson, *Practical Fragments*, 2011, http://practicalfragments.blogspot.com/2011/08/first-fragment-based-drug-approved.html.
12. D. Erlanson, *Practical Fragments*, 2010, http://practicalfragments.blogspot.com/2010/02/surface-plasmon-resonance-spr.html.
13. D. Erlanson, *Practical Fragments*, 2010, http://practicalfragments.blogspot.com/2010/02/isothermal-titration-calorimetry-itc.html.
14. D. Erlanson, *Practical Fragments*, 2013, http://practicalfragments.blogspot.com/2013/07/fragments-vs-tankyrases-dsf-shines-again.html.
15. V. Vivat, *Practical Fragments*, 2011, http://practicalfragments.blogspot.com/2011/05/native-ms-turning-up-voltage.html.
16. D. Erlanson, *Practical Fragments*, 2009, http://practicalfragments.blogspot.com/2009/05/more-on-docking-fragments-and-sampling.html.
17. D. Erlanson, *Practical Fragments*, 2011, http://practicalfragments.blogspot.com/2011/03/weak-affinity-chromatography-wac.html.
18. D. Erlanson, *Practical Fragments*, 2010, http://practicalfragments.blogspot.com/2010/09/fragments-vs-membrane-proteins-with.html.
19. D. Erlanson, *Practical Fragments*, 2012, http://practicalfragments.blogspot.com/2012/10/microscale-thermophoresis-mst.html.
20. D. Erlanson, *Practical Fragments*, 2012, http://practicalfragments.blogspot.com/2012/06/capillary-electrophoresis.html.
21. D. Erlanson, *Practical Fragments*, 2013, http://practicalfragments.blogspot.com/2013/06/inhibition-in-solution-assay-isa-on.html.
22. D. Erlanson, *Practical Fragments*, 2011, http://practicalfragments.blogspot.com/2011/09/ultrafiltration-to-filter-fragments.html.
23. D. Erlanson, *Practical Fragments*, 2012, http://practicalfragments.blogspot.com/2012/01/enthalpy-arrays.html.

24. D. Erlanson, *Practical Fragments*, 2009, http://practicalfragments.blogspot.com/2009/08/avoiding-will-o-wisps-aggregation.html.
25. D. Erlanson, *Practical Fragments*, 2010, http://practicalfragments.blogspot.com/2010/04/getting-misled-by-nmr-iloe-artifacts.html.
26. D. Erlanson, *Practical Fragments*, 2009, http://practicalfragments.blogspot.com/2009/10/looks-can-be-deceiving-getting-misled.html.
27. J. B. Baell and G. A. Holloway, *J. Med. Chem.*, 2010, **53**, 2719.
28. D. Erlanson, *Practical Fragments*, 2010, http://practicalfragments.blogspot.com/2010/10/small-but-painful.html.
29. D. Erlanson, *Practical Fragments*, 2012, http://practicalfragments.blogspot.com/2012/04/universal-fragments.html.
30. D. Erlanson, *Practical Fragments*, 2013, http://practicalfragments.blogspot.com/2013/02/fragmenting-natural-products-sometimes.html.
31. D. Lowe, *In the Pipeline*, 2015, http://www.corante.com/pipeline/.
32. E. R. Zartler, *Practical Fragments*, 2012, http://practicalfragments.blogspot.com/2012/04/comments.html.
33. D. Erlanson, *Practical Fragments*, 2010, http://practicalfragments.blogspot.com/2010/06/poll-results-academiaindustry.html.
34. D. Erlanson, *Practical Fragments*, 2014, http://practicalfragments.blogspot.com/2014/01/poll-results-affiliation-fragment.html.
35. D. Erlanson, *Practical Fragments*, 2011, http://practicalfragments.blogspot.com/2011/10/poll-results-fragment-screening-methods.html.
36. A. L. Hopkins, C. R. Groom and A. Alex, *Drug Discovery Today*, 2004, **9**, 430.
37. M. M. Hann and G. M. Keseru, *Nat. Rev. Drug Discovery*, 2012, **11**, 355.
38. D. Erlanson, *Practical Fragments*, 2011, http://practicalfragments.blogspot.com/2011/08/ligand-efficiency-metrics-poll-results.html.
39. P. N. Mortenson and C. W. Murray, *J. Comput.-Aided Mol. Des.*, 2011, **25**, 663.
40. M. D. Shultz, *Bioorg. Med. Chem. Lett.*, 2013, **23**, 5980.
41. E. R. Zartler, *Practical Fragments*, 2013, http://practicalfragments.blogspot.com/2013/03/poll-results-hac-vs-mw.html.
42. M. M. Hann, A. R. Leach and G. Harper, *J. Chem. Inf. Comput. Sci.*, 2001, **41**, 856.
43. B. C. Doak, C. J. Morton, J. S. Simpson and M. J. Scanlon, *Aust. J. Chem.*, 2013, **66**, 1465.
44. M. Congreve, R. Carr, C. Murray and H. Jhoti, *Drug Discovery Today*, 2003, **8**, 876.
45. D. Erlanson, *Practical Fragments*, 2012, http://practicalfragments.blogspot.com/2012/06/poll-results-how-big-are-your-fragments.html.
46. D. Erlanson, *Practical Fragments*, 2013, http://practicalfragments.blogspot.com/2013/06/poll-results-how-small-are-your.html.
47. D. Erlanson and E. R. Zartler, *Practical Fragments*, 2012, http://practicalfragments.blogspot.com/2012/04/seventh-annual-fragment-based-drug.html.

48. D. Erlanson and R. Law, *Practical Fragments*, 2012, http://practicalfragments.blogspot.com/2012/02/molecular-medicine-tri-con-2012.html.
49. D. Erlanson, *Practical Fragments*, 2010, http://practicalfragments.blogspot.com/2010/10/fbld-2010.html.
50. P. W. Kenny, *Molecular Design*, 2015, http://fbdd-lit.blogspot.com/.
51. *LinkedIn Group: Fragment-based Drug Discovery*, 2015, http://www.linkedin.com/groups?gid=121172&trk=myg_ugrp_ovr.
52. B. J. Davis and D. A. Erlanson, *Bioorg. Med. Chem. Lett.*, 2013, **23**, 2844.
53. E. R. Zartler, *Quantum Tessera Consulting*, 2015, http://www.quantumtessera.com/.
54. C. Swain, *Cambridge MedChem Consulting*, 2015, http://www.cambridgemedchemconsulting.com/.
55. C. Swain, *Cambridge MedChem Consulting*, 2013, http://www.cambridgemedchemconsulting.com/resources/hit_identification/fragment_collections.html.
56. E. R. Zartler, *Practical Fragments*, 2013, http://practicalfragments.blogspot.com/2013/06/what-you-are-missing.html.
57. B. Petersson, *The Economist*, 2007, http://www.economist.com/blogs/theinbox/2007/07/the_perils_of_prediction_june.
58. H. M. Bik and M. C. Goldstein, *PLoS Biol.*, 2013, **11**, e1001535.

Personal Essay: Adventures and Experiences in Fragment-Based Drug Discovery

MARTIN J. DRYSDALE*[a]

[a]Beatson Institute Drug Discovery Programme, Beatson Institute for Cancer Research, Garscube Estate, Switchback Road, Bearsden, Glasgow, G61 1BD, UK
*E-mail: m.drysdale@beatson.gla.ac.uk

1 In the Beginning

At the start of my Industrial post-doc at the now defunct Parke–Davis Neuroscience Research Centre in Cambridge, UK, I was presented with the challenge (opportunity?) of synthesising a set of acid mimics of the then pre-clinical development candidate PD134308 aka CI-988 (Figure 1).

CI-988 was a selective cholecystokinin B antagonist being developed as a potential anxiolytic[1,2] and it was thought that the pharmacokinetic properties might be improved by modification of the carboxylic acid. Whilst this was achieved with some degree of success[3] the synthesis of compounds such as **1** was laborious to say the least. It required a multistep synthesis of the intermediate **2** followed by a coupling reaction to install the acid mimic which was itself often a 3–6 step sequence (Figure 2A).[2,3]

Whilst in the first flush of enthusiasm of a post-doc, the synthesis of **3** was not a problem. However after a few run-throughs to scale up *N*-*t*-Boc-**2**

RSC Drug Discovery Series No. 47
Fragment-Based Drug Discovery
Edited by Steven Howard and Chris Abell

Published by the Royal Society of Chemistry, www.rsc.org

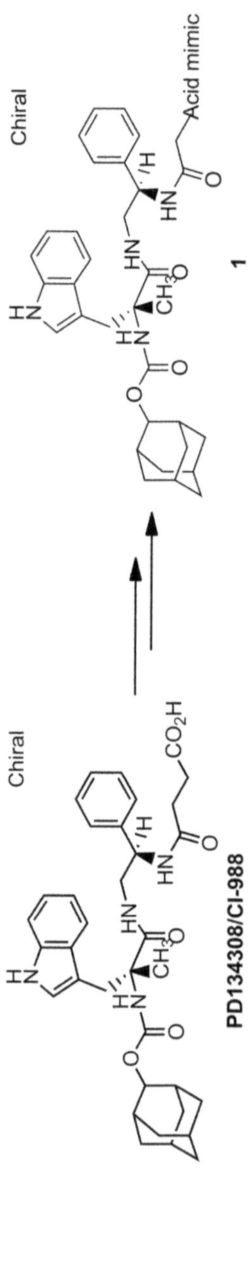

Figure 1 CI-988 and generic acid mimic structure.

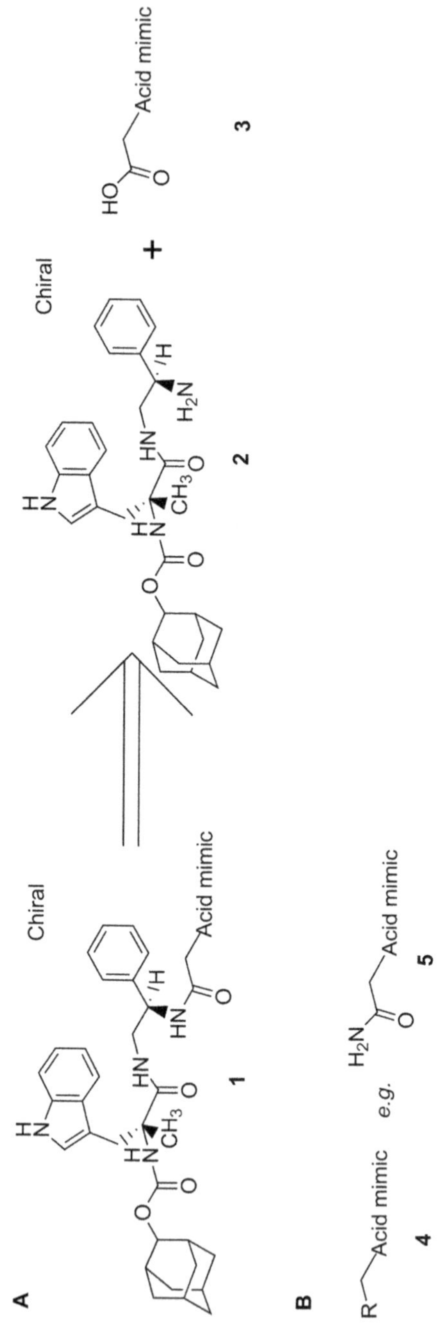

Figure 2 (A) Strategy towards acid mimics of CI-988. (B) Potential fragments of the acid mimics.

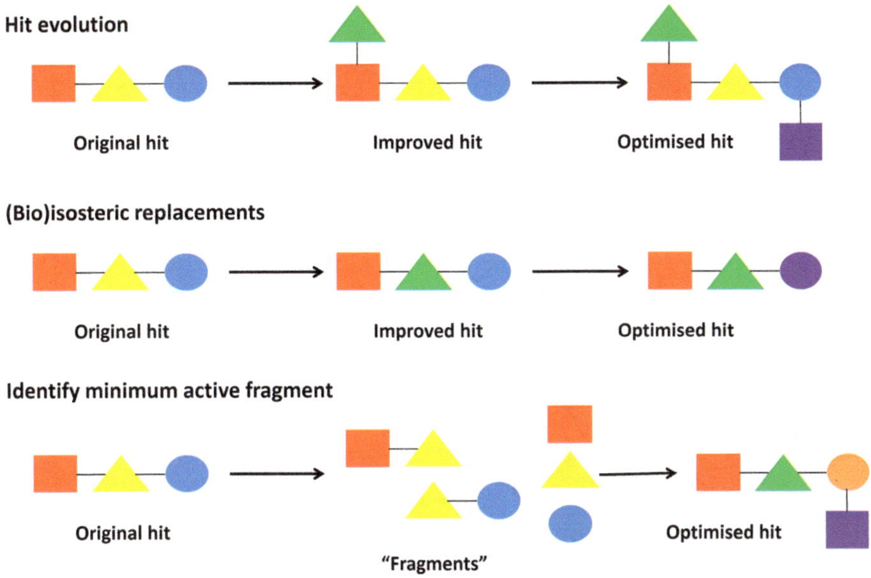

Figure 3 Representation of the mind of the medicinal chemist. Adapted from ref. 4.

and many more inactive analogues than active ones[3] it did occur to me that if there was a way of testing compounds (fragments?) such as **4** or **5** (Figure 2B) and only incorporating the active mimics into final compounds then that might be more efficient.

2 We are all Brought up to be Fragment Practitioners

I would contend that medicinal chemists are pre-disposed to be fragment-based practitioners. There are a number of ways in which hits from high-throughput screens (HTS) are interrogated from the medicinal chemist's point of view with a view to progression through hits-to-lead (H2L) and lead optimisation (LO) (Figure 3). But I would contend that amongst the first questions asked after validated hits are identified is: "What is the minimum active substructure of the hit?"

The limitation in terms of determining the minimum active fragment was the assay used in the first instance. For a typical HTS biochemical assay the lowest level of affinity which could be reasonably detected was about 30 μM. Thus that was the lowest level of affinity which could be ascribed to such a fragment. A nice example of this sort of practice comes from the AstraZeneca group at Charnwood (Figure 4). They identified compound **6** from an HTS looking for $P2X_7$ receptor antagonists.[5] Removal of one of the side-chain amides gave a minimum active fragment **7**[5] which was over an order of magnitude more potent than its parent **6** and with much improved

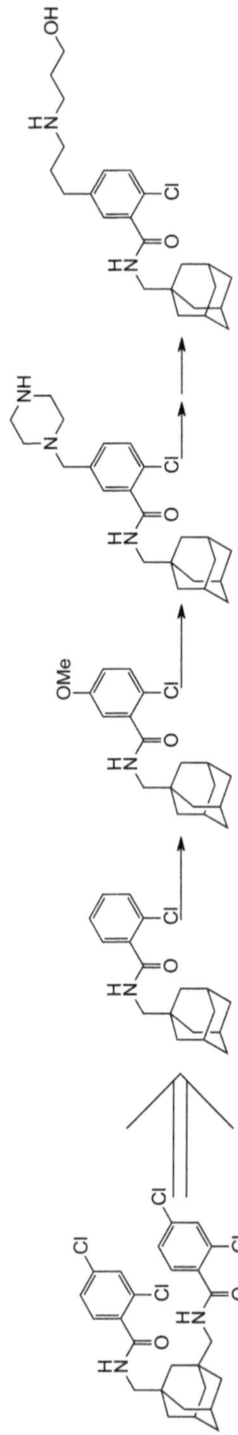

Figure 4 Discovery of AZ9056 from an HTS hit to a clinical development candidate.

physico-chemical properties as an added bonus. Subsequent optimisation (Figure 4) through compounds **8** and **9**[6] ultimately led to the development compound AZD9056[7,8] which has been the subject of multiple Phase II trials in Rheumatoid Arthritis (RA), Chronic Obstructive Pulmonary Disease (COPD) and Crohn's disease.[9]

3 Setting up Fragment Screening at RiboTargets/Vernalis

Like many others, we at RiboTargets/Vernalis were inspired by the publications from the Abbott laboratories.[10,11] We wanted to use NMR methods to carry out fragment screening, to a large extent because we had a 600 MHz NMR machine in-house, a legacy from our NMR structural work on RNA. Our approach was different however, driven by what we perceived as challenges both technically and cost-wise with the production of large quantities of labelled protein required to reproduce the Abbott structure–activity relationship (SAR) experiments by the NMR method (SAR-by-NMR). We decided to use 1D ligand-based observation methods and focussed on the triumvirate of STD,[12] WaterLOGSY[13] and T_2 relaxation-filtered 1D NMR.[14] We found that the use of the three techniques gave a much more reliable output from the NMR screens than using any one or even two techniques alone or in combination. The process we originally adopted was called the *SeeDs* approach (Figure 5). SeeDs stands for Structural exploitation of experimental Drug start-points, being coined after a competition to come up with a catchy acronym, and the overall approach has been described in a detailed review.[15]

There are several points to note in this review from the title itself onwards. Entitled "The SeeDs approach: Integrating Fragments into Drug Discovery", this reflects our original treatise that fragment-based approaches are not standalone techniques but a (valuable) addition to the overall drug discovery process. Though hits from a fragment screen can be optimised when there are no other available start-points from the literature, HTS or computational screens, they are most powerful if coupled with the validated output from other screening methodologies. This leads nicely to the notion of what is a validated hit. The hits from the original SeeDs process went directly into protein X-ray crystallography with no additional orthogonal validation. This is almost unthinkable now for Vernalis and indeed the Vernalis process, as described most recently, validates the original NMR hits by surface plasmon resonance (SPR) and HSQC NMR using ^{15}N-labelled protein as a minimum before hits are prioritised for crystallography. Such a thorough validation of hits is now more the norm and certainly something we subscribe to. Finally there is the process of using the competition step in the initial screening process. We were the first group to implement this step[15] and it seems an obvious one now. Briefly, the fragment library was cocktailed in mixtures of 10 and protein was added. After equilibration of

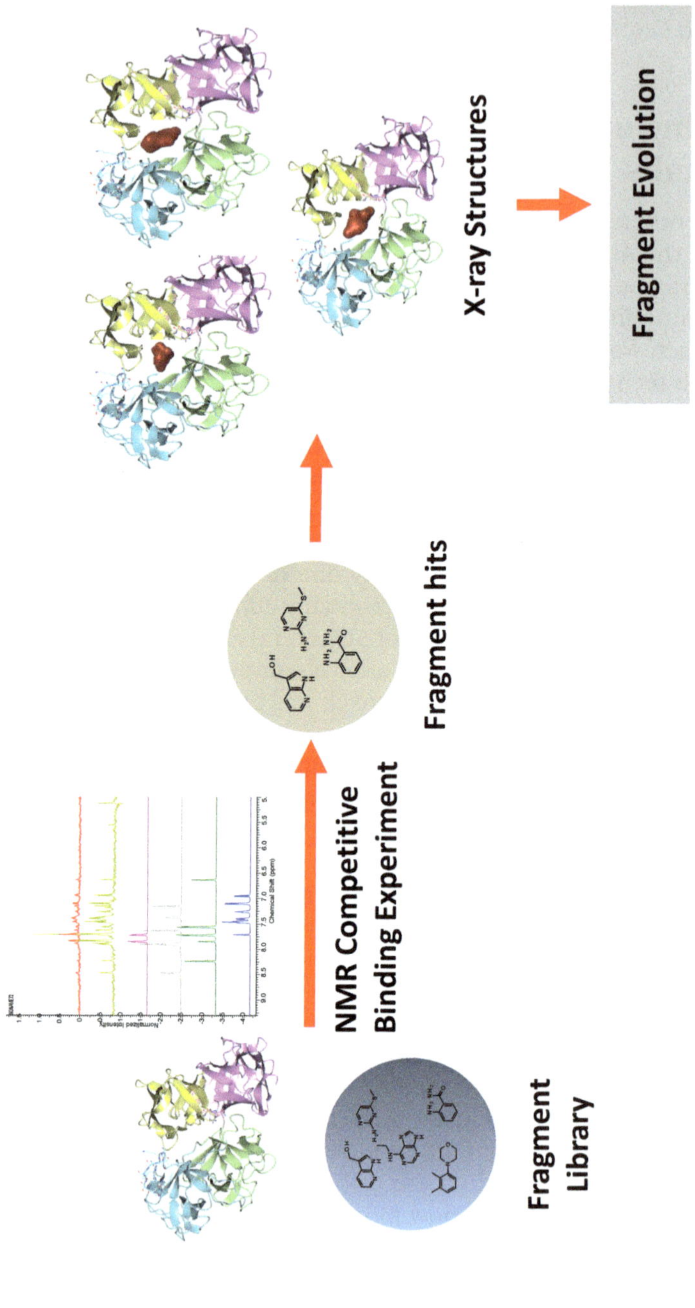

Figure 5 Outline of the original SeeDs approach to fragment screening and evolution.

the fragment cocktail a competitor ligand whose binding site is known was added to saturate that binding site. Looking at the fragments whose signals change on addition of the competitor focuses in on those which are most likely to bind to the same site.

3.1 Targeting Hsp90

Assuming that you want fragments binding at the same site as the competitor ligand, using the NMR competition step described above and in detail in ref. 15, is an excellent fragment screening approach. It worked spectacularly well with Hsp90 where 17 hits were identified as positives in all three NMR techniques mentioned above, 15 of these giving crystal structures from soaking experiments. The main chemotypes that provided crystal structures are exemplified by structures **10–14** in Figure 6. This was from an initial fragment library of 719 members representing an overall 2% hit rate based on the final output from X-ray crystallography.

The Hsp90 programme at RiboTargets and then Vernalis was originally a successful collaboration with the Institute for Cancer Research (ICR) Drug Discovery Group in Sutton. It latterly became a collaboration with Novartis and delivered AUY922[16,17] an intravenous candidate for cancer currently in Phase II clinical development, and BEP800 an orally bioavailable back-up compound (Figure 6).[18,19]

What was astounding from our Hsp90 fragment screen was both the diversity of the hit matter and also the fact that the core fragment chemotypes

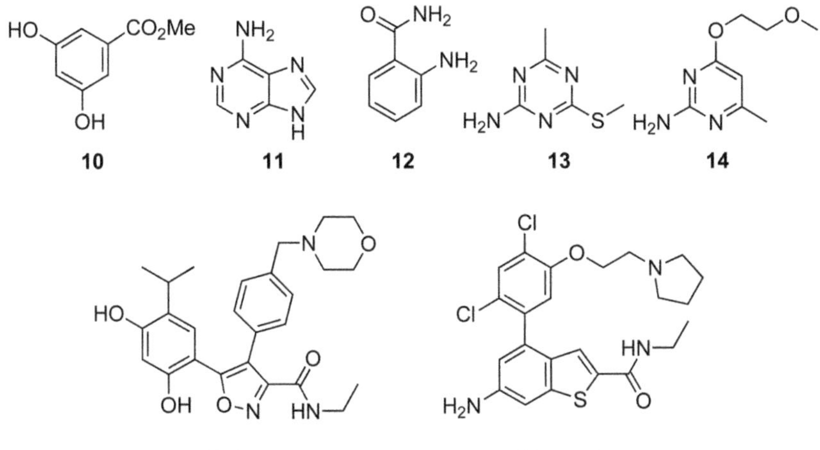

10 **11** **12** **13** **14**

AUY922 **BEP800**

Figure 6 Fragment chemotypes (**10–14**) identified at Vernalis targeting Hsp90. Structures of the Phase II development candidate AUY922[16] and pre-clinical development candidate BEP800[18] targeting Hsp90, both derived from campaigns utilising fragment-based methods as an integral part of hit identification.

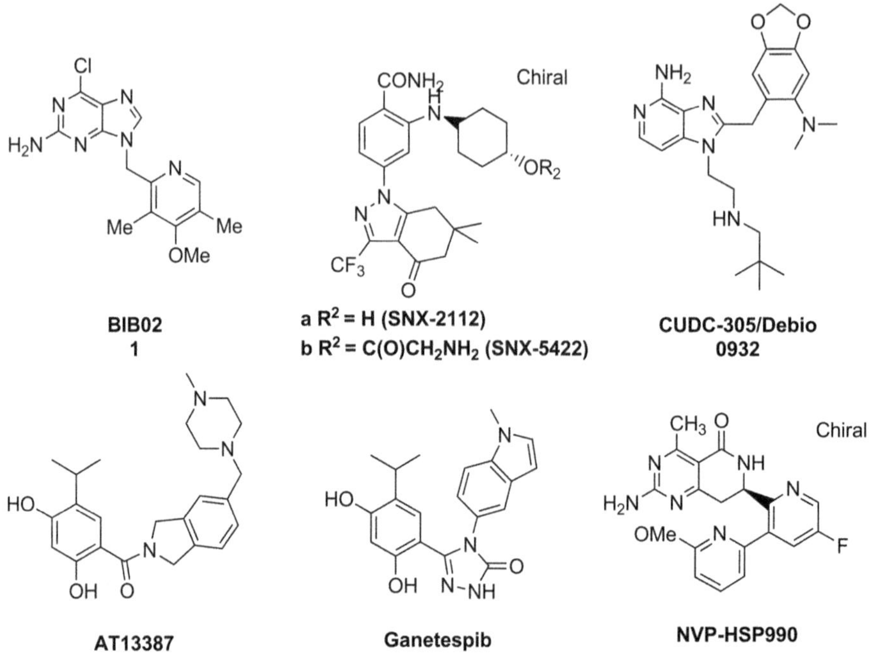

Figure 7 Published structures of Hsp90 clinical development candidates BIB021,[21] SNX-2122/SNX-5422,[22] CUDC-305/Debio 0932,[23] AT13387,[24] Ganetespib,[25] and NVP-HSP990.[26]

(**10–14**) are represented in pretty much all the published clinical Hsp90 structures from other groups (Figure 7).[20] The exception is CUDC-305/Debio 0932, a compound clearly inspired by the purine substrate (ATP) and compounds such as BIIB021 (Figure 7).

Thus a well-selected, well-curated fragment library can supply ample diversity of hits with the potential to be evolved into clinical candidates.

Whilst clinical candidates are exciting and are of course a pre-requisite as a stepping stone to a drug, there are now a number of drugs in clinical trials which originate from fragment-based methods.[27] Arguably the most important breakthrough for fragment-based methods was the approval of Zelboraf® (Vemurafenib) for the treatment of V600E mutant B-Raf melanoma.[28,29] Derived from the 7-azaindole fragment using biochemical screening, which is a very effective fragment screening technology for kinase targets, Zelboraf® was evolved using structure-based drug discovery (Figure 8).[28,29] Real validation is the delivery of a drug to the clinic whether that is validation of a target or a technology. Thus the development of Zelboraf® was vital in the recognition that fragment-based methods have a place in the armoury of the drug-discovery scientist of the future.

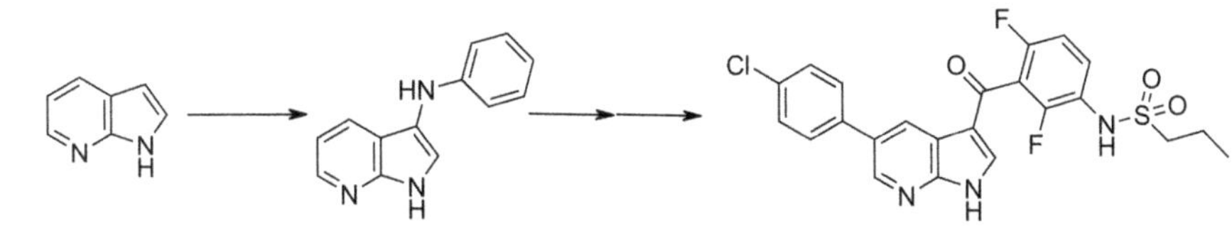

7-Azaindole
Pim-1 IC_{50} >200 μM
Pim-1 LE <0.56

15
Pim-1 IC_{50} ~100 μM
Pim-1 LE ~0.34

Zelboraf (Vemurafenib)
B-Raf^{V600E} IC_{50} = 0.031
μM B-Raf^{V600E} LE = 0.31

Figure 8 Evolution of Zelboraf® for the treatment of V600E mutant B-Raf melanoma.

4 The Future of Fragment-Based Methods – The Future is Now?

I think it is now fair to say that for a number of target classes, where protein production is reasonably straightforward and crystallography is facile, fragment methods will deliver hits from which structural information can be obtained. Though nothing is guaranteed in drug discovery, it is likely that these fragments can be optimised into nM leads. The success of these as pre-clinical and clinical candidates is down to the skills of drug-discovery scientists, the validity of the target and perhaps a modest dose of good fortune. And therefore the next set of challenges awaits fragment-based practitioners. For me this is reflected in the fact that what was once my last slide in a fragment presentation is now effectively my first Figure 9).

The exciting thing is that pretty much all of these are being addressed in one way or another.

4.1 Challenging Targets

4.1.1 β-Site Amyloid Precursor Protein Cleaving Enzyme 1

β-Site amyloid precursor protein cleaving enzyme 1 (β-secretase 1, BACE1) is involved in the formation of amyloid β peptide (Aβ) whose deposition is thought to be critical in the pathogenesis of Alzheimer's Disease (AD). BACE1 has been the subject of enormous efforts in the pharmaceutical industry with little success in delivering clinical candidates using traditional medicinal chemistry approaches.[30] Recently three companies have published on BACE1 inhibitors derived from fragment screening which have entered clinical development (Figure 10) with the most advanced, MK-8931 from Merck being reported to be in Phase III trials.

4.1.2 G Protein-Coupled Receptors

G protein-coupled receptor (GPCR) targeted drugs are amongst the most successful areas of drug discovery and have seen a renaissance recently[34,35] through the new high-quality GPCR crystal structures that have become

What Next for Fragment Screening ?

- Difficult/challenging targets
  - PPIs, transcription factors, allosteric sites, others
- Improvement in screening methods – Higher Throughput
  - NMR, SPR, X-ray, HCS, ITC, DSF, MS,........
- Progression in the absence of crystal structures
  - Using NMR to build models, computational methods
- Target prioritisation
- New, novel, different fragments
- ....

Figure 9 What next for fragment screening?

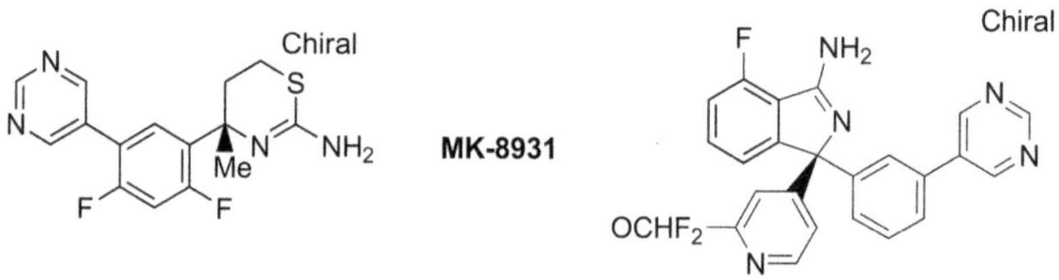

LY2811376 AZD3839

Figure 10 Structure of clinical candidates targeting β-secretase 1 (BACE1). LY2811376 – Lilly;[31] MK-8931 – Merck (structure not known);[32] AZD3839 – AstraZeneca.[33]

available. This has led directly to the enabling of fragment screening through accessing stabilised protein.[36] In a recent collaborative report from Heptares Therapeutics, Biosensor Tools and the Medical Research Council Laboratory of Molecular Biology, SPR screening of a stabilised β1-adrenergic receptor led to fragment hits for which receptor-bound crystal structures were then solved.[37] This topic is covered in much more detail in Chapter 5.

4.1.3 Protein–Protein Interactions

Protein–protein interactions (PPIs) are considered to be critically important for the progression of numerous disease states and are also considered to be amongst the most challenging targets for drug discovery. Whilst there are many cases where HTS campaigns have failed to give useful start-points for H2L and LO campaigns, fragment methods have been successful in this space. Perhaps the most impressive example so far is the discovery by the group at Abbott of the Bcl-xL and Bcl-2 inhibitors ABT-737 and Navitoclax (ABT-263) using their SAR-by-NMR technique to identify fragment start-points which they subsequently linked and optimised in a fantastic example of high-quality structure-guided optimisation.[38,39]

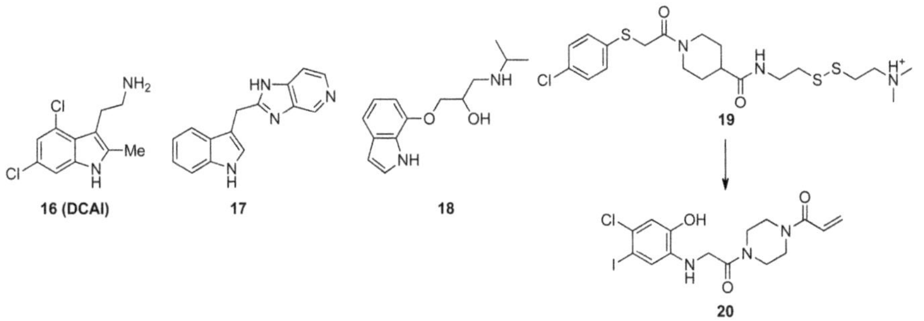

Figure 11 Reversible fragment hits targeting RAS from Genentech (**16**),[40] Vanderbilt (**17**),[41] and Novartis (**18**).[42] Compound **19** is an irreversible fragment that binds to oncogenic $KRAS^{G12C}$. Compound **20** is evolved from **19** and is also an irreversible inhibitor which binds to oncogenic $KRAS^{G12C}$.[44]

More recently, reports have been published on fragment-based hits targeting rat sarcoma (RAS), the holy grail of cancer targets that have for many years been considered undruggable. Reports from Genentech,[40] Vanderbilt University[41] and Novartis[42] have reported on reversibly binding fragments (Figure 11). Using the disulphide fragment screening approach called *tethering*[43] workers at UCSF identified the irreversible inhibitor based on **19** that specifically binds to oncogenic $KRAS^{G12C}$ (ref. 44) (Figure 11). Compound **19** was subsequently modified to the irreversible ligand **20** which was shown to block $KRAS^{G12C}$ interactions, decrease viability and increase apoptosis of lung cancer cell lines harbouring the $KRAS^{G12C}$ mutation.[44]

4.1.4 Allosteric Site Inhibitors

Allosteric site inhibitors are an attractive option for a number of reasons. They may offer selectivity advantages across family members with otherwise similar orthosteric binding sites such as the ATP site in kinases,[45,46] they may provide binding features for small molecules which are better associated with drug-like functionality and they may also induce conformations which lead to activation or inactivation, an important feature for targeting PPIs. Recently the group at Novartis identified farnesyl pyrophosphate synthase (FPPS) inhibitors using NMR fragment-based approaches which target an allosteric site on FPPS (Figure 12).[47] The group identified fragments (**21–24**) as binders in the NMR screen and in a series of elegant NMR experiments showed that these fragments were not competitive with the known FPPS bisphosphonate inhibitor Zoledronic acid. Crystal structures of **21–24** were obtained and all the fragments were shown to bind to an allosteric site on FPPS with fragment **24** being crystallised in a ternary complex with Zoledronic acid. Initial optimisation of these fragments led to compounds **25** and **26** with significantly improved potency (Figure 12).

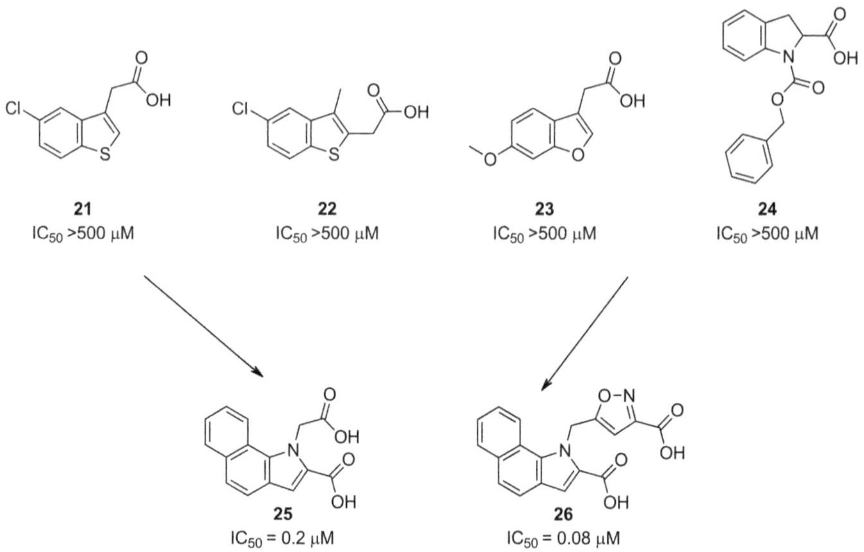

Figure 12 Allosteric binders to FPPS. Fragment hits (**21–24**) and optimised analogues (**25** and **26**).[47]

Workers at Astex published a recent paper on an X-ray crystallographic fragment screen targeting the full-length HCV NS3–NS4a protein.[48] The fragment series in the paper bound to an allosteric site in the protein at the interface between the helicase and protease domain. Structural and other biophysical data, including isothermal titration calorimetry (ITC) suggested that binding of the compounds in this site reduced the conformational flexibility of the protein leading to functional inhibition. Interestingly it was not until the compounds were significantly optimised that they became functionally active. Improvements were driven by biophysical methods requiring significant patience on the part of the group.[48] Allosteric site inhibitors are the subject of a more extensive discussion in Chapter 7.

4.2 Progress in the Absence of Crystal Structures

Whilst crystal structures are considered the most desirable structural data, there are occasions with some targets when it is difficult or even impossible to obtain them with fragments. This does not make the target a bad target, just more challenging, and we need to find ways to move forwards. NMR-guided structural information can be used as an alternative in conjunction with computational models. Two recent fragment screening examples of this have been published both using the original SAR-by-NMR method, and both targeting the anti-apoptotic protein Mcl-1.[49,50] In the first paper from the Vanderbilt group[49] two classes of fragments were identified represented by the benzothiophene **27** and compound **28**. Nuclear Overhauser effect (NOE)-derived distance restraints were acquired for **27** and **28** using doubly labelled

^{15}N-,^{13}C-Mcl-1 protein, and using these restraints the fragments docked into a Mcl-1/Bim BH3 peptide complex. Though both classes of fragments were observed to bind to different regions of Mcl-1, in each case the carboxylic acids were modelled as interacting with the same R263 residue. Merging of the two chemotypes using these docking models resulted in the synthesis of the optimised hit **29**. Compound **29** was much more potent as measured biochemically in a fluorescence polarisation anisotropy (FPA) assay (0.32 μM compared to 131 μM and 60 μM for **27** and **28** respectively). More gratifyingly, the crystal structure of **29** was obtained in complex with Mcl-1 and showed that the optimised hit **29** occupied both pockets occupied separately by the models for fragments **27** and **28** with the carboxylic acid interacting as expected with R263 (Figure 13). In a second paper, by workers at AbbVie, several fragment hits were identified from the NMR screen which had measurable activity in a fluorescence polarisation (FP) assay.[50] The sulphonamide **30** and salicylic acid **33** fragments were selected for further study. For **30**, intermolecular NOEs observed between the compound and Mcl-1 were used to guide *in silico* docking of the compound into the protein. Again, the carboxylic acid was observed to make a key interaction with R263 as was seen by the Vanderbilt group in their study. Based on the docked model the fragment was elaborated to give **31** for which a crystal structure was obtained. The acid of **31** binds as expected to R263 and the naphthyl group induces a conformational change in the protein to accommodate this moiety. Further optimisation led to the lead structure **32** with an FP IC_{50} = 0.03 μM. For the salicylic acid fragment **33**, this was simply docked into the BH3 peptide groove using the electrostatic interaction between the acid and R263 as the only restraint. Using this model **33** was evolved into **34** for which a crystal structure was again obtained. The crystal structure of **34** was as expected from the docked model of **33** (Figure 13).

4.3 New, Novel Different Fragments

From the very first fragment library we put together at RiboTargets[15] it was clear that most of our fragments were pretty flat and 2D. This is not surprising since most fragment molecules that are available for purchase are fairly flat by most methods of analysis. Given that many drugs are derived from natural products and that drugs themselves tend to be more shapely, it seems sensible to look to see whether there are opportunities to add more shape or three-dimensionality to fragments. There is much talk in this space but little in the way of publications with a couple of exceptions. Damian Young, then of Harvard now at the Centre for Drug Discovery at Baylor College of Medicine, has published on the synthesis of 3D fragments using the principles of diversity orientated synthesis (DOS) (Figure 14).[51]

The other published effort in this area is from the 3D Fragment Library consortium.[52] This is a UK-based group of eight not-for-profit drug-discovery groups who are trying to access fragment libraries with enhanced 3D characteristics and subsequently evaluate them in a range of fragment screens using a variety of screening methodologies.[53]

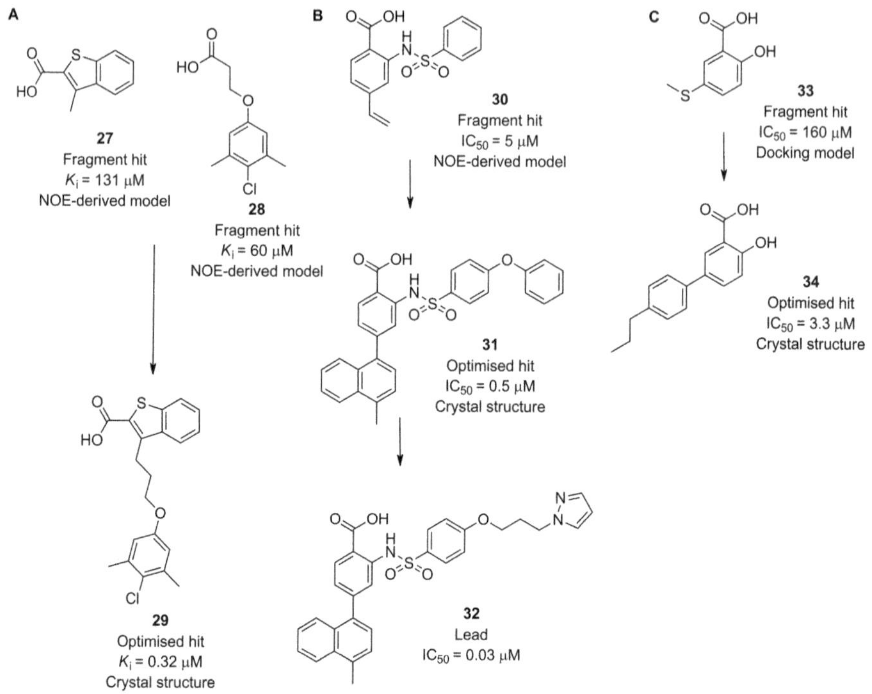

Figure 13 (A) Fragment merging targeting Mcl-1.[49] NOE-Guided models of **30** and **31** were generated and used to merge fragments leading to **32**. A crystal structure of **32** bound to Mcl-1 validated the NMR-guided docking. Fragment growing in Mcl-1[50] using (B) NOE-derived models to evolve fragment **33**, and (C) Computational docking to evolve fragment **36**. In both (B) (compound **34**) and (C) (compound **37**) crystal structures of evolved compounds in complex with Mcl-1 were obtained.

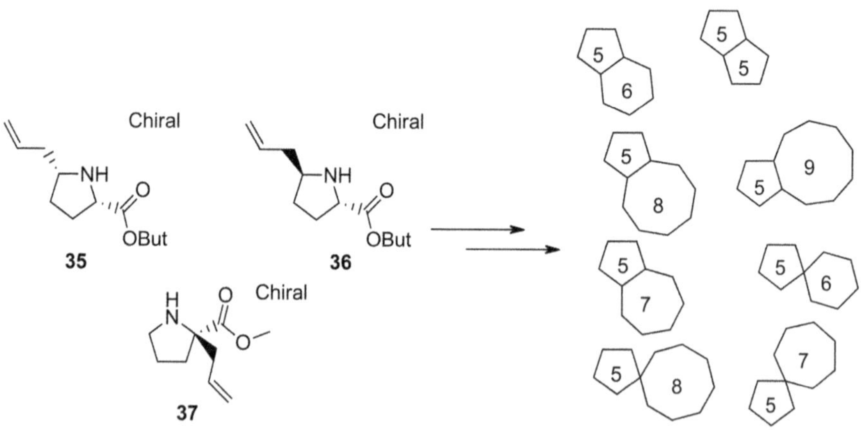

Figure 14 Building blocks and schematic for the preparation of 3D fragments.[51] Note that the fused and spirocycles can contain heteroatoms.

5 Fragment Discovery Outside Industry

My most recent experiences with fragment-based drug discovery have been outside of the commercial field; at the Beatson Institute for Cancer Research in Glasgow. Compared to HTS methods in commercial laboratories, fragment-based screening platforms are relatively lower in cost and require less initial capital investment. This 'relatively' low barrier to entry means that small not-for-profit and academic laboratories are well placed to operate in this field and make exciting and useful contributions.[44,54] This is summarised nicely in the review from the Abell group in Cambridge[55] which says "Unlike high-throughput screening libraries, fragment libraries are commercially available and are within the budget of academic research groups." There is certainly a difference in culture outside of the commercial world but the main difference is one of resource, especially when considering big pharma. I think this is seen vividly in the strategy for fragment library design published by Pfizer;[56] one which is replicated in many large organisations. Of particular value in the pharma industry setting is access to large corporate collections. For example, in the Pfizer paper, fragment libraries are designed such that library members have multiple close analogues as well as other near-neighbours that are invaluable for rapid exploitation of hits. Whilst the rest of us can mine commercial chemical space I think the corporate collections are a huge asset. However the most important fundamental aspect of fragment screening relates to rigour and quality. If you stick to these principals there is no real difference between carrying out fragment-based screening in a large pharmaceutical company, biotech company, not-for-profit organisation or academic laboratory. The recent *Nature Protocols* paper[57] from the Abell laboratory describing "A three-stage biophysical screening cascade for fragment-based drug discovery" has the robustness and rigour required for the quality standards needed to do fragment-based drug discovery properly, regardless of the setting.

6 Summary

From being the methodology of last resort and the home of the failed HTS project, fragment-based drug discovery has come a long way. Now embedded within the drug discovery community we can expect it to continue to punch above its weight well into the future.

References

1. J. Hughes, P. Boden, B. Costall, A. Domeney, E. Kelly, D. C. Horwell, J. C. Hunter, R. D. Pinnock and G. N. Woodruff, *Proc. Natl. Acad. Sci. U. S. A.*, 1990, **87**, 6728.
2. D. C. Horwell, J. Hughes, J. C. Hunter, M. C. Pritchard, R. S. Richardson, E. Roberts and G. N. Woodruff, *J. Med. Chem.*, 1991, **34**, 404.

3. M. J. Drysdale, M. C. Pritchard and D. C. Horwell, *J. Med. Chem.*, 1992, **35**, 2573.
4. G. M. Keseru and G. M. Makara, *Drug Discovery Today*, 2006, **11**, 741.
5. A. Baxter, J. Bent, K. Bowers, M. Braddock, S. Brough, M. Fagura, M. Lawson, T. McInally, M. Mortimore, M. Robertson, R. Weaver and P. Webborn, *Biorg. Med. Chem. Lett.*, 2003, **13**, 4047.
6. M. Furber, L. Alcaraz, J. E. Bent, A. Beyerbach, K. Bowers, M. Braddock, M. V. Caffrey, D. Cladingboel, J. Collington, D. K. Donald, M. Fagura, F. Ince, E. C. Kinchin, C. Laurent, M. Lawson, T. J. Luker, M. M. P. Mortimore, A. D. Pimm, R. J. Riley, N. Roberts, M. Robertson, J. Theaker, P. V. Thorne, R. Weaver, P. Webborn and P. Willis, *J. Med. Chem.*, 2007, **50**, 5882.
7. S. D. Guile, L. Alcaraz, T. N. Birkinshaw, K. C. Bowers, M. R. Ebden, M. Furber and M. J. Stocks, *J. Med. Chem.*, 2009, **52**, 3123.
8. (a) A. Teasdale, S. Fenner, A. Ray, A. Ford and A. Phillips, *Org. Process Res. Dev.*, 2010, **14**, 943; (b) A. Teasdale, D. Elder, S.-J. Chang, S. Wang, R. Thompson, N. Benz and I. H. Sanchez Flores, *Org. Process Res. Dev.*, 2013, **17**, 221.
9. National Center for Advancing Translational Sciences (NCATS) at the National Institute of Health website. http://www.ncats.nih.gov/files/AZD9056.pdf, March 2014.
10. S. B. Shuker, P. J. Hajduk, R. P. Meadows and S. W. Fesik, *Science*, 1996, **274**, 1531.
11. P. J. Hajduk, G. Sheppard, D. G. Nettesheim, E. T. Olejniczak, S. B. Shuker, R. P. Meadows, D. H. Steinman, G. M. Carrera, P. A. Marcotte, J. Severin, K. Walter, H. Smith, E. Gubbins, R. Simmer, T. F. Holzman, D. W. Morgan, S. K. Davidsen, J. B. Summers and S. W. Fesik, *J. Am. Chem. Soc.*, 1997, **119**, 5818.
12. M. Mayer and B. Meyer, *Angew. Chem., Int. Ed. Engl.*, 1999, **38**, 1784.
13. C. Dalvit, P. Pevarello, M. Tato, M. Veronesi, A. Vulpetti and M. Sundstrom, *J. Biomol. NMR*, 2000, **18**, 65.
14. P. J. Hajduk, E. T. Olejniczak and S. W. Fesik, *J. Am. Chem. Soc.*, 1997, **119**, 12257.
15. R. E. Hubbard, B. Davis, I. Chen and M. J. Drysdale, *Curr. Top. Med. Chem.*, 2007, **7**, 1568.
16. P. A. Brough, W. Aherne, X. Barril, J. Borgognoni, K. Boxall, J. E. Cansfield, K.-M. J. Cheung, I. Collins, N. G. M. Davies, M. J. Drysdale, B. Dymock, S. A. Eccles, H. Finch, A. Fink, A. Hayes, R. Howes, R. E. Hubbard, K. James, A. M. Jordan, A. Lockie, V. Martins, A. Massey, T. P. Matthews, E. McDonald, C. J. Northfield, L. H. Pearl, C. Prodromou, S. Ray, F. I. Raynaud, S. D. Roughley, S. Y. Sharp, A. Surgenor, D. L. Walmsley, P. Webb, M. Wood, P. Workman and L. Wright, *J. Med. Chem.*, 2008, **51**, 196.
17. S. A. Eccles, A. Massey, F. Raynaud, S. Y. Sharp, G. Box, M. Valenti, L. Patterson, S. Gowan, F. Boxall, W. Aherne, M. Rowlands, A. Hayes, V. Martins, F. Urban, K. Boxall, C. Prodromou, L. Pearl, K. James, T. P. Matthews, K.-M. Cheung, A. Kalusa, K. Jones, E. McDonald, X. Barril, P. Brough,

J. E. Cansfield, B. Dymock, M. Drysdale, H. Finch, R. Howes, R. Hubbard, A. Surgenor, P. Webb, L. M. Wood, L. Wright and P. Workman, *Cancer Res.*, 2008, **68**, 2850.
18. P. A. Brough, X. Barril, J. Borgognoni, P. Chene, N. G. M. Davies, B. Davis, M. J. Drysdale, B. Dymock, S. A. Eccles, C. Garcia-Echeverria, C. Fromont, A. Hayes, R. E. Hubbard, A. M. Jordan, M. Rugaard Jensen, A. Massey, A. Merrett, A. Padfield, R. Parsons, T. Radimerski, F. I. Raynaud, A. Robertson, S. D. Roughley, J. Schoepfer, H. Simmonite, S. Y. Sharp, A. Surgenor, M. Valenti, S. Walls, P. Webb, M. Wood, P. Workman and L. Wright, *J. Med. Chem.*, 2009, **52**, 4794.
19. A. J. Massey, J. Schoepfer, P. A. Brough, J. Brueggen, P. Chène, M. J. Drysdale, U. Pfaar, T. Radimerski, S. Ruetz, A. Schweitzer, M. Wood, C. Garcia-Echeverria and M. Rugaard Jensen, *Mol. Cancer Ther.*, 2010, **9**, 906.
20. S. D. Roughley and R. E. Hubbard, *J. Med. Chem.*, 2011, **54**, 3989.
21. S. R. Kasibhatla, K. Hong, M. A. Biamonte, D. J. Busch, P. L. Karjian, J. L. Sensintaffar, A. Kamal, R. E. Lough, J. Brekken, K. Lundgren, R. Grecko, G. A. Timony, Y. Ran, R. Mansfield, L. C. Fritz, E. Ulm, F. J. Burrows and M. F. Boehm, *J. Med. Chem.*, 2007, **50**, 2767.
22. K. H. Huang, J. M. Veal, R. P. Fadden, J. W. Rice, J. Eaves, J.-P. Strachan, A. F. Barabasz, B. E. Foley, T. E. Barta, W. Ma, M. A. Silinski, M. Hu, J. M. Partridge, A. Scott, L. G. DuBois, T. Freed, P. M. Steed, A. J. Ommen, E. D. Smith, P. F. Hughes, A. R. Woodward, G. J. Hanson, W. S. McCall, C. J. Markworth, L. Hinkley, M. Jenks, L. Geng, M. Lewis, J. Otto, B. Pronk, K. Verleysen and S. E. Hall, *J. Med. Chem.*, 2009, **52**, 4288.
23. R. Bao, C. J. Lai, H. Qu, D. Wang, L. Yin, B. Zifcak, R. Atoyan, J. Wang, M. Samson, J. Forrester, S. DellaRocca, G. X. Xu, X. Tao, H. X. Zhai, X. Cai and C. Qian, *Clin. Cancer Res.*, 2009, **15**, 4046.
24. A. J. Woodhead, H. Angove, M. G. Carr, G. Chessari, M. Congreve, J. E. Coyle, J. Cosme, B. Graham, P. J. Day, R. Downham, L. Fazal, R. Feltell, E. Figueroa, M. Frederickson, J. Lewis, R. McMenamin, C. W. Murray, M. A. O'Brien, L. Parra, S. Patel, T. Phillips, D. C. Rees, S. Rich, D.-M. Smith, G. Trewartha, M. Vinkovic, B. Williams and A. J.-A. Woolford, *J. Med. Chem.*, 2010, **53**, 5956.
25. W. Ying, Z. Du, L. Sun, K. P. Foley, D. A. Proia, R. K. Blackman, D. Zhou, T. Inoue, N. Tatsuta, J. Sang, S. Ye, J. Acquaviva, L. Shin Ogawa, Y. Wada, J. Barsoum and K. Koya, *Mol. Cancer Ther.*, 2012, **11**, 475.
26. D. L. Menezes, P. Taverna, M. R. Jensen, T. Abrams, D. Stuart, G. K. Yu, D. Duhl, T. Machajewski, W. R. Sellers, N. K. Pryer and Z. Gao, *Cancer Ther.*, 2012, **11**, 730.
27. Practical Fragments Website, http://practicalfragments.blogspot.com/2013/01/fragments-in-clinic-2013-edition.html, March 2014.
28. J. Tsai, J. T. Lee, W. Wang, J. Zhang, H. Cho, S. Mamo, R. Bremer, S. Gillette, J. Kong, N. K. Haass, K. Sproesser, L. Li, K. S. M. Smalley, D. Fong, Y.-L. Zhu, A. Marimuthu, H. Nguyen, B. Lam, J. Liu, I. Cheung, J. Rice, Y. Suzuki, C. Luu, C. Settachatgul, R. Shellooe, J. Cantwell, S.-H. Kim, J. Schlessinger, K. Y. J. Zhang, B. L. West, B. Powell, G. Habets,

C. Zhang, P. N. Ibrahim, P. Hirth, D. R. Artis, M. Herlyn and G. Bollag, *Proc. Natl. Acad. Sci. U.S.A.*, 2008, **105**, 3041.
29. G. Bollag, P. Hirth, J. Tsai, J. Zhang, P. N. Ibrahim, H. Cho, W. Spevak, C. Zhang, Y. Zhang, G. Habets, E. A. Burton, B. Wong, G. Tsang, B. L. West, B. Powell, R. Shellooe, A. Marimuthu, H. Nguyen, K. Y. Zhang, D. R. Artis, J. Schlessinger, F. Su, B. Higgins, R. Iyer, K. D'Andrea, A. Koehler, M. Stumm, P. S. Lin, R. J. Lee, J. Grippo, I. Puzanov, K. B. Kim, A. Ribas, G. A. McArthur, J. A. Sosman, P. B. Chapman, K. T. Flaherty, X. Xu, K. L. Nathanson and K. Nolop, *Nature*, 2010, **467**, 596.
30. S. J. Stachel, *Drug Dev. Res.*, 2009, **70**, 101.
31. P. C. May, R. A. Dean, S. L. Lowe, F. Martenyi, S. M. Sheehan, L. N. Boggs, S. A. Monk, B. M. Mathes, D. J. Mergott, B. M. Watson, S. L. Stout, D. E. Timm, E.-s. LaBell, C. R. Gonzales, M. Nakano, S. S. Jhee, M. Yen, L. Ereshefsky, T. D. Lindstrom, D. O. Calligaro, P. J. Cocke, D. G. Hall, S. Friedrich, M. Citron and J. E. Audia, *J. Neurosci.*, 2011, **31**, 16507.
32. D. F. Wyss, Y.-S. Wang, H. L. Eaton, C. Strickland, J. H. Voigt, Z. Zhu and A. W. Stamford, *Top. Curr. Chem.*, 2012, **317**, 83.
33. F. Jeppsson, S. Eketjall, J. Janson, S. Karlstrom, S. Gustavsson, L.-L. Olsson, A.-C. Radesater, B. Ploeger, G. Cebers, K. Kolmodin, B.-M. Swahn, S. von Berg, T. Bueters and J. Falting, *J. Biol. Chem.*, 2012, **287**, 41245.
34. M. A. Hanson and R. C. Stevens, *Structure*, 2009, **17**, 8.
35. J. A. Salon, D. T. Lodowski and K. Palczewski, *Pharmacol. Rev.*, 2011, **63**, 901.
36. M. Congreve, R. L. Rich, D. G. Myszka, F. Figaroa, G. Siegal and F. H. Marshall, *Methods in Enzymology*, ed. L. C. Kuo, Academic Press, Burlington, MA, U.S.A, 2011, vol. 493, pp. 115–136.
37. J. A. Christopher, J. Brown, A. S. Doré, J. C. Errey, M. Koglin, F. H. Marshall, D. G. Myszka, R. L. Rich, C. G. Tate, B. Tehan, T. Warne and M. Congrave, *J. Med. Chem.*, 2013, **56**, 3446.
38. T. Oltersdorf, S. W. Elmore, A. R. Shoemaker, R. C. Armstrong, D. J. Augeri, D. J. Augeri, B. A. Belli, M. Bruncko, T. L. Deckwerth, J. Dinges, P. J. Hajduk, M. K. Joseph, S. Kitada, S. J. Korsmeyer, A. R. Kunzer, A. Letai, C. Li, M. J. Mitten, D. G. Nettesheim, S.-C. Ng, P. M. Nimmer, J. M. O'Connor, A. Oleksijew, A. M. Petros, J. C. Reed, W. Shen, S. K. Tahir, C. B. Thompson, K. J. Tomaselli, B. Wang, M. D. Wendt, H. Zhang, S. W. Fesik and S. H. Rosenberg, *Nature*, 2005, **435**, 677.
39. C. M. Park, M. Bruncko, J. Adickes, J. Bauch, H. Ding, A. Kunzer, K. C. Marsh, P. Nimmer, A. R. Shoemaker, X. Song, S. K. Tahir, C. Tse, X. Wang, M. D. Wendt, X. Yang, H. Zhang, S. W. Fesik, S. H. Rosenberg and S. W. Elmore, *J. Med. Chem.*, 2008, **51**, 6902.
40. T. Maurer, L. S. Garrenton, A. Oh, K. Pitts, D. J. Anderson, N. J. Skelton, B. P. Fauber, B. Pan, S. Malek, D. Stokoe, M. J. C. Ludlam, K. K. Bowman, J. Wu, A. M. Giannetti, M. A. Starovasnik, I. Mellman, P. K. Jackson, J. Rudolph, W. Wang and G. Fang, *Proc. Natl. Acad. Sci. U.S.A.*, 2012, **109**, 5299.
41. Q. Sun, J. P. Burke, J. Phan, M. C. Burns, E. T. Olejniczak, A. G. Waterson, T. Lee, O. W. Rossanese and S. W. Fesik, *Angew. Chem., Int. Ed.*, 2012, **51**, 6140.

42. J. M. Jansen, W. Jahnke, S. Fong, L. Tandeske, C. Wartchow, K. Pfister, T. Zavorotinskaya, A. Blechschmidt, D. Bussiere, Y. Dai, M. Dillon, J. Dove, E. Fang, D. Farley, J.-M. Florent, J. Fuller, S. Gokhin, A. Gossert, M. Hekmat-Nejad, C. Henry, J. Klopp, B. Lenahan, A. Lingel, A. Meyer, J. Narberes, G. Pardee, G. Paris, S. Ramurthy, P. Renhowe, S. Rieffel, K. Shoemaker, S. Subramanian, T. Tsang, S. Widger, A. Widmer, I. Zaror and S. Hardy, *RAS Oncogenes: From biology to therapy*, 2014, poster presentation.
43. D. A. Erlanson, A. C. Braisted, D. R. Raphael, M. Randal, R. M. Stroud, E. M. Gordon and J. A. Wells, *Proc. Natl. Acad. Sci.*, 2000, **97**, 9367.
44. J. M. Ostrem, U. Peters, M. L. Sos, J. A. Wells and K. M. Shokat, *Nature*, 2013, **503**, 548.
45. J. A. Lewis, E. P. Lebois and C. W. Lindsley, *Curr. Opin. Chem. Biol.*, 2008, **12**, 269.
46. L. Skora, J. Mestan, D. Fabbro, W. Jahnke and S. Grzesiek, *Proc. Natl. Acad. Sci. U. S. A.*, 2013, **110**, E4437.
47. W. Jahnke, J.-M. Rondeau, S. Cotesta, A. Marzinzik, X. Pellé, M. Geiser, A. Strauss, M. Götte, F. Bitsch, R. Hemmig, C. Henry, S. Lehmann, J. F. Glickman, T. P. Roddy, S. J. Stout and J. R. Green, *Nat. Chem. Biol.*, 2010, **6**, 660.
48. S. M. Saalau-Bethell, A. J. Woodhead, G. Chessari, M. G. Carr, J. Coyle, B. Graham, S. D. Hiscock, C. W. Murray, P. Pathuri, S. J. Rich, C. J. Richardson, P. A. Williams and H. Jhoti, *Nat. Chem. Biol.*, 2012, **8**, 920.
49. A. Friberg, D. Vigil, B. Zhao, R. N. Daniels, J. P. Burke, P. M. Garcia-Barrantes, D. Camper, B. A. Chauder, T. Lee, E. T. Olejniczak and S. W. Fesik, *J. Med. Chem.*, 2013, **56**, 15.
50. A. M. Petros, S. L. Swann, D. Song, K. Swinger, C. Park, H. Zhang, M. D. Wendt, A. R. Kunzer, A. J. Souers and C. Sun, *Bioorg. Med. Chem.*, 2014, **24**, 1484.
51. A. W. Hung, A. Ramek, Y. Wang, T. Kaya, J. A. Wilson, P. A. Clemons and D. W. Young, *Proc. Natl. Acad. Sci. U. S. A.*, 2011, **108**, 6799.
52. 3DFragment Consortium Website, http://www.3dfrag.org/, March 2014.
53. A. D. Morley, A. Pugliese, K. Birchall, J. Bower, P. Brennan, N. Brown, T. Chapman, M. Drysdale, I. H. Gilbert, S. Hoelder, A. Jordan, S. V. Ley, A. Merritt, D. Miller, M. E. Swarbrick and P. G. Wyatt, *Drug Discovery Today*, 2013, **18**, 1222.
54. D. E. Scott, M. T. Ehebauer, T. Pukala, M. Marsh, T. L. Blundell, A. R. Venkitaraman, C. Abell and M. Hyvçnen, *ChemBioChem*, 2013, **14**, 332.
55. D. E. Scott, A. G. Coyne, S. A. Hudson and C. Abell, *Biochemistry*, 2012, **51**, 4990.
56. W. F. Lau, J. M. Withka, D. Hepworth, T. V. Magee, Y. J. Du, G. A. Bakken, M. D. Miller, Z. S. Hendsch, V. Thanabal, S. A. Kolodziej, L. Xing, Q. Hu, L. S. Narasimhan, R. Love, M. E. Charlton, S. Hughes, W. P. van Hoorn and J. E. Mills, *J. Comput.-Aided Mol. Des.*, 2011, **25**, 621.
57. E. H. Mashalidis, P. Sledz, S. Lang and C. Abell, *Nat. Protoc.*, 2013, **8**, 2309.

CHAPTER 1

Different Flavours of Fragments

CHRIS ABELL*[a] AND CLAUDIO DAGOSTIN[a]

[a]Department of Chemistry, University of Cambridge, Lensfield Road, Cambridge, CB2 1EW, UK
*E-mail: ca26@cam.ac.uk

1.1 Fragment-Based Drug Discovery

Fragment-based drug discovery is becoming a powerful technology in the arsenal of the pharmaceutical companies to aid discovery of new small-molecule therapeutics, establish the druggability of biological targets, and discover alternative inhibition sites on already established ones.[1,2] With the market approval of Zelboraf® (Vemurafenib – the first drug discovered *via* fragment-based drug discovery[3]) the approach has a measure of validation and interest in the field continues to grow.

Fragments are small molecules that may become parts of a larger molecule, but in some cases were fragments of known drugs, that have been used as starting points to find new inhibitors for different biological targets. A fragment is a small, typically aromatic, organic molecule of molecular weight <250 Da, which is very soluble and chemically stable. The use of such small molecules at the beginning of a medicinal chemistry program allows an effective exploration of chemical space and a more rational design of the final molecule through iterations guided by structural knowledge of the site of binding. In the best cases, the approach can offer a less expensive and faster route to potent and better quality lead compounds compared

RSC Drug Discovery Series No. 47
Fragment-Based Drug Discovery
Edited by Steven Howard and Chris Abell

Published by the Royal Society of Chemistry, www.rsc.org

with more traditional approaches such as high-throughput screening (HTS) of higher molecular weight compounds.

1.2 Different Types of Fragments

Views on the composition of fragment libraries have evolved over time. Some libraries were assembled by taking known drugs and dissecting them into fragments[4] or by computationally finding diverse fragments.[5] It was apparent from the start that solubility would be very important because of the concentrations used for screening (often tens of mM). Another key criterion was lack of reactive groups[6] although current interest in covalent inhibitors is beginning to impact on thinking here.[7] Fragment libraries, although generally diverse, were mainly composed of flat aromatic or heteroaromatics. Astex introduced the *Rule of 3*[8] to map across to the *Lipinski Rule of 5*, and soon after, commercial Rule of 3 compliant libraries became available.

There remain different views on the best size for a fragment library. Some organisations have libraries of over 20 000 fragments whereas others use small libraries (for a recent study of an NMR library see ref. 9). We typically screen 800–1300 compounds (generally by thermal shift assay), from which we identify 20–100 fragments for further screening by ligand-based NMR. Structure–activity relationship (SAR) analysis is then conducted around the NMR hits by buying or making similar compounds. At Astex, similar sized libraries are screened by X-ray crystallography and NMR spectroscopy.[10] In large pharmaceutical companies there has been a tendency to screen larger fragment libraries, *e.g.*, by surface plasmon resonance (SPR); an advantage being that SARs will emerge from the initial screen. A consideration in deciding how many fragments to screen is the expected hit rate. This can be very high (20%+) in the case of a kinase screen or vanishing low when looking for compounds binding to a surface site to disrupt a protein–protein interaction.[11]

Fragment libraries have been designed based on natural products *e.g.*, "fragments of life" libraries including highly soluble metabolites, natural product derivatives and biaryl molecules designed to mimic the architectural motifs of proteins (α, β and γ turns).[12] This library was tested against leukotriene A4 hydrolase by screening about 200 fragments using X-ray crystallography. Thirteen hits were obtained of which 11 bound to the active site, while 2 biaryls were found binding in a shallow pocket on the surface of the protein.

In another report, the Waldmann group reported a chemoinformatic analysis of more than 180 000 natural product structures that was used to obtain about 750 000 fragments of different M_w and properties rich in sp^3 centres.[13] This virtual library was then filtered according to an HTS filter and Rule of 3 derived parameters and re-ordered in 2000 clusters. Part of the library (193 fragments) was tested on p38α MAP kinase using biochemical assays, leading to 12 hits being identified with potencies between 6 and 0.4 mM and ligand efficiencies (LEs) ranging from 0.19 to 0.28. Most of the hits predictably bound to the hinge region of the kinase ATP site, but interestingly there were some that bound to an allosteric site, as a novel class of type-III inhibitors.

The library was also tested on a less tractable class of targets, the tyrosine phosphatases, including *Mycobacterium tuberculosis* protein tyrosine phosphatase A and B. The best hits had an IC_{50} of <100 μM and LEs of almost 0.50. However, despite the growing interest in 3D fragments there is no compelling evidence yet that this will result in more or better hits.[14]

1.3 How We Identify Fragments

Given their size, fragments generally bind very weakly (100 μM to 10 mM) to biological targets. Consequently, there is a particular emphasis in the biophysical techniques to detect binding.[15] These include NMR (ligand- and protein-based), X-ray crystallography, SPR, isothermal titration calorimetry (ITC), thermal shift, and native mass spectrometry. These different approaches mean that the start of a fragment-based drug discovery (FBDD) project can be quite different in different labs, but the end game is generally similar. The screening approach used is partly dictated by available facilities, and partly by organisational constraints, although in all the cases, a single technique is rarely used in isolation. Typically a combination of techniques is employed in order to minimise false positives and false negatives and thus provide reliable chemical starting points for subsequent design and synthesis.

Virtual screening using *in silico* docking methods has been used, but usually in support of subsequent biochemical and biophysical assays. The different binding modes of simple fragments may only exhibit small differences in free energy of binding compared to larger compounds, so the inaccuracy inherent in scoring functions does not allow a reliable ranking of fragment-binding modes.[16]

In our lab, the first screen generally employs the Sypro Orange fluorescence-based thermal shift screen (Figure 1.1);[17] where the selected hits are those giving the largest increase in thermal stability of the protein. This offers a quick way to identify 50–100 compounds to proceed on to NMR screening. However, the shifts are generally small and the screen should be repeated to improve the reliability of the hits. This is a very simple application of this screen, and it is clear that in more complex systems, *e.g.*, when studying protein–protein interactions, good hits can give even significant negative shifts. A recent study on Checkpoint kinase 2 used the AlphaScreen kinase biochemical assay and a thermal shift screen.[18] A good correlation was found for compounds which were hits in both these orthogonal screens. Such high-concentration biochemical screens can give a high frequency of false positives so it is important to do a counter-screen to check for interfering compounds.

In many companies, SPR has become the primary screening technique to identifying fragment hits. It has advantages of ease of setting up the assay, relatively high throughput, and requires low amounts of protein compared to other techniques (*e.g.*, NMR and X-ray) used. The methodology is likely to become more popular as the cost of the instrumentation decreases.[19] A later chapter by Giannetti is dedicated to this approach.

NMR techniques[1,20,21] are among the most powerful and widely used to detect binding of fragments to target proteins. Ligand-based and protein-based

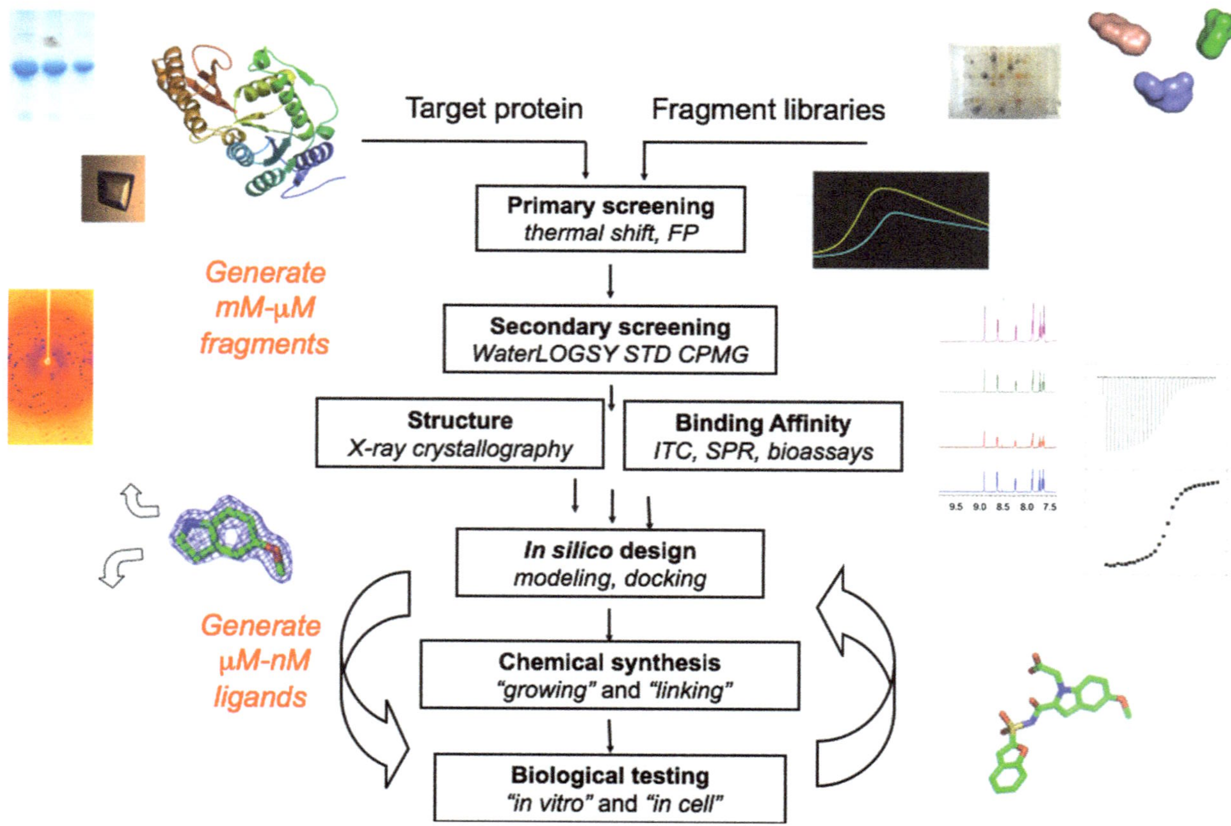

Figure 1.1 Fragment screening cascade.

methods each have advantages and limitations, and can complement each other. However it is the ligand-based methods that have found most general application. In our lab we use a combination of three such techniques (described in detail in Krimm's chapter): WaterLOGSY, STD and CPMG. WaterLOGSY detects binding through transfer of magnetisation from the water molecules to the ligand bound to the protein. STD uses a pulse to saturate signals from methyl groups in the protein. The disturbance in magnetisation is then transferred through an NOE to the bound ligand, with the resulting difference spectrum only showing a peak of the ligand when this binds to the protein target. Finally CPMG, is a relaxation-time-edited NMR experiment that exploits differences in transverse relaxation time (T_2). Proteins (and bound ligands) have a small T_2 while free ligands have a large T_2. Thus monitoring T_2, binding can be detected when the signal of the ligand decreases.

In a given screen, some fragments show binding by all three techniques (*e.g.*, Figure 1.2), some by only two, some by one. We progress preferentially those giving a positive outcome from all three experiments. An example of the power of fragment screening by NMR was in a study on Pin1. After a million-compound HTS failed to give any hits, an NMR screen of 1200 fragments using the three 1D ligand-based techniques identified 5 hits from 4 different series and yielded a lead compound of 260 nM IC_{50}.[22] Astex used a multistage NMR screen

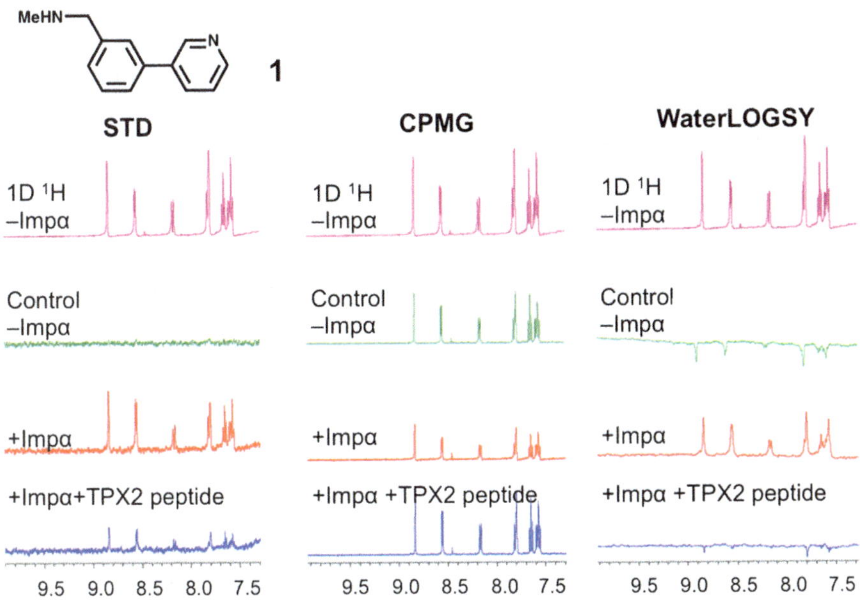

Figure 1.2 Ligand-observed NMR techniques STD, CPMG and WaterLOGSY illustrating the binding of fragment 1 to importin-α (Impα). NMR 1D spectrum of fragment 1 (magenta). Control –Impα (green): fragment in the absence of the protein. +Impα (red): fragment in the presence of the protein, changes in signal indicating binding. +Impα +TPX2 peptide (blue): displacement study by addition of the TPX2 peptide.

against HSP90 to select 125 compounds from a 1600 fragment library for X-ray crystallography. Fragments that bound to the protein were initially identified in a WaterLOGSY experiment. The presence of 200 μM ADP made it possible to identify those compounds that bound to the nucleotide site from their effect on the ADP signal. When Mg^{2+} was then added, this increased the affinity of HSP90 for ADP 20-fold, and led to displacement of those fragments.[23]

There are more sophisticated NMR techniques that can give additional information. For example, we have used the inter-ligand Overhauser effect (ILOE). This can detect two fragments binding simultaneously to the protein, and in close proximity to each other, through the analysis of their negatives NOEs. This is particularly useful if the intention is to increase potency through linking or merging fragments binding in the same cavity.[24]

Fragment screening using protein-based NMR is a powerful approach that has been used to great effect in some laboratories.[1] To be most effective, it does require a pre-assignment of the protein NMR spectrum, involving multiply-labelled protein. It is generally applied to proteins of molecular weight below 30 kDa. An elegant example using this technique has been described recently where two scaffolds were identified to bind in different regions of the Mcl-1 protein through 2D ^{1}H–^{15}N-HMQC. Using a merging approach, Mcl-1 inhibitors were developed with an equilibrium dissociation constant (K_D) < 100 nM.[25]

Having confirmed fragment binding by NMR, we usually attempt to quantify the interaction using ITC. Specifically we are interested in the K_D and stoichiometry, paying relatively little attention to the enthalpy and entropy of binding – which are generally difficult to explain (although that does not stop some authors attempting to do so). The approach does require a significant amount of protein, and someone skilled in the art. Even then, some proteins or chemical series are not amenable to ITC, often due to low heats of binding. From the K_D we derive the LE.[26] The LE continues to be a useful metric although increasingly surpassed by parameters that account for lipophilicity.[27] If the fragment library has been assembled thoughtfully, there should also be analogues of the fragment that are commercially available that can be used to rapidly scope out SARs.

The use of X-ray crystallography for fragment screening, in spite of its early use *e.g.*, by Astex[28] has not been adopted generally. This was initially a surprise, as putting an X-ray fragment screen early in a screening cascade is arguably the most powerful way to drive the process. There may be many reasons for the reluctance to adopt this approach, some of which are probably as much to do with organisational issues, as they are to do with science. Other issues will be around the cost of X-ray equipment and associated robotics, and the specialised skills required to crystallise new proteins in a form suitable for fragment screening. In our academic group we try to incorporate X-ray crystallography as soon as we can, but in practice this is usually after we have hits validated by NMR spectroscopy.

For us, it is vital that we do get X-ray structures of our fragments bound to the target protein. This is the major go/no go gateway for project progression. We consider that structural information on the precise mode of fragment binding is needed in order to inform the subsequent chemical elaboration of the fragment. Several projects have been dropped at this point, because either we were not able to get suitable, soakable crystals of the target protein, or because even with these in hand, we could not identify fragments bound to these structures. This consideration highlights one of the key issues for success in this area within academic teams; the need for very strong collaborations between different disciplines. This includes biophysical scientists, structural biologists and chemists, with each project team having a major investment in the project. Getting the first structure of a fragment–protein complex is often the most difficult aspect of the project but is always a significant event for the team.

Ideally the chemist would like to have crystal structures of several fragments bound to the protein of interest, and generally also fragments bound at several sites (even if this can complicate the understanding of the K_D values). Many early studies focused on protein kinases, which led to some level of predictability about the site where fragments would bind.[29] At the other end of the spectrum, our studies on fragment binding to cytochrome P450s have identified many different binding sites ranging from the heme iron across to the protein surface (Figure 1.3).[30] Where and how fragments bind has a major influence on the approach the chemists can take to the subsequent chemical elaboration. The early papers on FBDD against kinases mainly described fragment growing from a common fragment-binding site (usually the ATP binding site). The availability of structures of fragments bound at several sites in a protein gives options for fragment merging,[31] linking or growing.[32]

Fragment growing has been the major method of fragment elaboration described in the literature, perhaps because of the initial focus on protein kinases. A good fragment should offer immediate sites for elaboration. Initial elaboration should be incremental and not too ambitious, and should

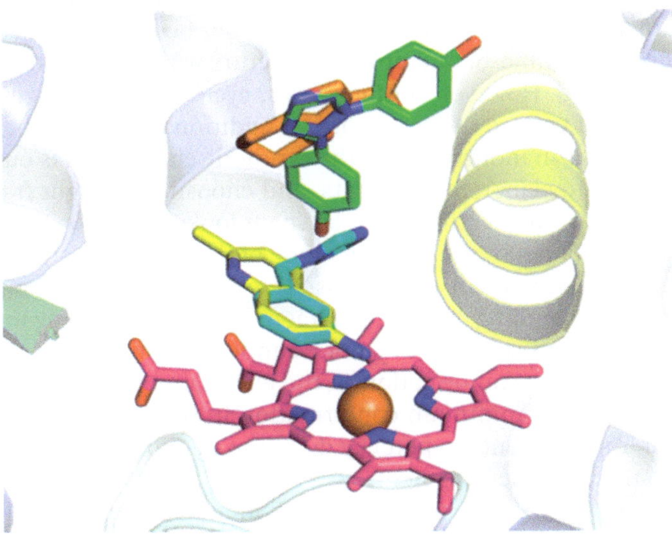

Figure 1.3 Overlay of fragments bound into the active site of *M. tuberculosis* CYP121.[30]

be guided by appropriate structural information and docking studies. Most fragments located by X-ray crystallography will have LEs of 0.3 or greater, and an aspirational aim should be to maintain this level of LE throughout the iterative synthetic cycles. This is challenging, and sometimes seen as too demanding, *e.g.*, adding a phenyl ring requires an approximate 20-fold increase in potency in order to maintain an LE = 0.3. Such incremental elaboration is perhaps not helped if the synthetic chemistry is done through outsourcing when multiple analogues may be ordered at once.

In elaborating a fragment, the aim is to pick up new interactions *e.g.*, π-stacking interactions, hydrogen bonds *etc.*, as effectively as possible using any information that can be inferred from SARs around the fragment, while also avoiding obvious clashes with the protein surface. Fragment elaboration is an iterative process. Successive molecules are designed and synthesised, and their structure bound to the target protein is solved. Looking at a new co-crystal structure may do nothing more than confirm the predicted binding, but in some cases the binding may lead to changes in the protein conformation. In this case, new opportunities for structure-based design that were not obvious in the previous structures, may present themselves.

The iterative nature of fragment elaboration is potentially a strength of the process. It gives the chemist multiple opportunities to tailor the properties of the molecules being assembled. Fragments are typically small and highly water-soluble, and as such represent a good starting point in terms of physico-chemical properties. The temptation at this stage is to seek a quick increase in binding by adding mainly hydrophobic groups. This should be avoided as it leads to molecules with low solubility, less selectivity, and, ultimately, poor drug-like properties. The elaboration process does depend upon close co-operation between the chemist and structural biologist. Good cross-discipline relationships are essential to minimise cycle times and ensure momentum within the project. In our best cases, structural biology can obtain a ligand-bound protein structure within 24 hours of a ligand being synthesised. However in other cases, cycle time will be governed by the frequency of trips to the synchrotron. Additional delay can be imposed on this cycle time in industry when the synthetic elaboration is outsourced.

It is not easy to elaborate fragments, and choosing the right fragment to elaborate is important, so time must be spent looking closely at what opportunities are presented by the binding mode of the fragment, and combining that with some assessment of synthetic tractability. If there is a perceived problem with the binding mode of the fragment, then correcting that should lead to quick increases in potency. For example, affinity has been gained by stabilising the active conformation of the elaborated fragment, ensuring it is bound close to its minimum conformation in free solution.[23,31] However if expected improvements in potency do not materialise, it may be better to start again from another fragment.

Sometimes, extensive synthesis aimed at adding an extra hydrogen-bonding interaction does not reward the chemist with a significant increase in potency. Conversely, innocuous changes can lead to significant jumps in potency. Our inability to build in potency in an entirely predictable way, in part reflects the limitations of the models we are using *e.g.*, the target protein is not a rigid structure, it is a dynamic and in a way that can change as ligands evolve. A study of 98 pairs of structures (apoprotein and cognate ligand complexes), showed that 57 of them had conformational differences.[33]

Often structural information will reveal that several fragments can bind to the same part of the protein. These overlapping fragments can suggest new merged structures. We have used fragment merging extensively in our development of inhibitors of cytochrome P450 inhibitors, discussed in some detail in the later chapter on antibacterial compounds.[30,31] Finally, fragment linking is intellectually appealing, as it appears to offer access to the binding energy of two fragments that should theoretically be enough to give nM potency in the linked molecule. We have had projects where some level of additivity can be achieved. In an early study on thrombin inhibitors, an X-ray screen identified a chlorophenyl fragment that bound in the S1 pocket (IC_{50} = 330 μM). This was linked to a novel sulphonamide ligand (IC_{50} = 12 μM), that straddled the S2–S4 sites, through a single methylene group, to give a 1.4 nM inhibitor (Figure 1.4).[34]

In other study on pantothenate synthetase, linking a 1 mM fragment (**5**) and a 1100 μM fragment (**1**) gave a 2 μM inhibitor (**8**) (Figure 1.5).[32] Although

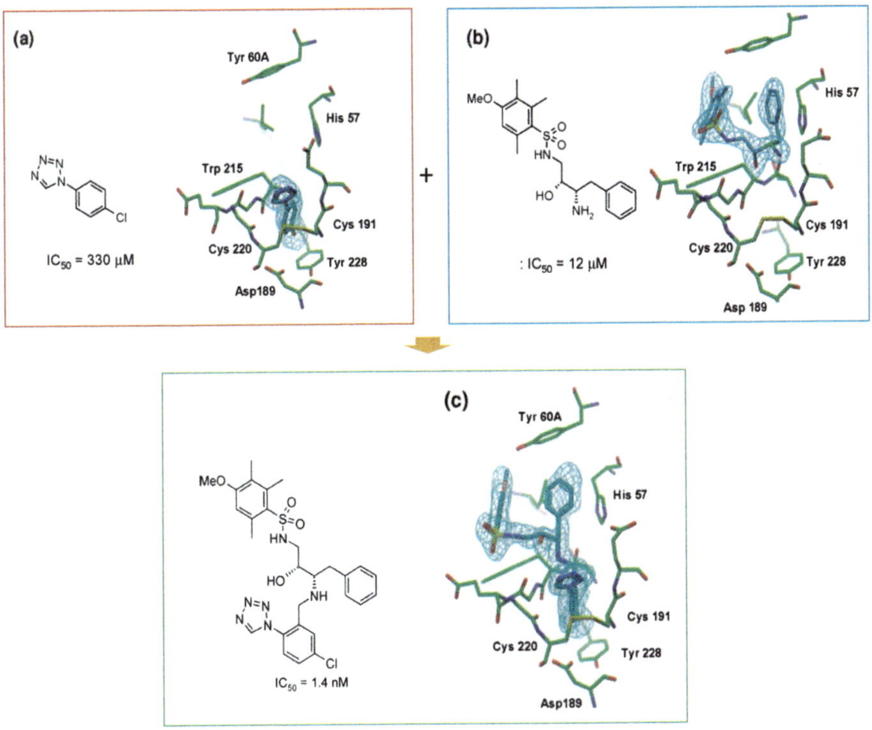

Figure 1.4 Thrombin inhibitors derived from fragment linking. (a) Chlorobenzene fragment (IC_{50} = 330 μM) bound into the S1 pocket of thrombin, (b) sulfonamide (IC_{50} = 12 μM) straddling S2–S4 sites, (c) linked compound (IC_{50} = 1.4 nM).[34]

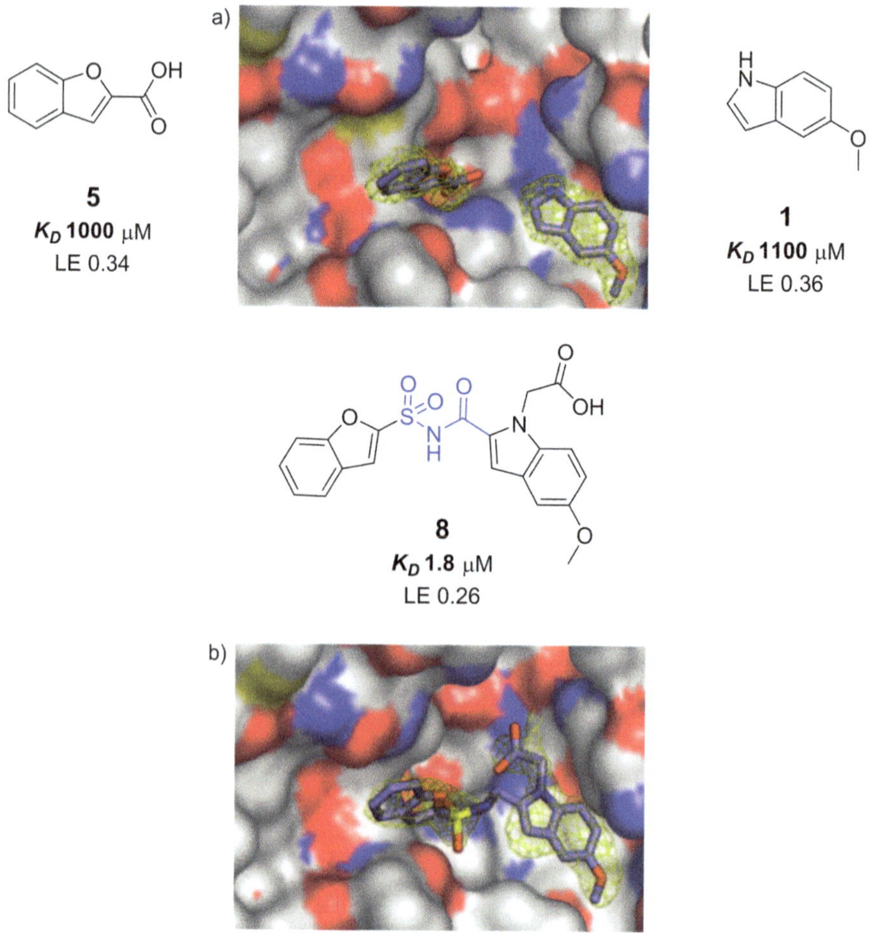

Figure 1.5 Fragment linking to develop a pantothenate synthetase inhibitor. (a) Benzofurane **5** (K_D = 1 mM) and indole **1** (K_D = 1100 μM) bind simultaneously at the active site, (b) they have been linked to give **8** (K_D = 1.8 μM).[32]

this looks impressive at first sight, it actually gives a product with lower LE than the original fragments. Interestingly, group analysis[35] of this and related inhibitors showed that the part of the molecule derived from fragment **5** (originally an acid, but now incorporated into an acyl sulphonamide) was not making a great contribution to binding. This is presumably because the carboxylate group that has been modified was the source of most of the original binding.

1.4 Selectivity

One question with fragment-based approaches is when is selectivity to be expected? We have addressed this issue in our studies on cytochrome P450s (CYPs) from *M. tuberculosis*. There are 20 CYPs in *M. tuberculosis* compared to none in *E. coli*. Some are involved in key metabolic processes such as steroid degradation, whereas the role of others is not understood, or unknown. A key point is that the active sites of all the CYPs are dominated by a heme group and the chemistry they catalyse is similar. It was therefore a major surprise when similar fragment screens of CYP121 and CYP125 identified 26 hits for the former and 9 for the latter, with only 1 hit in common.[30] Another surprise from the crystallography was that fragments bound to CYP121 at several different places in the active site, and effectively defined a trajectory from solution to the heme iron (Figure 1.3).

1.5 Applications of Fragment-Based Methods

1.5.1 Assigning Function

Encouraged by the selectivity observed with the fragment screening of CYPs (see above), we wondered if fragment screening could be used to provide clues about the natural substrate of the functionally uncharacterised *M. tuberculosis* CYP126.[36] In 2011 it was reported that there was no functional data for at least a quarter of gene products in *M. tuberculosis*[37] and it is clear new approaches to functional annotation are needed.

Our approach was to first assemble a relevant fragment library. There were many possible ways to do this, but we chose to base ours on ligands found in the many crystal structures of CYPs in the Protein Data Bank. In principle this would enable us to cross-reference hits back to structural information. The composition of the resulting 63-member biofragment library was relatively rich in sp^3 centres. Screening the tailored library gave a far higher hit rate (14%) than screening our drug-like fragment library (4%). Fragments were identified that bound with substrate-like (type I) and inhibitor-like (type II) binding modes. A chemical-fingerprint-based model was built from these hits and used to screen a virtual *M. tuberculosis* metabolome. The outcome was the identification of a chlorophenol hit (K_D = 150 μM) binding in a substrate-like manner. CYPs are known to mediate the de-halogenation of such compounds.[38]

1.5.2 Assessing Ligandability

To develop a small-molecule drug against a particular protein, it is obvious that there needs to be a site where such a molecule can bind. The hit rates in fragment screening can give us information about how likely it is to identify and exploit such a site. In a study by AstraZeneca, 36 different programs distributed across several target classes were assessed.[39] A fragment screen and a HTS were run for each program, and a ligandability score was assigned on the basis of the number of hits, their affinity and their diversity. Of the 12 programs that scored ‘low’ in terms of ligandability, none was successful in finding a hit through HTS screening. However, four programs that were failed in HTS were progressed to the ‘hit-to-lead’ phase on the basis of fragment screening.

1.5.3 Speeding Up Chemical Validation

We are interested in developing inhibitors of enzymes in *M. tuberculosis*. This is extremely challenging, primarily because it is not clear what good targets are, and this is further complicated by the difficulty in getting compounds across the thick lipidic membrane of the bacterium. Our original assumption was that we would need to make sub μM enzyme inhibitors before we would see any activity against cells. Indeed our low μM inhibitors of pantothenate synthetase showed no activity against the wild-type strain.[40] However when the same compounds were tested against an *M. tuberculosis* strain where the level of the target enzyme was selectively attenuated, the compounds did kill tuberculosis in what was shown to be an on-target effect. This use of a conditional mutant strain combined with a fragment approach led to chemical validation of the target.

1.6 Targeting Protein–Protein Interactions and Beyond

The hope is that FBDD can be used on targets where HTS has struggled. An obvious example is drugging protein–protein interactions (PPIs).[41] PPIs represent a widely heterogeneous range of targets. Some will prove very difficult to disrupt with small molecules. But other PPIs, *e.g.*, where there is a preformed groove to bind an α-helix, appear tractable. At present it is assumed that it may be necessary to make slightly larger (>500 Da) and less ligand-efficient molecules but this remains to be fully explored. A major challenge is posed by surface malleability; changes in the surface topography can be very dramatic. Computer simulations of Bcl-xL have been used to show that pockets can appear on the nanosecond timescale on surfaces that appear relatively featureless.[42] We have used a benzene probe in molecular dynamics simulations to expand the accessible conformational space sampled on a protein surface, enabling us to see an otherwise cryptic pocket. Based on the benzene sampling we were then able to design a novel ligand and subsequently confirm its binding by X-ray crystallography.[43]

Our first significant foray into disrupting PPIs was an attempt to disrupt the interaction between the tumour suppressor peptide BRCA2 and the recombination enzyme RAD51.[11] The key interaction site on RAD51 is also part of the oligomerisation site for RAD51 self-assembly. It features two pockets, one binds the aromatic side-chain of a phenylalanine, the other the methyl group of an alanine in an FXXA tetrapeptide. The technical hurdles to this project were significant, not least the fact that RAD51 is oligomeric and it has not proved possible to crystallise an unliganded monomeric form. This forced us to use a surrogate system RadA, the orthologue present in the thermophilic organism *Pyrococcus furiosus*. RadA was then systematically humanised by replacing six amino acids in the area of the interaction between the two proteins, generating a mutant called MAYSAM RadA. A collection of 1249 fragments was screened against this protein in a thermal shift experiment.

This led to the identification of only two hits, both indoles. One of these compounds was then used as a competitive probe in a NMR screen, which identified other scaffolds, all of which bound into the Phe pocket (Figure 1.6).

Other studies have been less successful. Despite extensive fragment screening against the polo-box domain (PBD) of polo-like Kinase 1, no fragments were detected. Interestingly we even failed to detect binding of short phosphothreonine peptides. However, by considering the crystal packing interactions in crystals structures of the PBD, it was possible to identify a novel surface pocket.[44] It was possible to probe binding into this pocket by delivering molecules tethered onto binding peptides.[45] A preliminary indication of binding in the pocket was obtained by comparing the thermal shift of the acylated peptide against the protein and against a double mutant where the pocket had been removed.

Studies to develop inhibitors of HCV protease led to an interesting example of disrupting an intramolecular PPI, and uncovered a previously unrecognized allosteric control mechanism. X-Ray crystallographic fragment screening of the bifunctional HCV NS3 protein, that includes the protease and helicase domains, identified a conserved allosteric site at the interface

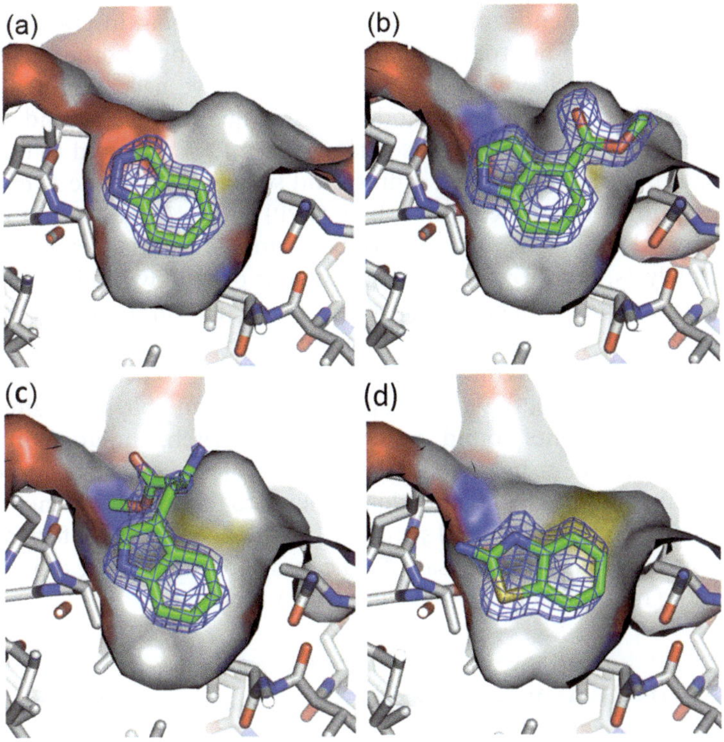

Figure 1.6 Side-view through the Phe pocket of X-ray structures of fragments bound to MAYSAM RadA: (a) indazole; (b) 4-methylester indole; (c) L-methylester tryptophan; and (d) 2-aminobenzothiazole.[11]

of the two domains. Initial weakly binding fragments were optimised to fill the pocket and so stabilise a closed auto-inhibited conformational state. These compounds exhibited potent antiviral activity. Several lines of evidence supported the proposed mechanism of action confirming that compounds inhibited the protease activity by an allosteric mechanism. In particular, sequencing of viral resistant replicons, generated by applying selective pressure form the compound, identified mutations in the interfacial allosteric site.[46]

Finally two important and quite distinct applications: in an interesting play on thermal stability, Fersht screened for ligands that bind to a surface cavity formed by the oncogenic Y220C mutation in the tumour suppressor p53, with the aim of stabilising and thereby rescuing the protein.[47] Perhaps the most challenging example of using a fragment approach against a protein target is in screening against GPCRs, where breakthroughs in the structural biology of these systems has opened them up to FBDD.[48] This is exciting work in progress.

1.7 Targeting Oligonucleotides

Fragment-based methods have primarily been focused on drug discovery, but as the approach depends on the biophysics of a fragment interacting with a macromolecule, the approach clearly has applications in other areas, *e.g.*, ligands for oligonucleotides. At Cambridge we have used fragment-based approaches to identify novel ligands for DNA quadruplexes[49] and also for riboswitches.[50] Riboswitches are sequences of mRNA that fold to bind specific small molecules with high affinity and selectivity, *e.g.*, the *E. coli* ThiM riboswitch binds thiamine diphosphate (TPP) with a K_D of 200 nM but thiamine much less well, K_D 1.5 μM. We performed an initial screen by equilibrium dialysis, monitoring the displacement of weakly binding thiamine from the riboswitch using cocktails of fragments. This approach worked surprisingly well, identifying cocktails that were then easily de-convoluted. The fragment hits were then confirmed by NMR and ITC. Some were remarkably ligand efficient (LE > 0.5). We have subsequently been able to obtain crystal structures of fragments bound to the ThiM riboswitch, and are using these to design more potent ligands. These crystal structures have also provided important insight into the ligand requirements to fold the apo structure into a much more compact liganded structure (Figure 1.7).[51] The conformation of G72, which unfolds to compensate for the loss of binding to the diphosphate group of the natural ligand, was a complete surprise.

1.8 Concluding Remarks

Our academic studies on fragment-based methods have been wide ranging and intellectually stimulating. We have developed our own approach, with its intrinsic assumptions and methodological idiosyncrasies. To the uninitiated, the key features of fragment-based methods are front-loaded; how to make

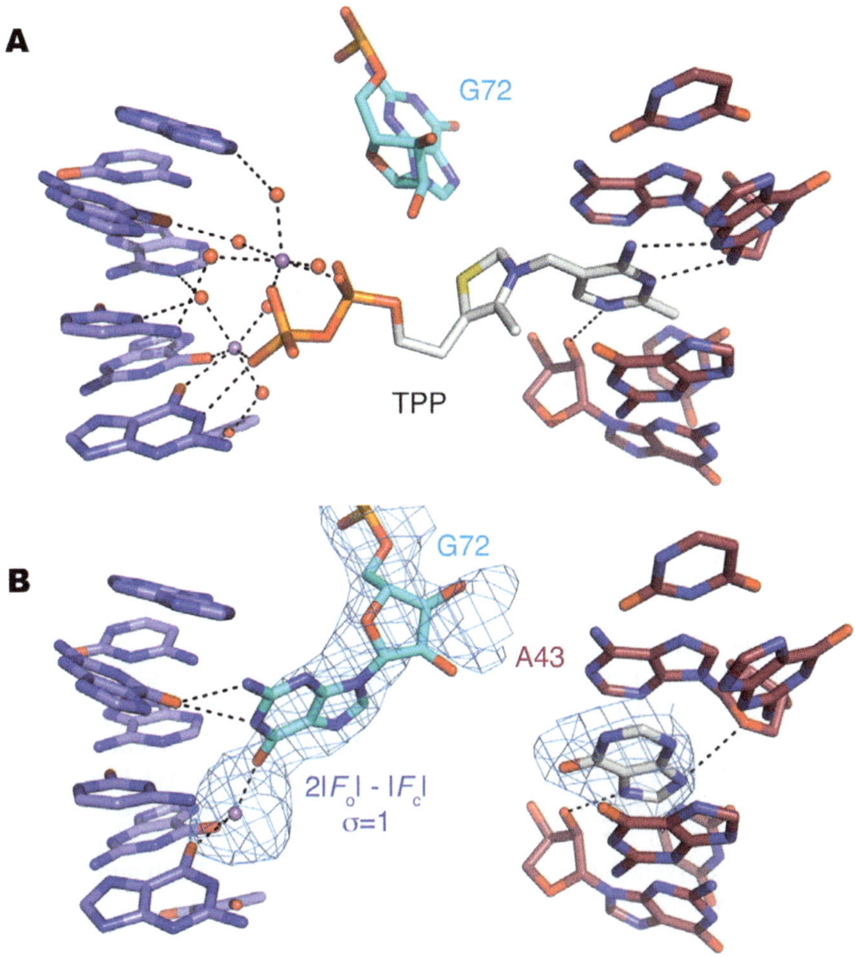

Figure 1.7 (A) X-ray crystal structure of thiamine diphosphate (TPP) complexed with the ThiM riboswitch. Metal ions are shown in purple and water molecules in red. (B) Structure of a purine fragment (grey) complexed with the ThiM riboswitch. Note the large conformational shift of G72.[51]

the library, how to do the screening? These are important questions, but for the practitioner, they are less important than might be imagined. The library and the screening have only to deliver a handful of hits, whose binding can be characterised, ideally by X-ray crystallography. The main challenges are downstream. Which fragments to elaborate? How to assess the importance of potential interactions, and then translate these into a synthetic strategy? In the commercial world there are issues about how to prioritise fragment-based approaches so they are not a method of last resort, and to ensure that they are executed by experienced practitioners. It is an approach that is intellectually satisfying and can deliver, but as always, detail is important.

References

1. S. B. Shuker, P. J. Hajduk, R. P. Meadows and S. W. Fesik, *Science*, 1996, **274**, 1531–1534.
2. T. L. Blundell, H. Jhoti and C. Abell, *Nat. Rev. Drug Discovery*, 2002, **1**, 45–54.
3. G. Bollag, J. Tsai, J. Zhang, C. Zhang, P. Ibrahim, K. Nolop and P. Hirth, *Nat. Rev. Drug Discovery*, 2012, **11**, 873–886.
4. J. Fejzo, C. A. Lepre, J. W. Peng, G. W. Bemis, M. A. Ajay Murcko and J. M. Moore, *Chem. Biol.*, 1999, **6**, 755–769.
5. N. Blomberg, D. A. Cosgrove, P. W. Kenny and K. Kolmodin, *J. Comput.-Aided Mol. Des.*, 2009, **23**, 513–525.
6. J. B. Baell, *Future Med. Chem.*, 2010, **2**, 1529–1546.
7. S. G. Kathman, Z. Xu and A. V. Statsyuk, *J. Med. Chem.*, 2014, **57**, 4969–4974.
8. M. Congreve, R. Carr, C. Murray and H. Jhoti, *Drug Discovery Today*, 2003, **8**, 876–877.
9. B. C. Doak, A. C. J. Morton, J. S. Simpson and M. J. Scanlon, *Aust. J. Chem.*, 2013, **66**, 1465–1472.
10. C. W. Murray, M. L. Verdonk and D. C. Rees, *Trends Pharmacol. Sci.*, 2012, **33**, 224–232.
11. D. E. Scott, M. T. Ehebauer, T. Pukala, M. Marsh, T. L. Blundell, A. R. Venkitaraman, C. Abell and M. Hyvonen, *ChemBioChem*, 2013, **14**, 332–342.
12. D. R. Davies, B. Mamat, O. T. Magnusson, J. Christensen, M. H. Haraldsson, R. Mishra, B. Pease, E. Hansen, J. Singh, D. Zembower, H. Kim, A. S. Kiselyov, A. B. Burgin, M. E. Gurney and L. J. Stewart, *J. Med. Chem.*, 2009, **52**, 4694–4715.
13. B. Over, S. Wetzel, C. Grutter, Y. Nakai, S. Renner, D. Rauh and H. Waldmann, *Nat. Chem.*, 2013, **5**, 21–28.
14. H. Jhoti, G. Williams, D. C. Rees and C. W. Murray, *Nat. Rev. Drug Discovery*, 2013, **12**, 644.
15. E. H. Mashalidis, P. Śledź, S. Lang and C. Abell, *Nat. Protoc.*, 2013, **8**, 2309–2324.
16. M. L. Verdonk, I. Giangreco, R. J. Hall, O. Korb, P. N. Mortenson and C. W. Murray, *J. Med. Chem.*, 2011, **54**, 5422–5431.
17. M. C. Lo, A. Aulabaugh, G. Jin, R. Cowling, J. Bard, M. Malamas and G. Ellestad, *Anal. Biochem.*, 2004, **332**, 153–159.
18. M. C. Silva-Santisteban, I. M. Westwood, K. Boxall, N. Brown, S. Peacock, C. McAndrew, E. Barrie, M. Richards, A. Mirza, A. W. Oliver, R. Burke, S. Hoelder, K. Jones, G. W. Aherne, J. Blagg, I. Collins, M. D. Garrett and R. L. M. van Montfort, *PLoS One*, 2013, **8**, e65689.
19. I. Navratilova and A. L. Hopkins, *ACS Med. Chem. Lett.*, 2010, **1**, 44–48.
20. B. Meyer and T. Peters, *Angew. Chem., Int. Ed.*, 2003, **42**, 864–890.
21. S. Vanwetswinkel, R. J. Heetebrij, J. van Duynhoven, J. G. Hollander, D. V. Filippov, P. J. Hajduk and G. Siegal, *Chem. Biol.*, 2005, **12**, 207–216.

22. A. J. Potter, S. Ray, L. Gueritz, C. L. Nunns, C. J. Bryant, S. F. Scrace, N. Matassova, L. Baker, P. Dokurno, D. A. Robinson, A. E. Surgenor, B. Davis, J. B. Murray, C. M. Richardson and J. D. Moore, *Bioorg. Med. Chem. Lett.*, 2010, **20**, 586–590.
23. C. W. Murray, M. G. Carr, O. Callaghan, G. Chessari, M. Congreve, S. Cowan, J. E. Coyle, R. Downham, E. Figueroa, M. Frederickson, B. Graham, R. McMenamin, M. A. O'Brien, S. Patel, T. R. Phillips, G. Williams, A. J. Woodhead and A. J.-A. Woolford, *J. Med. Chem.*, 2010, **53**, 5942–5955.
24. P. Śledź, H. L. Silvestre, A. W. Hung, A. Ciulli, T. L. Blundell and C. Abell, *J. Am. Chem. Soc.*, 2010, **132**, 4544–4545.
25. A. Friberg, D. Vigil, B. Zhao, R. N. Daniels, J. P. Burke, P. M. Garcia-Barrantes, D. Camper, B. A. Chauder, T. Lee, E. T. Olejniczak and S. W. Fesik, *J. Med. Chem.*, 2013, **56**, 15–30.
26. A. L. Hopkins, C. R. Groom and A. Alex, *Drug Discovery Today*, 2004, **9**, 430–431.
27. A. L. Hopkins, G. M. Keserü, P. D. Leeson, D. C. Rees and C. H. Reynolds, *Nat. Rev. Drug Discovery*, 2014, **13**, 105–121.
28. M. J. Hartshorn, C. W. Murray, A. Cleasby, M. Frederickson, I. J. Tickle and H. Jhoti, *J. Med. Chem.*, 2005, **48**, 403–413.
29. R. Urich, G. Wishart, M. Kiczun, A. Richters, N. Tidten-Luksch, D. Rauh, B. Sherborne, P. G. Wyatt and R. Brenk, *Chem. Biol.*, 2013, **8**, 1044–1052.
30. S. A. Hudson, K. J. McLean, S. Surade, Y.-Q. Yang, D. Leys, A. Ciulli, A. W. Munro and C. Abell, *Angew. Chem., Int. Ed.*, 2012, **51**, 9311–9316.
31. S. A. Hudson, S. Surade, A. G. Coyne, K. J. McLean, D. Leys, A. W. Munro and C. Abell, *ChemMedChem*, 2013, **8**, 1451–1456.
32. A. W. Hung, H. L. Silvestre, S. Wen, A. Ciulli, T. L. Blundell and C. Abell, *Angew. Chem., Int. Ed.*, 2009, **48**, 8452–8456.
33. K. Gunasekaran and R. Nussinov, *J. Mol. Biol.*, 2007, **365**, 257–273.
34. N. Howard, C. Abell, W. Blakemore, R. Carr, G. Chessari, M. Congreve, S. Howard, H. Jhoti, C. W. Murray, L. C. A. Seavers and R. L. M. van Montfort, *J. Med. Chem.*, 2006, **49**, 1346–1355.
35. M. L. Verdonk and D. C. Rees, *ChemMedChem*, 2008, **3**, 1179–1180.
36. S. A. Hudson, E. H. Mashalidis, A. Bender, K. J. McLean, A. W. Munro and C. Abell, *ChemBioChem*, 2014, **15**, 549–555.
37. J. M. Lew, A. Kapopoulou, L. M. Jones and S. T. Cole, *Tuberculosis*, 2011, **91**, 1–7.
38. J. C. Hackett, T. T. Sanan and C. M. Hadad, *Biochemistry*, 2007, **46**, 5924–5940.
39. F. N. B. Edfeldt, R. H. A. Folmer and A. L. Breeze, *Drug Discovery Today*, 2011, **16**, 284–287.
40. G. L. Abrahams, A. Kumar, S. Savvi, A. W. Hung, S. Wen, C. Abell, C. E. Barry, D. R. Sherman, H. I. M. Boshoff and V. Mizrahi, *Chem. Biol.*, 2012, **19**, 844–854.
41. J. A. Wells and C. L. McClendon, *Nature*, 2007, **450**, 1001–1009.
42. S. Eyrisch and V. Helms, *J. Med. Chem.*, 2007, **50**, 3457–3464.

43. Y. S. Tan, P. Śledź, S. Lang, D. R. Spring, C. Abell and R. B. Best, *Angew. Chem., Int. Ed.*, 2012, **51**, 10078–10081.
44. P. Śledź, C. J. Stubbs, S. Lang, Y. Yang, G. J. McKenzie, A. R. Venkitaraman, M. Hyvönen and C. Abell, *Angew. Chem., Int. Ed.*, 2011, **50**, 4003–4006.
45. P. Śledź, S. Lang, C. J. Stubbs and C. Abell, *Angew. Chem., Int. Ed.*, 2012, **51**, 7680–7683.
46. S. M. Saalau-Bethell, A. J. Woodhead, G. Chessari, M. G. Carr, J. Coyle, B. Graham, S. D. Hiscock, C. W. Murray, P. Pathuri, S. J. Rich, C. J. Richardson, P. A. Williams and H. Jhoti, *Nat. Chem. Biol.*, 2012, **8**, 920–925.
47. N. Basse, J. L. Kaar, G. Settanni, A. C. Joerger, T. J. Rutherford and A. R. Fersht, *Chem. Biol.*, 2010, **17**, 46–56.
48. J. A. Christopher, J. Brown, A. S. Doré, J. C. Errey, M. Koglin, F. H. Marshall, D. G. Myszka, R. L. Rich, C. G. Tate, B. Tehan, T. Warne and M. Congreve, *J. Med. Chem.*, 2013, **56**, 3446–3455.
49. H. R. Nasiri, N. M. Bell, K. I. E. McLuckie, J. Husby, C. Abell, S. Neidle and S. Balasubramanian, *Chem. Commun.*, 2014, **50**, 1704–1707.
50. L. Chen, E. Cressina, F. J. Leeper, A. G. Smith and C. Abell, *ACS Chem. Biol.*, 2010, **5**, 355–358.
51. K. D. Warner, P. Homan, K. M. Weeks, A. G. Smith, C. Abell and A. R. Ferré-D'Amaré, *Chem. Biol.*, 2014, **21**, 591–595.

CHAPTER 2

Getting the Most Value from Your Screens: Advances in Hardware, Software, and Methodologies to Enhance Surface Plasmon Resonance Based Fragment Screening and Hit-to-Lead Support

ANTHONY M. GIANNETTI*[a], HOUSTON N. GILBERT[b], DONALD P. HUDDLER[c], MAC REITER[d], CHRIS STRANDE[d], KEITH E. PITTS[a], AND BRANDON J. BRAVO[a]

[a]Department of Biochemical and Cellular Pharmacology, Genentech, Inc., 1 DNA Way, South San Francisco, CA 94080, USA; [b]Department of Biostatistics, Genentech, Inc., 1 DNA Way, South San Francisco, CA 94080, USA; [c]Molecular Discovery Research, GlaxoSmithKline, 1250 South Collegeville Road, Collegeville, PA 19426, Philadelphia, USA; [d]SensiQ Technologies Inc., 800 Research Parkway, Suite 100, Oklahoma City, OK 73104, USA
*E-mail: giannetti.tony@gmail.com

RSC Drug Discovery Series No. 47
Fragment-Based Drug Discovery
Edited by Steven Howard and Chris Abell

Published by the Royal Society of Chemistry, www.rsc.org

2.1 Introduction

The use of fragment-based lead discovery (FBLD) as a tool for the identification of hit matter in the pursuit of small-molecule pharmaceutical drug discovery has been steadily increasing since the mid 1990s.[1] The successful application of FBLD requires the tight integration of biophysical, structural, biochemical, computational, and medicinal chemistry groups to identify and progress fragment hits towards lead-like compounds in a timeframe compatible with both the organizational and project goals. Application of FBLD has proven successful in generating clinical candidates and, at the time of this writing, the marketed drug Vemurafinib.[2] The success of FBLD approaches within particular organizations tends to be directly proportional to their ability to maintain a focus on progressing the frequently low-affinity hits into series of compounds with robust biochemical and cellular potencies. With improvements in hit-finding technologies for fragment-based screening (FBS) and high-throughput screening (HTS) platforms, and an increasing body of published chemical probes and inhibitors, the modern medicinal chemist can frequently face a deluge of potential starting points. For kinase targets especially, the ability to morph known inhibitors from the academic literature, patents, and internal assets, means the output from random screening approaches may significantly lag the imaginations of the chemists, thus potentially blunting the impact of fragment-derived knowledge on the course of a program.

FBS is designed to deliver hits with very different properties than traditional HTS. Weak binding, in particular, presents numerous issues since discerning between weak and non-specific binding can be challenging and mute hit confidence. When hit affinities are close to the detection assay's limits then the generation of structure–activity relationships (SARs) can be challenging. Substitutions that reduce affinity may not be quantifiable if the affinity exceeds the particular assay's limit. Most technologies applied to FBS generally observe binding rather than direct inhibition creating questions regarding the relevance of the binding site to modulating target activity. Confidence in hits increases if their activity can be demonstrated in a functional assay, but frequently this is not possible with many of the initial hits. Additionally, fragment libraries are purposefully designed to be small in number and chemically diverse in support of the advantage of FBS in sampling large segments of chemical space.[3] Indeed, groups that perform fragment and traditional HTS campaigns against the same target often find concordance among chemotypes in both hit sets and can use the information to guide the selection of both early fragment and lead-like chemical series. However, the size and diversity of fragment libraries limits the sizes of hit clusters, increases the number of singletons, and produces few or no SARs in the hit set, all of which can mute confidence in crystallography and chemistry for deploying the time and resources required to advance the hits. This is especially true in highly constrained organizations, or in environments where FBLD has not been previously applied or met limited success.

While every facet of FBLD is evolving, and the advances in the different specialties are synergizing, this chapter will focus primarily on recent advances in surface plasmon resonance (SPR) technologies and techniques that can increase hit confidence, shorten timelines, and allow the screening groups to provide richer datasets and more comprehensive analysis than previously possible. The elevation of the technology to the robustness and standards of more established screening technologies has increased the adoption, usage, and impact of the technology and its information. Much of this chapter will build on the modern experimental best practices that have been published elsewhere.[4–12] In particular we will focus on new tools to automate and standardize screening data reduction and interpretation, present a strategy for high-throughput competition assays including more rapid and unbiased data analysis procedures, and then discuss fundamental advances in the configuration of the SPR experiment that open up potentially new applications for biosensors. These new applications can give screening groups more flexibility in developing screening cascades that provide more of the types of data that increase the confidence and understanding of hits earlier and help bridge gaps between FBLD and traditional lead discovery methods.

2.2 SPR Technology as Applied to FBS

The SPR assay, as implemented by a variety of instruments from different vendors, is extremely flexible and it is critical for the screening groups to be familiar with as many assay variations and case studies as possible in order to adapt their FBS strategy to their own organization, project's goals and infrastructure.[4,11,13–26] There are no prescribed recipes for a successful screen and the full suite of applications, tools, and technologies needs to be judiciously applied according to the project goals. In addition we caution against heavy reliance on a single technique. While SPR has numerous advantages,[27] each kind of biochemical and biophysical technique has additional advantages and offers unique perspectives on molecular interactions. A flexible strategy for combining multiple techniques is encouraged. For example, while the competition assays described below can rapidly build confidence that some fragment hits are binding in a site of interest, it may make sense in some circumstances to use NMR-shift maps instead of, or in combination with, the SPR approach depending on project needs, time, resources, available instrumentation and expertise. There are no "one size fits all" recommendations and this chapter seeks to highlight the pool of tools and resources available for SPR labs to integrate into their larger organization.

FBS benefits from several advantages offered by SPR. The throughput of modern SPR instrumentation is more limited than most biochemical assay formats, but is often significantly faster than most other biophysical techniques that are also suitable for FBS. In addition, virtually all SPR instruments contain multiple experimental surfaces on the sensor allowing for target multiplexing. FBS groups frequently capitalize on this by screening targets for multiple programs simultaneously, or screening variants of the same targets such as mutants,[10,23,25,28] inactive forms of the protein such as zymogens,[19] or albumin and/or carbonic anhydrase as surrogates for non-specific binding.[24,29,30] Multiplexing is especially useful if *a priori* guidelines on desired selectivity and specificity profiles are available from the project team. We applied this in the case of Bruton's tyrosine kinase (BTK) where BTK was immobilized on two surfaces, and then one activated by a long exposure to high concentrations of ATP. This allowed the simultaneous characterization of the effects of activation on the affinity and kinetics of compound binding.[31]

Every SPR assay comes with a pre-requisite of immobilization of one of the interacting species on the surface. We typically promote protein-immobilized strategies due to the large number of coupling strategies and chip chemistries available from most SPR-hardware vendors that allow significantly flexibility in adapting the protein to the assay platform. We find most soluble targets can be robustly immobilized using either carefully controlled amine-coupling, *in vitro* biotinylation, or *in vivo* biotinylation using engineered biotinylation sequences. Nickel-coated chips can give most labs rapid access to immobilization of proteins engineered with poly-histidine tags, as is extremely common in modern protein production. Advances in chip types and methodologies have resulted in a recent increase in the number of membrane proteins accessible by SPR assays and even fragment screens.[10] Every chip type has its advantages and disadvantages and each must be considered in turn. Streptavidin, for example, often exhibits more non-specific binding to fragments than neutravidin, and antibody-capture systems are not always able to achieve high enough target densities for fragment work. We usually try at least two coupling strategies and compare the results for increased confidence.

SPR instrumentation is highly quantitative and rapidly yields affinity information using very little sample. This is especially useful in FBLD where various ligand efficiency metrics are used to help identify the most valuable hits.[32] Most ligand efficiency calculations rely on a known binding affinity. Affinity can be obtained from NMR or ITC experiments, but those are typically lower throughput and consume protein on the order of milligrams to tens of milligrams rather than the micrograms to low milligram quantities that an SPR screen may require.

Historically, the time-resolved SPR experiment has added value to molecular analysis through the direct observation and quantitation of binding kinetics. Kinetic data can be used to understand binding mechanisms, discover interactions with unusually slow-on/slow-off rates,[33,34] or derive half-life information when the dwell time on target is of significant interest.[31,35] Fragment interactions are purposely weak and tested at high concentrations. Thus, most interactions equilibrate faster than the 50–100 ms sampling rate of most SPR instruments and curvature in the association and dissociation phases is rarely observed. There are instances where weak binders will exhibit resolvable kinetics that are more typical of sub μM interactions, but rarely is the significance of this readily apparent. It could be a sign of a differentiated binding site, multiple-distinct points of contact, interaction of a site affected by slow conformational changes within the protein itself, or any number of mechanisms. We are not suggesting dismissing unusually slow fragment kinetics as an interesting feature when observed, but care should be taken to prevent over interpretation of their significance. Both hits with, and without, observable kinetics have lead to fragment-derived inhibitors. While we do observe a handful of kinetically resolved hits in most screens (see Figure 8.9 in ref. 4 for examples of fragment hits with similar potencies but different kinetics on related targets) kinetic features are frequently associated with higher affinity fragments with equilibrium dissociation constant (K_D) values typically tighter than 10 μM. However, when clean single exponential association and dissociation is observed it increases the confidence of the fragment being a true and specific binder.

For FBS the time-resolved binding signal offered by SPR is most useful in assessing if the binding signal is consistent with a fragment binder (*i.e.*, a square wave) or if it has features consistent with promiscuous binders, which are compounds that aggregate and bind non-specifically to proteins.[5,36–39] The dependency of promiscuous binding not just on compound identity, but also target protein, and buffer conditions, as well as the ability for some promiscuous binders to resist detergent, means that this kind of non-specific interaction is likely to occur at some level in all screens.[5] The SPR experiment is exquisitely sensitive to the presence of promiscuous binders. These aggregated structures have molecular weights in the millions and produce enormous SPR signals far out of the range expected for fragments. Aggregates also diffuse slowly and their binding tends to exhibit unusually slow apparent kinetics. The poly-valent nature of the aggregates often results in extremely tight binding to the surface, but their low concentration in solution generally means an insignificant fraction of the surface is fouled on exposure to these compounds, especially for short times. While rigorous solubility qualification of compounds before inclusion in a fragment collection significantly reduces the number of promiscuous binders, it will never completely eliminate it. Thus, the technology allows the SPR screening groups to remove this class of false positives from their own primary screening data without additional counter-screens. The types of promiscuous binders typically observed in SPR experiments have been cataloged and classified, and a set of several hundred has been published and annotated.[5] The assay can also be extended to the characterization of hits from other screening approaches, as discussed in Section 2.5.1. Testing of HTS, fragment, and literature compounds for promiscuous binding is critical to prevent the waste of resources in trying to establish SARs for non-specific inhibition.

2.3 The Basic Screening Cascade

SPR-based screening largely follows an experimental progression similar to that of HTS (Figure 2.1). Generally each fragment from the library is flowed, one at a time, over an immobilized target on the sensor chip and the binding

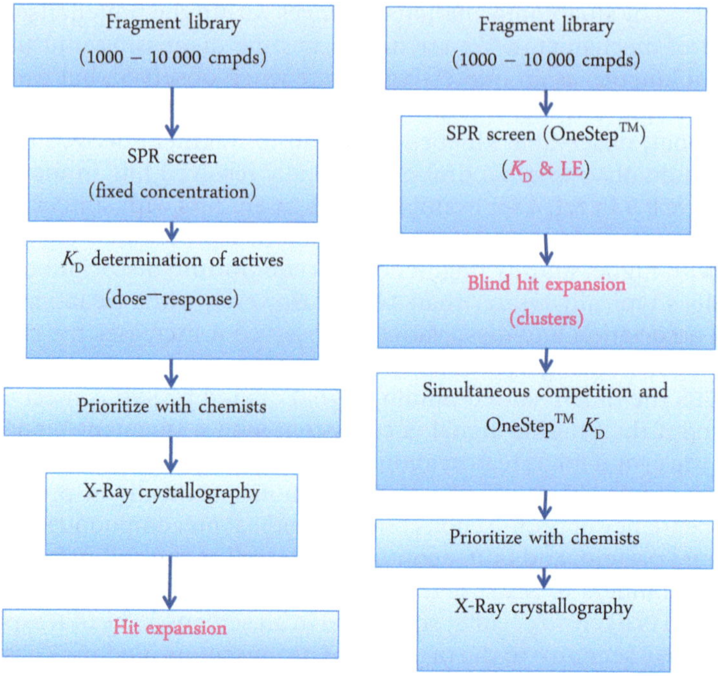

Figure 2.1 Typical SPR-based fragment screening cascade (left). A variation (right) is shown incorporating some of the hardware and methodological changes described in this chapter. Screening by OneStep™ allows for K_D and LE to be assigned at the time of screening. Additional compounds are chosen by hit expansion and added to the competition step to incorporate binding site information and determine K_D values for the original and expanded compounds simultaneously.

levels at the end of the injection are recorded. As most SPR devices contain multiple flow cells, several targets, or the on-target and relevant off-targets, may also be screened in parallel if buffer conditions support binding activity and sufficient stability for all immobilized species.[25] After the complete library is screened, the data are assembled and corrected for systematic and non-systematic drift, baseline adjusted, and normalized to a global scale such as fractional-binding. False positives are removed by inspecting each binding profile, or *sensorgram*, to assess it for features consistent with promiscuous binders.[5] This process can be tedious and, depending on software, can take hours to days if the inspection is done manually. Algorithms have been published that automate this interpretation and can identify the most frequent kinds of non-ideal behavior, though they may be challenged by non-specific interactions that give rise to features not previously observed.[40] Additionally, software improvements that allow for inspection of only the compounds whose binding level exceeds the hit threshold can reduce the number of inspected sensorgrams by 90–95%. Once the quality control is completed, binders *vs.* non-binders must be identified. Methods for this vary with some groups choosing a global statistical cut-off based on the assay noise, such as three-times the standard deviation,[4] or all binders greater than 5 RU (response or resonance units; the linear scale on which SPR data are measured).[25] Internally we refer to the compounds selected at this stage as *actives* to differentiate them from the more thoroughly qualified hits that will be released to the team.

Once the actives are chosen the screening group will typically re-order them from compound management and perform dose–response experiments to determine the K_D of the interaction. During this phase additional false-positives are removed due to deviations from ideal behavior that can be more easily observed when analyzing multiple concentrations. Once the affinity data are compiled, additional information such as ligand efficiencies (LEs)[32] is added, and hits of interest for crystallography as well as hit expansion to derive additional SARs, are selected in collaboration with the structural biology, computational chemistry, and medicinal chemistry groups. Numerous variations on this strategy exist. For example, hits may be analyzed by additional biophysical or biochemical techniques and competition-binding formats to complement the initial SPR data.

2.4 Advances in Data Reduction and Active Selection in Primary Screening

SPR instrumentation vendors have been steadily improving hardware throughput to further capitalize on SPR's combined advantages in speed and data richness. Various combinations of increased numbers of flow cells, parallelized injections, concentration-gradient injections, and more rapid injection protocols, have increased throughput several fold over the state-of-the-art at the turn of the century. However, data reduction and analysis software has not generally kept pace and increasingly, significant amounts of user time have been needed to perform repetitive and tedious tasks on increasingly large and complex datasets. To systematically select hits one must apply corrections to the screening data to account for drift, chip-to-chip and instrument-to-instrument variability, and possibly protein-to-protein variability in the case of off-target screens. The actives from the screen are then chosen from an analysis of the global data. A protocol describing one procedure for this has been previously detailed.[4] However, this approach could not account well for non-systematic effects, especially drifting baselines, and the actives-selection procedure based on standard deviations is not necessarily valid for single-point screening data. Additionally, we lacked sophisticated tools to apply the approach, and data was slowly handled manually within Excel templates, or in a semi-automated fashion using customized workflows within IDBS's Biobook suite. Processing a screen with these tools takes many hours to days to properly assemble the data and select binders. The highly manual process introduces multiple points where human-to-human preferences and variance can interfere with objective interpretation. The software solutions for managing screening data available to HTS groups are typically plate-based and have limited utility in managing issues that arise from time-dependent, rather than position-dependent, effects. As an additional complication, our lab now hosts instruments from multiple vendors and increasingly we split screens across diverse platforms to enhance speed, requiring that the data reduction software accommodate the output from different hardware. We report here an enhanced statistical approach for detecting actives in order to bring additional formalism to the analysis of the data and accelerate and nearly automate the entire process while still allowing the experimenter the flexibility to account for any potential uniqueness inherent to an individual screening campaign.

First, the method had to account for the observation that not all proteins are perfectly stable and the response signal of a control analyte injected into the system at equal concentrations tends to decline over the course of processing a single 384-well plate, which typically takes 18–24 hours (Figure 2.2, blue dots, 20 screening plates combined). Fragments injected at later times would therefore produce a response with an intensity inversely proportional to their run order. Second, corrections for decay in binding capacity do not address baseline drift, which is system behavior we also wish to correct. Other features of an ideal screening method would allow responses to a control analyte to remain consistent both within and across multiple runs, whether performed on different sensor surfaces, possibly by different experimenters, on multiple days, on different machines, and increasingly, on SPR platforms from different vendors. Finally, the method would be reproducible in the

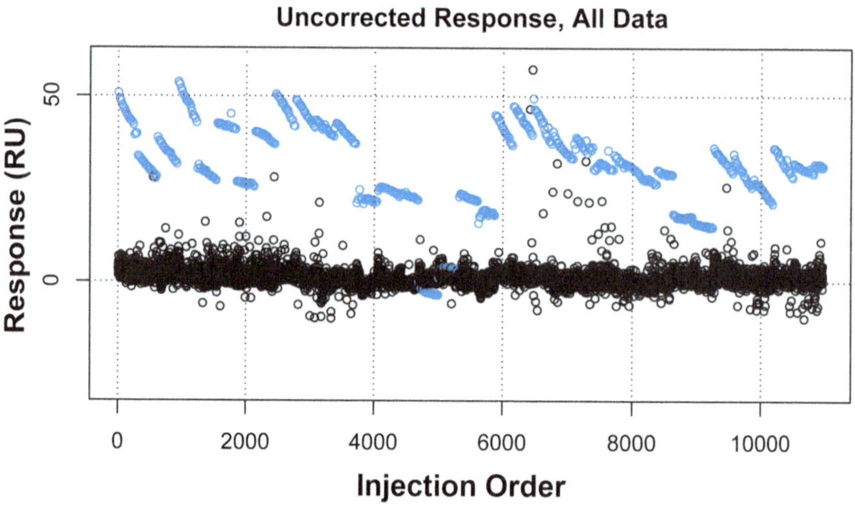

Figure 2.2 All fragment (black) and control replicate (blue) binding responses for an SPR-screen of a *ca.* 6500-member fragment library. Data were collected across multiple surfaces, chips, and instruments. Chip changes are demarcated by gaps between groups of controls. Daily decay of the controls is apparent along with fluctuations in the shape and magnitude of the baseline. Two plates in the middle show controls with no signal overlaid on the fragment background. This was due to a bad batch of control. The plates were rerun but the negative runs were maintained in the dataset to demonstrate that they fail subsequent statistical tests (see Figure 2.4A).

sense that two screeners, given the same data, would generate the same list of hits by specifying a few, well-defined parameters, rather than relying on subjective criteria when visualizing the raw data.

2.4.1 Robust Drift Correction by Local Estimation

Visualizing the data from multiple runs across multiple sensor surfaces reveals several trends in the data (Figure 2.2). Within a plate, the response signal of a positive control injected regularly into the system declined over time (blue dots, Figures 2.2 and 2.3A). Chip-to-chip variability also causes fluctuations in

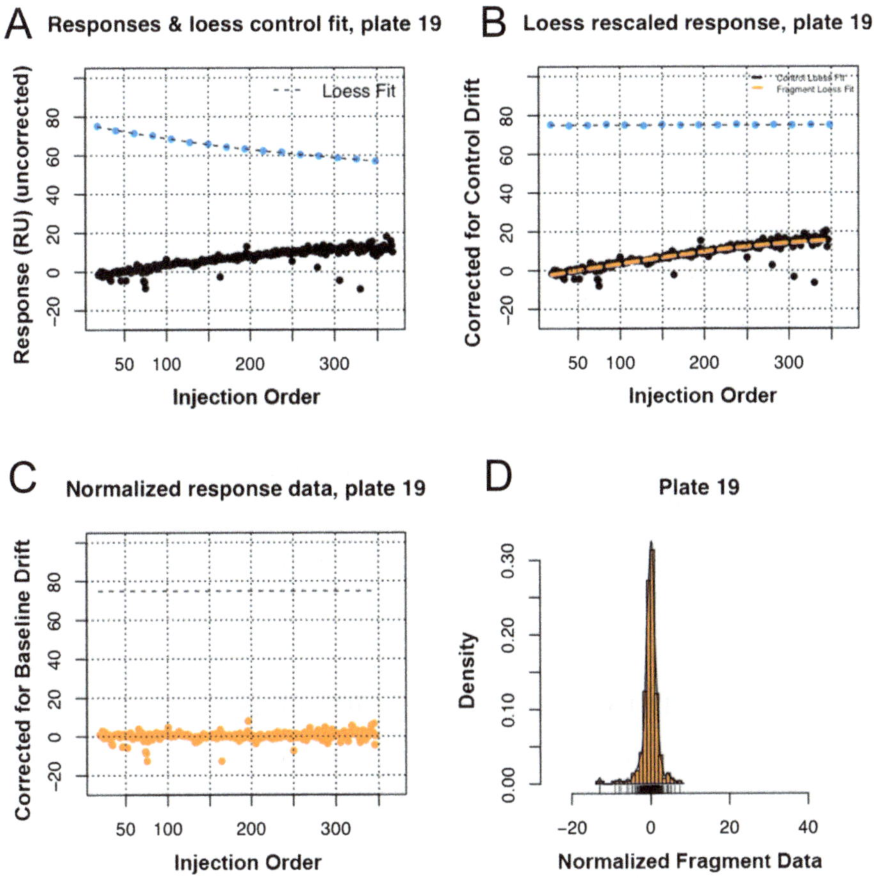

Figure 2.3 The late-binding response data from a single plate with a low hit rate is shown. A LOESS fit through the controls is shown by the black dashed line (A). After correcting the controls and fragments for the decay model, a LOESS fit (B, orange line) is determined and the baseline adjusted applied to yield a flat baseline (C) with noise evenly distributed around zero. Application of the normal–exponential convolution model on this very low-hitting plate yields an expected normal distribution whose spread represents the noise of the assay (D).

the maximal intensity of the signal. This occurs due to variation in the coupling density (Figure 2.2: each grouping of blue dots represents a separate chip). All of these experimental artifacts should be accounted for prior to hit selection.

Approximately 20 replicates are collected at regular intervals for every 300–400 fragments (Figures 2.2 and 2.3). Previous approaches to fit the decline in control binding using cubic-polynomial regression models[4] had the drawback of specifying a particular model to carry out the correction that may or may not be appropriate for the data. Therefore, we sought a more robust approach for our third-generation screening strategies.

Local regression, or locally weighted regression (LOESS), is a non-parametric alternative that does not require an *a priori* specified model.[41–44] Local regression applies classical least-squares techniques over a sliding window of smaller subsets of the data. The fits over the smaller subsets are aggregated to form a *smoothed fit* over the data and the normalization becomes model-independent. In our application, the positive control fragment response values trend monotonically downwards, with some minor fluctuations (Figure 2.3A, blue dots). We fit a local regression model to the control response data using the corresponding injection cycle number as the time variable. This results in a fit that robustly captures the change in the surface's binding capacity over time (Figure 2.3B). Robust fits to data were generally obtained using the default parameters to the LOESS function in the R environment for statistical computing (R Core Team, 2013, http://www.r-project.org). Once fit, the fragment responses can be normalized across the plate to the LOESS model.

After correction for a run's time-dependent loss of surface activity there may still be other artifacts observable in the data. Typically, these present as baseline drift, whereby the family of non-binders, when plotted against run order, may trend downward or upward as shown in an extreme example in Figure 2.3. A second LOESS model is fit through the fragment response data after rejecting strong positive or negative outliers (Figure 2.3B, orange dashed line). The smoothed fit of this line is subsequently subtracted from the run-order corrected data to produce normalized values now centered close to zero (Figure 2.3C).

2.4.2 Method #1 for Hit Detection: Box Plots

A box plot or box-and-whisker plot is an exploratory graphical tool for displaying distributions of data collected on continuous variables.[45,46] The box is typically constructed from a five-number summary comprised of the minimum, 25th percentile, median, 75th percentile, and maximum of the observed data. The edges of the box are set at the 25th and 75th percentiles, spanning the *interquartile range* (IQR), with a line in the middle of the box drawn at the median. Whiskers extend from the edge of the box out to the maximum or minimum observed data point or to the distances calculated by the 25th percentile − $k \times$ IQR, or the 75th percentile + $k \times$ IQR, whichever is closer (Figure 2.4A). Data points falling outside the boundary of the whisker are considered outliers. In our default implementation, we find $k = 3$ to provide a good balance in filtering out noise from potential signal.

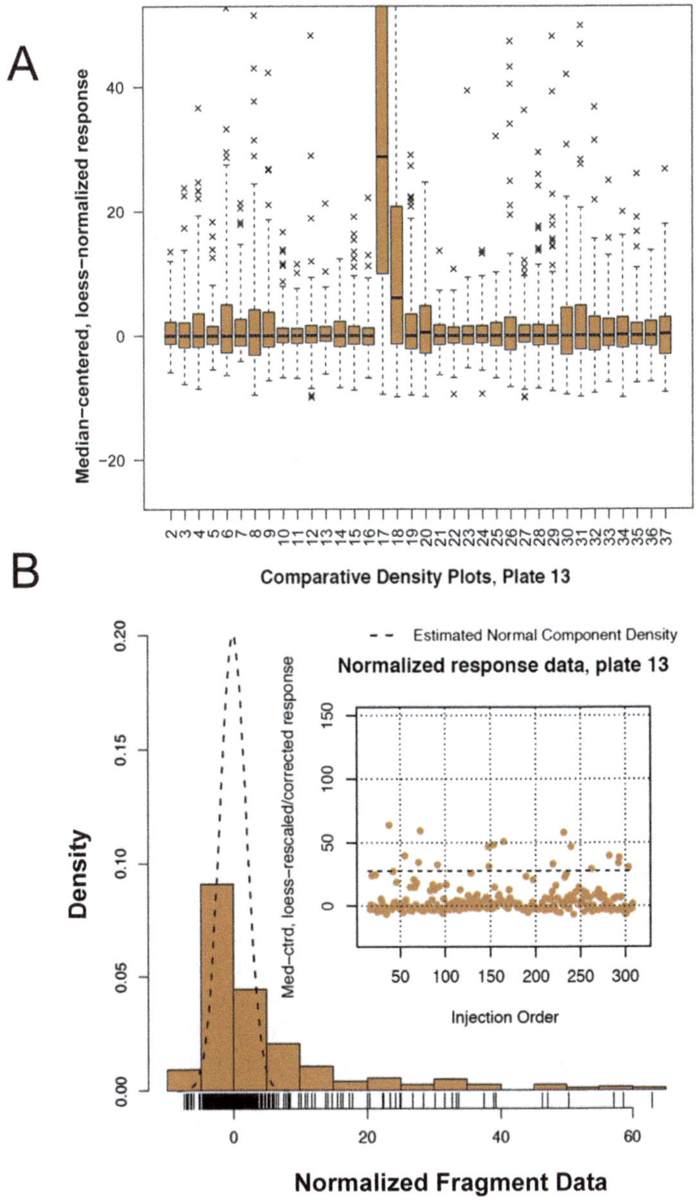

Figure 2.4 (A) A box-and-whisker plot for a screen of ~11 000 fragments allows a quick assessment of the plate-to-plate variability and apparent number of hits. Quality control is readily assessed and plates 17 and 18 are clearly outliers due to the lack of inclusion of the control compound. (B) Application of the normal-exponential convolution model showing the estimated normal density (dashed line). Individual compounds (short vertical lines) falling outside the noise are putative binders. The analysis derives from the data in the inset.

The box plot is also an easy-to-produce quality control tool for an experiment. In the example shown in Figure 2.4A, one can quickly glean the plate-to-plate variability throughout the experiment while also easily spotting plates that one may wish to discard or rerun. For example, the positive control was accidentally excluded for plates 17 and 18 in Figure 2.4A and they clearly return unacceptable statistical results.

2.4.3 Method #2 for Hit Detection: Normal–Exponential Convolution Model

Another method for detecting fragment–ligand binding interactions is borrowed from the bioinformatics literature for pre-processing (*i.e.*, signal background correction and normalization) of data generated by a variety of expression and mutation detection microarray platforms. In particular, a *normal–exponential convolution model* has been evaluated as a standard method for background correcting data in Affymetrix arrays.[47–50] The model has been implemented and is widely used, for example, in the Bioconductor project (http://www.bioconductor.org) open-source software packages, Affy[51] and Limma.[52]

The normal–exponential convolution model considers the data as a sum of normally distributed noise plus exponentially distributed signal.[47,48] The distribution of the response data can be thought of as a combination of both the normally distributed background noise, and an exponentially decreasing population of hits falling off at higher potency. Several methods exist for estimating the normal and the exponential distribution parameters of the convolution model.[47–49] We have implemented density estimation to derive the normal and exponential parameters of the convolution model (Figures 2.3D and 2.4B). The calculated estimate of the normal noise (dashed curve) may be compared to the histogram and distribution of the fragments (vertical dashes in Figures 2.3D and 2.4B). The mean and variance of the noise, obtained either within or across plates, are used to select a cut-off for identifying putative binders. In practice, we have used the 99.99th or even the 99.999th percentiles of the estimated normal distribution as a cut-off. Finally, both box plot and convolution model hit selection are run automatically on both the globally scaled dataset and plate-by-plate so the screener may compare methods for detecting hits by visualizing the fragments called by each method simultaneously (Figure 2.5).

2.4.4 Implementation and Automation in R

The standardization and automation of these procedures resulted in significant efficiency gains. Previous analyses could cover several days and generate results influenced by subjective decisions by the screener. With the new method and its implementation in the open-source R programming language, we are now able to generate numerical and graphical output for a screen comprising ~11 000 fragments in approximately 10–20 s using

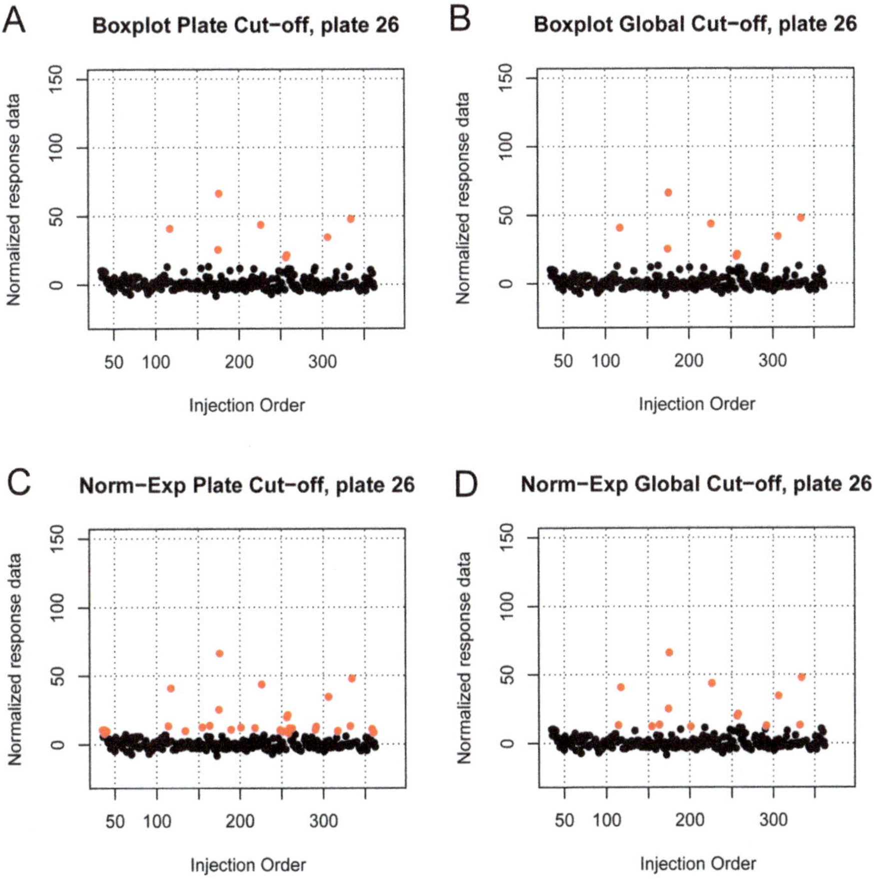

Figure 2.5 Hit selection as implemented within R-Applied LOESS for Picking Hits (RALPH). Hits are selected both globally and on a plate-by-plate basis by both box plot (A and B) or the normal-exponential convolution approaches (C and D). One can rapidly note that, as applied here, the selection approach in (C) may be too generous. The analyst is free to choose a hit set proposed by a particular method or from a concordance among them.

standard laptops (*ca.* 2014). The program automatically generates easy-to-read graphical reports in PDF format, exports the statistical analysis cut-offs, and produces a standardized text file tracking every step of the analysis in additional columns. Adjustment to parameters followed by rerunning is fast and allows the user to evaluate the sensitivity of the hit-selection process on more or less stringent criteria. R is supported on most computing platforms and can be run in a standalone version of R on a personal or lab computer, or integrated by cheminformatics groups as a module in Pipeline Pilot (Accelrys), Knime,[53] or other workflow solutions that accommodate R code.[54] Since it only requires a batch identifier to separate plates, a run order, sample name, fragment or control identifier, and a response value, the analysis becomes independent of the particular SPR platform and has been validated with data from instruments produced by Biacore and SensiQ.

2.5 High-Throughput Competition Experiments for Binding Site and MOA Studies

While SPR cannot directly observe the location of a fragment's binding site with the atomic precision of X-ray crystallography, or as directly as labeled NMR experiments, it can indirectly provide binding-site information through testing fragment hit binding to mutants, inactive or alternate forms of proteins, or in the presence of competitors. If a ligand of known binding site is available, then competition can be a preferred format since validating binding assays for mutants or inactive forms can be challenging if the positive control no longer works, thus making it hard to validate the integrity of the protein and assay. Increasingly more SPR-based screeners are integrating a competition step in their screening cascade to increase confidence and bias towards hits interacting at sites of interest. This is analogous to NMR screens where hits identified by ligand-detected methods are retested in the presence of a competitor to see if the binding signals are reduced.[55,56] The SPR approach described below leverages the low sample consumption and speed of the modern SPR instrumentation, and is easily multiplexed if multiple instruments are available.

Many examples of SPR-based competition experiments have been published[11,14,27,29,30] and this chapter will focus on formats where the protein is immobilized rather than the competitor. Immobilized ligand approaches, where protein is flowed over a known binder in the presence of putative fragments has been robustly demonstrated[55] but requires the development of additional compounds that can be tethered to the biosensor surface as well as validating additional assays. By keeping the protein immobilized, the original screening assay can be used with little modification to enhance project speed and assay/data alignment.

2.5.1 Configuring SPR-Competition Assays

In most published examples of SPR-based competition, the binding level of the fragment alone is assessed, followed by determining the binding level of the competitor alone. The two are then mixed and the new binding level of the mixture is recorded. If the binding level of the mixture is roughly the sum of the two independent events, then the fragment is considered to bind away from the competitor's site. If the binding level is the same as the competitor (or slightly reduced) then the fragment is considered competitive. In some cases the binding of the query compound can be enhanced by the competitor, revealing uncompetitive binding.[16,18] Data collection in this manner is resource-intensive as three injections per fragment are required. Additionally this puts significant constraints on the properties of the competitor. In addition to the general requirement that the competitor be soluble and stable in aqueous solution at 10–20× its K_D and have a known binding site (ideally by X-ray crystallography), the competitor would need to be reversible and dissociate completely to baseline in a short time so the surface can be re-used for testing the complex. We frequently find that known inhibitors with X-ray crystal structures early in a program are potent and exhibit long half-lives making them unsuitable for testing large numbers of fragments. Data from these three-injection screens are frequently displayed as busy bar graphs with three bars per compound representing the fragment, competitor, and fragment + competitor injections. This visualization can be tedious to analyze and is unsuitable for screens that produce a large number of potential binders (the example below contains ~800 fragment hits).

We sought to simplify the competition experiment and make it as analogous as possible to our single-point library screening assays, thus minimizing the amount of time and effort required to develop the competition format by leveraging the liquid handling and instrument protocols already established. We proposed instead to run the putative fragment binders twice, ideally on two machines in parallel though this is not required: once just as in the primary screening assay (Figure 2.6A), and once again where the competitor is present in all the buffers (Figure 2.6B). This is especially useful for less stable targets as compound binding frequently enhances protein stability. Set up this way, the experiment runs at the same speed as the primary screen. As a further advantage, slowly dissociating competitors, and slowly equilibrating competitors whose binding may take a long time to achieve equilibrium, need not be accommodated with different protocols because they are constantly present throughout the experiment. Additionally, the screener gets a second test of the putative binders from the primary screen in the (−)competitor experiment (Figure 2.6A) free from any interference or issues associated with being run alongside the non-binders and promiscuous binders from the primary screen. This helps filter out more false-positives earlier in the screen, reducing the workload on the downstream dose–response and K_D determination steps. For most screens the competition assessment for 300–1000 compounds can be completed in one to three runs depending on the hit rate and size of the screening library. We include the control compound in the (+)competitor condition, even though it is expected to be competed throughout the run, to verify both the integrity of the blockade and its maintenance throughout the run (Figure 2.6B, red line).

The report points from the (+)competitor and (−)competitor experiments are normalized and scaled using the LOESS methodology described above. The two datasets are graphed (Figure 2.6C). Both non-competitive and non-specific binders fall along the 1 : 1 line. As most screens contain at least a few of these, the exact slope of this line can be determined from the data in case it is affected by an experimental artifact such as imperfect scaling.

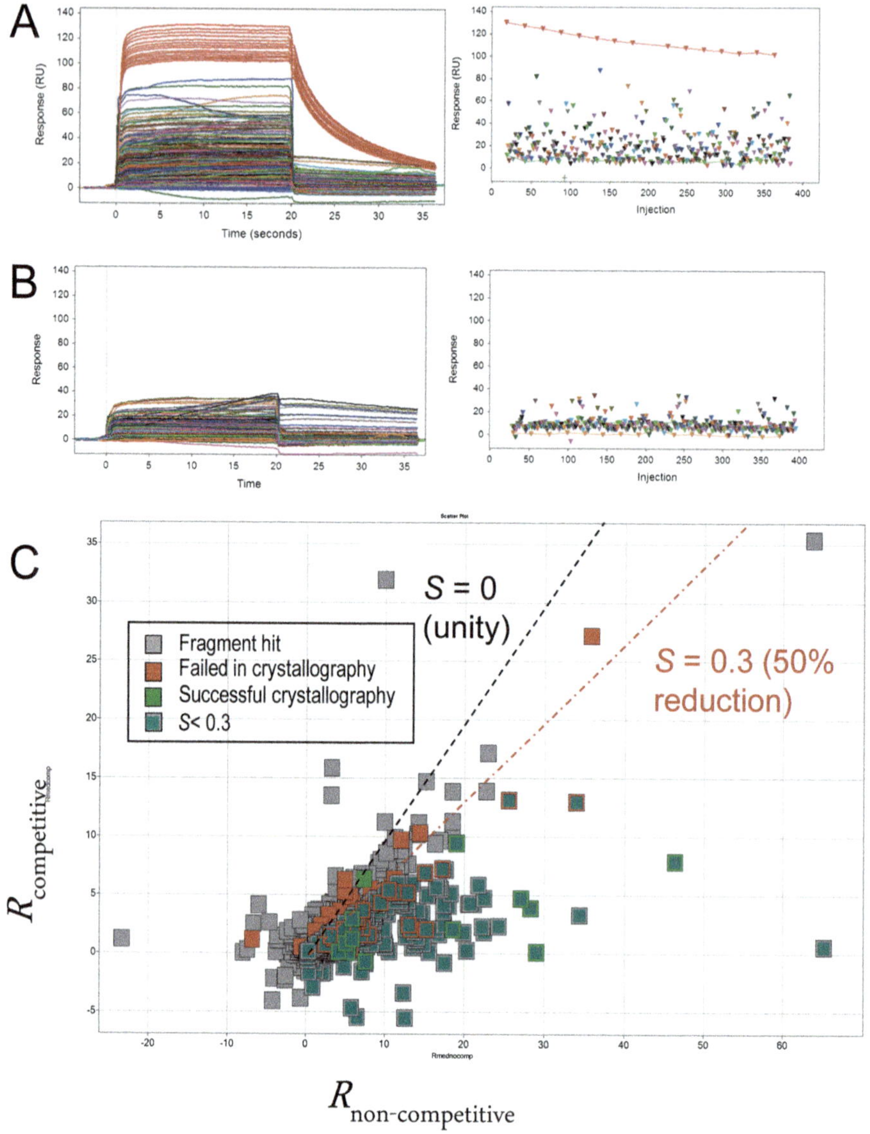

Figure 2.6 (A) Response *vs.* time (left) and response *vs.* injection number (right) for 320 actives from a primary SPR screen. The positive control compounds are shown in red. Compounds that do not bind (those that lie along the green baseline) are false positives from the primary screen and can be eliminated from further steps. In this case many non-binders are selective hits to a second target being co-screened (data not shown). (B) The same plate and protocol as in (A) but with competitor compound present in all the buffers. The suppression of the positive control (orange) to baseline levels is consistent with a complete blockade by the competitor. Some of the compounds retaining a binding signal are non-specific, others show evidence of promiscuous binding, and others

2.5.2 A Quantitative Specificity Metric for Selecting Competitive Compounds

Since the biosensor produces quantitative data, and not yes/no answers, a threshold must be set beyond which compounds are considered specific enough. This could vary from screen to screen depending on experimental or project specifics. Further complicating matters, some fragments contain a specific as well as a non-specific component of binding, or bind in more than one site at the concentrations tested. It is important to allow some tolerance for the convolution of specific and non-specific binding events to not prematurely reject a potentially valuable hit. To this end we developed a simple specificity metric. $S = (R_{\text{non-competitive}} - R_{\text{competitive}})/(R_{\text{non-competitive}} + R_{\text{competitive}})$ where the binding levels for each fragment are determined in the buffers with ($R_{\text{competitive}}$) or without ($R_{\text{non-competitive}}$) the competitor. A 50% binding reduction would yield an $S = 0.3$. Compounds with $S > 0.3$ (assuming a non-specific binder slope of 1 : 1) should represent compounds with a significant component of their binding in the site of interest. Using interactive visualization tools the screener can adjust the S cut-off to see the effects of the selection. If the number of actives is overwhelming to the throughput of the next step then the set can be trimmed to a desired size while simultaneously biasing towards the more competitive fragments.

We attempted to validate the choice of $S > 0.3$ as an initial selection criterion by performing retroactive competition analysis on a kinase program that, through multiple rounds of screening multiple libraries had yielded ~900 hits. These were collected and retested by the competition approach described above using staurosporine as a competitor (Figure 2.6). Prior to the competition experiment, significant crystallographic effort had been expended to determine structures. Fragments that did (green) or did not (red) yield occupied structures are shown. The $S = 0.3$ and unity lines are shown (Figure 2.6C). Exhaustive characterization of the compounds with $S < 0.3$ has not been carried out with orthogonal methods such as NMR so we cannot unambiguously determine that they are false positives. Some could be non-competitive binders to a site not easily amendable to co-crystallization with the current system. However, it is clear that had the structural effort been focused only on $S > 0.3$ then the co-crystallization efficiency would have certainly been increased by not diluting the sample pool with potentially less

---

may be specific with a small amount of non-specific binding necessitating calculation of S values to determine the competition threshold. (C) Graph showing the report-point data from (A) and (B) after processing with the LOESS approaches described in Section 2.4. The unity line is shown in black. Compounds with their binding levels suppressed by >50% in the presence of the competitor compound are represented by the dark-green squares and lie below the red line defined by $S = 0.3$. Red and bright-green squares represent compounds that were attempted in crystallography, and occupied structures were observed only for the bright-green samples.

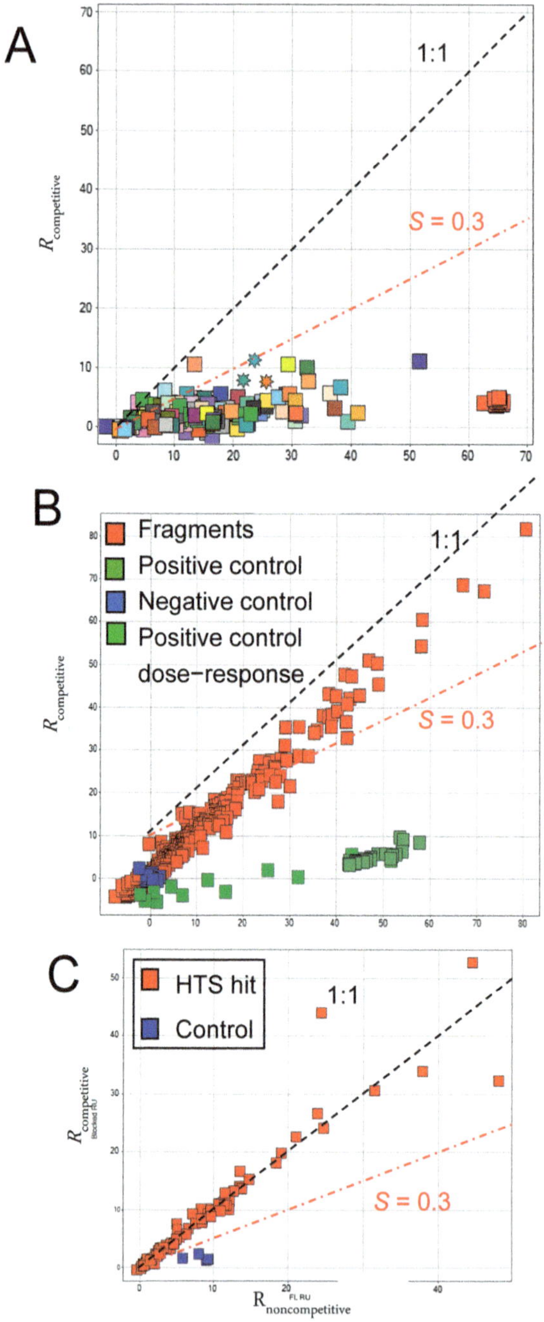
A
1:1
S = 0.3
$R_{competitive}$
B
Fragments
Positive control
Negative control
Positive control dose−response
1:1
S = 0.3
$R_{competitive}$
C
HTS hit
Control
1:1
S = 0.3
$R_{competitive}$
Blocked RU
$R_{noncompetitive}$
FL RU

specific hits or false positives. One compound with a K_D of 750 μM would have been eliminated from consideration if a strict $S > 0.3$ filter had been applied, but in a time- and resource-constrained environment some false-negatives will have to be tolerated.

Similar results were obtained for a screen of the NAMPT protein.[57] In that work 40 of 47 fragments competitive against a known inhibitor yielded co-crystal structures. We cannot definitively argue that application of the specificity filter prior to crystallography is uniquely responsible for this high success rate (*e.g.*, the NAMPT crystallographic system may be particularly amenable to fragment–structure determination), but none of the elucidated hits was localized outside the competitor site. The data also reduced the number of putative hits followed up. In a third screen (Figure 2.7A) all but one fragment showed significant competition and the competition step principally served to filter out weak or non-binding false positives from the primary screen prior to dose–response. In addition, the competition provides increased confidence in pursuing the remaining fragments as they are more certain to bind to the site of interest.

In another project this approach was applied to a screen seeking fragments competitive to a known inhibitor. The competition assay was developed and the competitor was shown to be effective in competing with other known active-site binders. However, not one fragment binder from the primary screen was shown to be competitive (Figure 2.7B), Interestingly the HTS hits for the same protein were profiled by both SPR and HTS competition assays and no competitive HTS hits were found (Figure 2.7C). Given the biochemical validation in multiple assays and SPR's ability to identify promiscuous binders/aggregators and other non-specific binding mechanisms, we hypothesized the presence of a second site. Combined with other data and the team's desire to focus on the known inhibitory site, the pursuit of hits from the random screening exercises was halted and the team focused on other leads, demonstrating the impact of the assay on a team's discovery strategy.

Non-competitive binders will be found around $S = 1$, however, they will be mixed in with compounds exhibiting non-specific binding. If non-competitive hits are desired, great care and caution should be taken to validate them with additional orthogonal methods. Compounds with an $S > 1$ would be uncompetitive and fragment binding would be dependent on the presence of

Figure 2.7 Competition diagrams as in Figure 2.6C with the 1:1 and $S = 0.3$ lines shown. (A) All the putative binders from a primary screen were tested in the presence of an active-site binder. The control (red squares on the right) compound as well as virtually all fragments (colored squares) show significant competition. Starred compounds appeared to be aggregators from a visual assessment. (B) In another program all the apparent hits from a fragment screen (red) lie along the 1:1 line with no fragment representatives below the $S = 0.3$ line. Only the control compounds (light- and dark-green squares) demonstrated competition with active-site blockade. (C) The same result in (B) was observed when testing HTS hits for the same target, suggesting the presence of a functional site distinct from the active site.

the competitor or co-factors included in the buffer. This was observed when characterizing HTS hits from a screen designed to identify inhibitors of lactate dehydrogenase (LDHA) that are uncompetitive with NADH.[16] During SPR testing, little to no binding was observed in a (−)NADH buffer. When NADH was supplemented at HTS assay concentrations, real and potent binding equivalent to the enzymatic IC_{50} was observed.[16] Similar experiments to determine Mechanism of Action (MOA) have also been performed for a protease, caspase-6, where SPR was able to demonstrate, alongside biochemical methods, fluorophore-dependent uncompetitive interactions.[18]

2.6 Hardware Improvements

The core SPR hardware has historically been supplied by Biacore Inc./GE Healthcare, since the introduction of the Biacore 1000 in 1995. Incremental improvements in hardware and software combined with expanding applications and user base lead to the introduction of the Biacore 3000 in 1999 and the Biacore S51 in 2001.[58] The S51 was the first SPR system specifically designed for detailed analysis of protein–small molecule interactions. With improved sensitivity, speed, and compatibility with industrial labware, the S51 enabled SPR-binding assays as a platform suitable for authentic hit identification in pharmaceutical discovery.[59] Continued improvements lead to the Biacore T100 and T200 instruments further enabling the measurement of weak binding events and rapid kinetics with small quantities of protein captured on the sensor surface. Parallelizing the core technology of the S51 yielded the Biacore A100 and Biacore 4000.[60] The Biacore 4000 aspirates 4 samples in parallel, rather than one per cycle, significantly increasing assay throughput. The Biacore 4000 and other various implementations of multiplexing including those offered by the Mass-1 (Sierra Sensors, 8 independent channels) and the ProteOn (Biorad)[61] with its 6 × 6 injection strategy, define the modern standard for fixed-concentration injection SPR hardware.

All of these instruments share a common design and operating principle: a multilayer integrated fluidic cartridge (IFC) with vacuum-actuated valves is used to construct the SPR detection flow cell while maintaining the shortest possible fluid path to minimize sample-plug diffusion en route to the SPR detector and prevent artifacts associated with exposing the surface to non-uniform sample concentration. However, the complex low-diameter channels and flexible valves within the IFC, makes it prone to clogging. The requirement for a fixed concentration throughout the injection and hardware requires that a significant fraction of screening time is spent observing more of the flat and quickly equilibrated binding response (20–30 s) than is strictly necessary (5–10 s) to obtain the binding level and perform sensorgram quality control.

2.6.1 SensiQ Pioneer Hardware and Novel Types of Injections

Recently, SensiQ Technologies Inc., has introduced an instrument that employs a different solution to the flow cell and fixed-concentration challenges. The Pioneer creates a flow cell by compressing the sensor chip in a manifold block. This simplified hardware design uses HPLC grade pumps and tubing for the flow system and a novel inlet/outlet flow cell to achieve individual flow-cell addressing.

The SensiQ Pioneer offers two novel types of injection protocol: FastStep™ and OneStep™. FastStep™ is an on-the-fly mixing approach where sample aspirated from a single well, at a concentration defining the top concentration of a dilution series, is mixed during the injection with running buffer at precisely controlled ratios to generate a series of discrete concentration steps. Figure 2.8A shows typical FastStep™ data for a fragment SAR hit-expansion compound. The combination of high flow rates, on-the-fly dilution, and rapid equilibration of fragment binding, places equilibrium affinity measurements over traditional pre-plated dose–response concentration series. The instrument can accommodate 2 deep 384-well plates allowing for the setup of many more samples in a single run. As fragment SARs progress to higher affinity molecules, kinetic models may be used to fit FastStep™ data. Rich *et al.* presented a detailed analysis of FastStep™, examples of typical assay parameters, and throughput comparisons to traditional fixed-concentration injection (FCI) experimental approaches.[62]

The second type of innovative injection protocol unique to the Pioneer is the OneStep™, or Taylor-dispersion Injection (diSPR™). In this experiment the fluid flow-path is intentionally lengthened using a coiled capillary tube that precedes the flow cell. Aspirated samples enter an already flowing column of buffer in the coil allowing for sample diffusion to create a continuous *ca.* four-log concentration gradient of the analyte prior to reaching the SPR-detection cell.[63] Therefore, the binding response is collected simultaneously with changing concentrations (Figure 2.8B). Quinn demonstrated that adding the diffusion component to SPR intermolecular interaction analysis: (1) does not introduce additional correlations between fitted parameters; (2) allows robust application of the two-compartment model to account for mass transport effects; (3) provides rapid and robust affinity measurements by fitting a Langmuir isotherm to the binding response generated from the continuous analyte concentration gradient; and (4) yields robust affinity and kinetic rate measurements from a single diSPR™ injection.[63,64] Like FastStep™ this requires only one well to provide affinity and kinetic information, but additionally produces a much wider and more finely sampled concentration range. The large number of data points and concentrations sampled can provide robust fits against even partial gradient response curves to aid in determining fragment K_D values where only partial dose–response curves are often collected due to solubility limitations of fragments relative to their potency.[4] The inability to collect fragment data at concentrations in the 5–20 mM range means incomplete dose–response curves are obtained. Frequently these can appear 'linear' due to sampling concentrations far from the K_D (can be too high or too low). The information content of the gradient data is much less sensitive to selection of the top concentration than with traditional 6–10 point dose–response data.[63,64] The increased data collection rate, curvature in the data due to the gradient, and wide gradient formed within the diffusion coil, serve to enhance

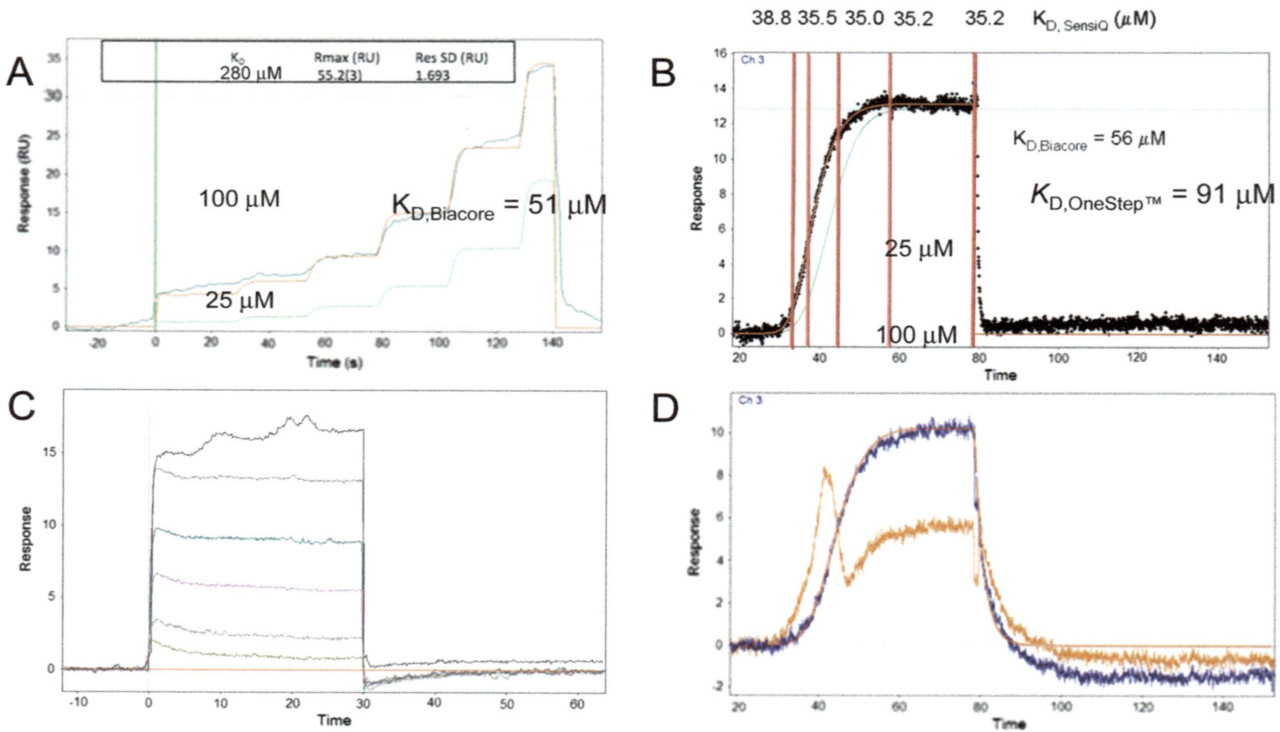

Figure 2.8 (A) FastStep™ injection obtained from a SensiQ Pioneer. The concentration of injected compound is raised six times (light-green curve tracking the instrument behavior) to produce the binding data (blue curve) that rapidly equilibrates to a new binding level before the concentration is increased again. A fit (orange curve) yields K_D information. (B) The OneStep™ injection up to 100 μM in (B) yields an affinity similar to that determined for the same interaction in a Biacore. Refitting the OneStep™-only data from the left side of the graph to the red bar using the locked R_{max} procedure yields robust K_D determinations. (C) Compounds with complex binding, such as the micro-aggregates create the jagged features at high concentration in the Biacore S51-derived data. Despite this, a binding constant can be assigned using lower concentration data. (D) The behavior of the compound in (C) is revealed during a OneStep™ injection by discontinuities in the data, such as the orange spike in (D) for the 100 μM injection. The spike results from separation of the heterogeneous populations during diffusion. Fitting to a single lower concentration injection (blue) yields a similar affinity as determined from the data in (C).

the locked R_{max} approach, which is currently the best practice for extracting reliable affinity information from incomplete dose-response curves.[4] Figure 2.8B shows the reproducibility of fits to less and less of the data from a gradient titration, while biasing to the low concentration data while returning a consistent K_D.

A second advantage of coupling the time and concentration domains in a diffusion injection is the resolution of soluble monomeric compounds from mixtures of complex aggregates. As in gel-filtration chromatography, the injected species will separate by size while passing through the diffusion coil, with their separation determined by their diffusion constants. Mixtures of species of different sizes will separate during diffusion creating discontinuities in the data. This ability further enhances SPR's already demonstrated application in readily detecting compounds that aggregate and interact through non-specific mechanisms.[4] Figure 2.8C shows data for a compound that aggregates at high concentration to create microaggregates that rapidly bind and disengage from the surface creating humps and spikes at the higher concentrations but much less at lower concentrations. The same interaction measured by OneStep™ shows a sharp spike in the 100 μM injection indicating the presence of aggregates in the mixture with the heavier component diffusing through around 40 s into the injection and about 10 s behind the monomer. At 25 μM the injection is smoother indicating a more homogenous solution that can be fit (red line) to yield a K_D similar to that determined by the Biacore S51 despite being sampled below the K_D. Combining the ability to characterize non-specific interactions by SPR with the ability to resolve samples that have heterogeneity due to concentration-sensitive aggregation allows the OneStep™ approaches to yields a higher percentage of authoritative and actionable data than traditional fixed-concentration injections.

2.6.2 Screening by Gradient Injections: Hardware and Software Integration

SPR screening best practices,[4] hit-selection statistics as described in this chapter, and general SPR assay optimization practices are applicable to OneStep™ screens and assays. OneStep™ provides the ability to both screen a library and measure the affinity of each hit in a single analyte injection. The SensiQ FE instrument (released July 2014), which was specifically designed for fragment discovery by SPR, incorporates higher sensitivity detectors with optimized injection protocols to yield the highest quality data as quickly as possible. Combining the screening step with a dose–response shortens the initial FBLD hit-identification cycle by several days: two separate experiments with their individual setup and plate logistics overhead are compressed into a single experiment. Screening campaigns have been performed this way and in one report,[65] a OneStep™ screen on a Pioneer compared with the same screen and compounds performed on a Biacore S51 showed a high concordance of identified hits. In addition, >90% of the K_D values determined by OneStep™ during primary screening matched those determined by the Biacore S51 using conventional dose–response approaches.[65] In our groups we have been able to make OneStep™ compatible with fragment screening. By shortening association and dissociation times dramatically compared to traditional fixed-concentration screening, running at 200 μL min^{-1} with a 10 s dissociation window and a sample volume that fills 50% of the diffusion loop, yields quality fragment screening data for one compound roughly every 75 s. For FBLD hit-expansion campaigns, a single OneStep™ injection of 100 μM is generally sufficient to determine binding and affinity. In high-throughput screening confirmations, we generally use a single injection of 10 μM.

In addition to the new experimental avenues offered by the Pioneer hardware, the control and analysis software were collaboratively developed with academic and industrial groups to better align it with real-world use cases.

Figure 2.9A and B show screen shots from Qdat of fragment screening data from a Pioneer FE collected either in the gradient OneStep™ mode or traditional fixed-concentration mode (Figure 2.9A and B, respectively). These figures show the hit-selection tab from Qdat (see below) incorporating automated LOESS and hit selection. Figure 2.9B shows the sliders for straightforward interaction with the data. Statistically identified hits are colored in green both in the trend plot and the sensorgram inset allowing the user to quickly check sensorgram quality. If there are more hits, individual ones can be isolated by clicking on the report points. The inset in Figure 2.9B shows the uncorrected data showing that, for this plate, there is little drift in either the control or the baseline, and needs minimal correction and normalization. The data in Figure 2.9C were collected on a target with a high degree of heterogeneity resulting in unstable control replicates and a highly scattered baseline indicative of both noise and a large number of weakly binding fragments. Application of the LOESS and hit selection within the Qdat software yields Figure 2.9D, which contains more noise and scatter than the data in Figure 2.9B, but is much more interpretable with higher confidence in the selected hits.

For example, the combination of the need for only one well per sample and reduced volume per well to generate a K_D reduces plating complexities. This makes the workflow for screening or rapid SAR follow up more compatible with HTS liquid-handling hardware and protocols and reduces SPR-specific customization regarding plate and compound handling. This should give easier access to internal HTS and corporate collections for rapid SAR follow on as well as HTS follow up for non-fragment discovery efforts. The workflow has been enhanced to highlight only samples with sufficient signal to be selected as hits. This eliminates more than 90% of the sensorgram quality control (compared to inspecting full plates). The screener can further enhance their assessment of affinity and data quality by automatically fitting the OneStep™ data and consider curve shape, model fits, and K_D assessment simultaneously from one experiment, rather than two or three.

The data analysis module for the SensiQ is called Qdat and builds upon the principles of the Scrubber platform (BioLogic Software, Campbell, Australia; http://www.biologic.com/au). In addition to further user-guided optimization

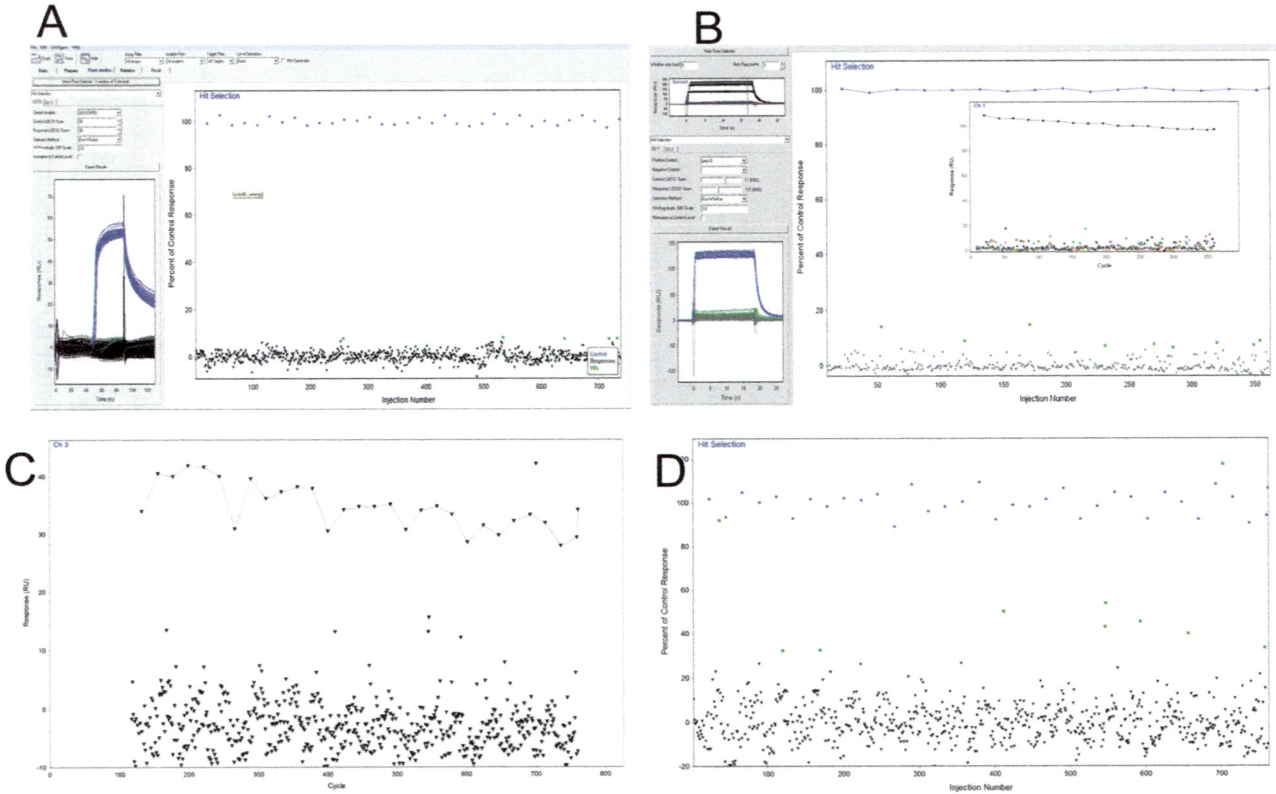

Figure 2.9 LOESS and hit selection applied in a data reduction workflow. Screenshots (A and B) of the Qdat software's hit-selection tab show screening data collected on a Pioneer FE. Fragment screening data are shown for experiments using the OneStep™ injection (A) or the traditional fixed-concentration injection mode (B). The inset in (B) is the non-corrected data and shows decay in the positive control replicates. Statistical hits are colored in green in both the trend plot and data insets for faster quality control. The uncorrected binding report-point data for a higher hit-rate target whose assay exhibits surface heterogeneity, variance in control binding, and a lot of weakly binding hits is shown in (C). Despite the greater noise and variability as compared to the data in (A), the application of the LOESS procedure (D) still significantly cleans up the presentation of the data and makes the hit threshold more clear.

of the data reduction workflow, the LOESS methods described in Section 2.4 have been incorporated directly into the data reduction. All control, baseline scaling, and hit selection are performed within an additional user-activated subtab. While the standalone R-code is sufficient to process a screen, the Qdat software allows it to be done as part of data reduction, enhancing the ability to select hits daily to feed into the hit-expansion workflow. In addition, the span parameters for the control and baseline LOESS fits have been updated with sliders allowing the user to quickly and interactively optimize the dataset. These workflow and automation enhancements further increase the speed and reliability of the screen.

These tools, combined with the decreased maintenance requirements and increased reliability resulting from the simplified flow system, the compatibility of the hardware with traditional fixed-concentration injections and applications, and the addition of the unique injection options described here, make the SensiQ hardware worthy of serious examination by FBLD practitioners.

2.7 Conclusions

As FBLD continues to develop we continue to seek ways to enhance this drug-discovery approach by providing richer project-relevant data earlier to give more confidence, support, and momentum to the fragment approach. By improving and automating data reduction and analysis the screens are further standardized and accelerated, allowing scientists more time for other efforts. Competition and other MOA experiments, enhanced with the proposed specificity metrics, allows binding site information and mechanism of action data to be collected more quickly, uniformly, and quantitatively. We have been combining these methods with the automation and liquid handling used by the HTS group to help unify the SPR and HTS workflows for greater compatibility. For example, by using acoustic dispensers to work from HTS source plates, SPR analysis of thousands of HTS hits concurrently with other HTS counter-screens allows for the option to incorporate biophysical assessments of binding, mechanism, and aggregation properties into the real-time HTS cascade rather than the post-hit triage.

Even more tantalizing are the prospects offered by the next generations of screening technology. Advances in throughput (such as with the Biacore 4000 or the 8-channel Mass-1) and richness of data such as K_D values from a single injection in the SensiQ allows for a re-imagining of the SPR-based fragment screening cascade as a whole. In particular, the ability to screen in dose–response mode or to get K_D information from single injections, allows us to provide more and richer information earlier to resolve some of the challenges in garnering FBLD traction and commitment. For example, if ligand efficiency information were available after the single-point screen then a focus on the higher LE molecules could be engaged at the beginning. We typically find that about half of the hits from an FBS have an LE <0.3 or below; the generally considered threshold for an interesting fragment.[32] These low-LE compounds could potentially be stripped away at this stage and competition and MOA experiments performed on a smaller subset. Similarly, hit expansion (also known as *analog by catalog* or *SAR by catalog*) activities are typically engaged after crystal structure determination, but generally no earlier than after K_D determination. Having K_D and LE information earlier would allow for the possibility of blindly choosing an expanded set of compounds from the internal collection to test alongside the initial binders in the competition and MOA experiments. If the K_D information can be determined by OneStep™ during the competition assessment, then the probability of producing SARs and delivering larger clusters to chemistry would give the fragments more traction and confidence at an earlier stage. Similarly, the LOESS approach and its adoption into data reduction software, allows for the actives to be selected on a daily basis. Hit expansion on those early sets could be performed during the primary screen and follow up compound plates appended to the end of the screen to fill out the SARs and clusters even before the MOA experiments.

With its current suite of applications and assay configurations, SPR technology continues to have a widening impact on early stage drug discovery. Expansion of the capabilities of the technology to help meet the demands of FBLD are helping deliver more detailed and project-relevant data earlier to teams not just for fragments, but also for hit matter from other sources such as HTS, focused libraries, computational subsets, DNA-encoded libraries, *etc.* By accelerating the competition studies, the technology can give assessments of mechanism with little additional effort beyond what is already required to produce the initial SPR assay. By adapting the methods to be compatible with the hardware and methodologies of HTS, the engagement of biophysics in lead discovery can continue to expand and help bridge and integrate the information gleaned from the different approaches to hit finding. In joining these efforts more closely over time, HTS and FBS can be seen less as separate and distinct efforts but integrated and complementary approaches that inform teams and provide more options for the development and enhancement of early lead series.

References

1. D. E. Scott, A. G. Coyne, S. A. Hudson and C. Abell, *Biochemistry*, 2012, **51**, 4990–5003.
2. M. J. Davis and J. Schlessinger, *J. Cell Biol.*, 2012, **199**, 15–19.
3. D. A. Erlanson, *Top. Curr. Chem.*, 2012, **317**, 1–32.
4. A. M. Giannetti, *Methods Enzymol.*, 2011, **493**, 169–218.
5. A. M. Giannetti, B. D. Koch and M. F. Browner, *J. Med. Chem.*, 2008, **51**, 574–580.
6. T. A. Morton and D. G. Myszka, *Methods Enzymol.*, 1998, **295**, 268–294.
7. D. G. Myszka, *Curr. Opin. Biotechnol.*, 1997, **8**, 50–57.
8. D. G. Myszka and R. L. Rich, *Pharm. Sci. Technol. Today*, 2000, **3**, 310–317.
9. I. Navratilova, M. Dioszegi and D. G. Myszka, *Anal. Biochem.*, 2006, **355**, 132–139.

10. M. Congreve, R. L. Rich, D. G. Myszka, F. Figaroa, G. Siegal and F. H. Marshall, *Methods Enzymol.*, 2011, **493**, 115–136.
11. S. Perspicace, D. Banner, J. Benz, F. Muller, D. Schlatter and W. Huber, *J. Biomol. Screening*, 2009, **14**, 337–349.
12. R. L. Rich and D. G. Myszka, *Curr. Opin. Biotechnol.*, 2000, **11**, 54–61.
13. W. Huber, S. Perspicace, J. Kohler, F. Muller and D. Schlatter, *Anal. Biochem.*, 2004, **333**, 280–288.
14. A. Kuglstatter, M. Stahl, J. U. Peters, W. Huber, M. Stihle, D. Schlatter, J. Benz, A. Ruf, D. Roth, T. Enderle and M. Hennig, *Bioorg. Med. Chem. Lett.*, 2008, **18**, 1304–1307.
15. S. S. Antonysamy, B. Aubol, J. Blaney, M. F. Browner, A. M. Giannetti, S. F. Harris, N. Hebert, J. Hendle, S. Hopkins, E. Jefferson, C. Kissinger, V. Leveque, D. Marciano, E. McGee, I. Najera, B. Nolan, M. Tomimoto, E. Torres and T. Wright, *Bioorg. Med. Chem. Lett.*, 2008, **18**, 2990–2995.
16. P. S. Dragovich, B. P. Fauber, L. B. Corson, C. Z. Ding, C. Eigenbrot, H. Ge, A. M. Giannetti, T. Hunsaker, S. Labadie, Y. Liu, S. Malek, B. Pan, D. Peterson, K. Pitts, H. E. Purkey, S. Sideris, M. Ultsch, E. VanderPorten, B. Wei, Q. Xu, I. Yen, Q. Yue, H. Zhang and X. Zhang, *Bioorg. Med. Chem. Lett.*, 2013, **23**, 3186–3194.
17. B. P. Fauber, P. S. Dragovich, J. Chen, L. B. Corson, C. Z. Ding, C. Eigenbrot, A. M. Giannetti, T. Hunsaker, S. Labadie, Y. Liu, S. Malek, D. Peterson, K. Pitts, S. Sideris, M. Ultsch, E. VanderPorten, J. Wang, B. Wei, I. Yen and Q. Yue, *Bioorg. Med. Chem. Lett.*, 2013, **23**, 5533–5539.
18. C. E. Heise, J. Murray, K. E. Augustyn, B. Bravo, P. Chugha, F. Cohen, A. M. Giannetti, P. Gibbons, R. N. Hannoush, B. R. Hearn, P. Jaishankar, C. Q. Ly, K. Shah, K. Stanger, M. Steffek, Y. Tang, X. Zhao, J. W. Lewcock, A. R. Renslo, J. Flygare and M. R. Arkin, *PLoS One*, 2012, **7**, e50864.
19. J. Murray, A. M. Giannetti, M. Steffek, P. Gibbons, B. R. Hearn, F. Cohen, C. Tam, C. Pozniak, B. Bravo, J. Lewcock, P. Jaishankar, C. Q. Ly, X. Zhao, Y. Tang, P. Chugha, M. R. Arkin, J. Flygare and A. R. Renslo, *ChemMedChem*, 2014, **9**, 73–77.
20. Z. K. Sweeney, S. Acharya, A. Briggs, J. P. Dunn, T. R. Elworthy, J. Fretland, A. M. Giannetti, G. Heilek, Y. Li, A. C. Kaiser, M. Martin, Y. D. Saito, M. Smith, J. M. Suh, S. Swallow, J. Wu, J. Q. Hang, A. S. Zhou and K. Klumpp, *Bioorg. Med. Chem. Lett.*, 2008, **18**, 4348–4351.
21. F. X. Talamas, G. Ao-Ieong, K. A. Brameld, E. Chin, J. de Vicente, J. P. Dunn, M. Ghate, A. M. Giannetti, S. F. Harris, S. S. Labadie, V. Leveque, J. Li, A. S. Lui, K. L. McCaleb, I. Najera, R. C. Schoenfeld, B. Wang and A. Wong, *J. Med. Chem.*, 2013, **56**, 3115–3119.
22. D. G. Teotico, K. Babaoglu, G. J. Rocklin, R. S. Ferreira, A. M. Giannetti and B. K. Shoichet, *Proc. Natl. Acad. Sci. U. S. A.*, 2009, **106**, 7455–7460.
23. M. Elinder, H. Nordstrom, M. Geitmann, M. Hamalainen, L. Vrang, B. Oberg and U. H. Danielson, *J. Biomol. Screening*, 2009, **14**, 395–403.
24. A. Frostell-Karlsson, A. Remaeus, H. Roos, K. Andersson, P. Borg, M. Hamalainen and R. Karlsson, *J. Med. Chem.*, 2000, **43**, 1986–1992.
25. M. D. Hamalainen, A. Zhukov, M. Ivarsson, T. Fex, J. Gottfries, R. Karlsson and M. Bjorsne, *J. Biomol. Screening*, 2008, **13**, 202–209.

26. H. Nordstrom, T. Gossas, M. Hamalainen, P. Kallblad, S. Nystrom, H. Wallberg and U. H. Danielson, *J. Med. Chem.*, 2008, **51**, 3449–3459.
27. M. Hennig, A. Ruf and W. Huber, *Top. Curr. Chem.*, 2012, **317**, 115–143.
28. M. Alterman, H. Sjobom, P. Safsten, P. O. Markgren, U. H. Danielson, M. Hamalainen, S. Lofas, J. Hulten, B. Classon, B. Samuelsson and A. Hallberg, *Eur. J. Pharm. Sci.*, 2001, **13**, 203–212.
29. Y. S. Day and D. G. Myszka, *J. Pharm. Sci.*, 2003, **92**, 333–343.
30. R. L. Rich, Y. S. Day, T. A. Morton and D. G. Myszka, *Anal. Biochem.*, 2001, **296**, 197–207.
31. J. A. Di Paolo, T. Huang, M. Balazs, J. Barbosa, K. H. Barck, B. J. Bravo, R. A. Carano, J. Darrow, D. R. Davies, L. E. DeForge, L. Diehl, R. Ferrando, S. L. Gallion, A. M. Giannetti, P. Gribling, V. Hurez, S. G. Hymowitz, R. Jones, J. E. Kropf, W. P. Lee, P. M. Maciejewski, S. A. Mitchell, H. Rong, B. L. Staker, J. A. Whitney, S. Yeh, W. B. Young, C. Yu, J. Zhang, K. Reif and K. S. Currie, *Nat. Chem. Biol.*, 2011, **7**, 41–50.
32. A. L. Hopkins, C. R. Groom and A. Alex, *Drug Discovery Today*, 2004, **9**, 430–431.
33. D. Casper, M. Bukhtiyarova and E. B. Springman, *Anal. Biochem.*, 2004, **325**, 126–136.
34. J. Q. Hang, Y. Yang, S. F. Harris, V. Leveque, H. J. Whittington, S. Rajyaguru, G. Ao-Ieong, M. F. McCown, A. Wong, A. M. Giannetti, S. Le Pogam, F. Talamas, N. Cammack, I. Najera and K. Klumpp, *J. Biol. Chem.*, 2009, **284**, 15517–15529.
35. E. C. Dueber, A. J. Schoeffler, A. Lingel, J. M. Elliott, A. V. Fedorova, A. M. Giannetti, K. Zobel, B. Maurer, E. Varfolomeev, P. Wu, H. J. Wallweber, S. G. Hymowitz, K. Deshayes, D. Vucic and W. J. Fairbrother, *Science*, 2011, **334**, 376–380.
36. B. Y. Feng and B. K. Shoichet, *Nat. Protoc.*, 2006, **1**, 550–553.
37. B. Y. Feng and B. K. Shoichet, *J. Med. Chem.*, 2006, **49**, 2151–2154.
38. S. L. McGovern, E. Caselli, N. Grigorieff and B. K. Shoichet, *J. Med. Chem.*, 2002, **45**, 1712–1722.
39. J. Seidler, S. L. McGovern, T. N. Doman and B. K. Shoichet, *J. Med. Chem.*, 2003, **46**, 4477–4486.
40. C. Kreatsoulas and K. Narayan, *Anal. Biochem.*, 2010, **402**, 179–184.
41. W. S. Cleveland, *J. Am. Stat. Assoc.*, 1979, **74**, 829–836.
42. W. S. Cleveland, *Am. Stat.*, 1981, **35**, 54.
43. W. S. Cleveland and S. J. Devlin, *J. Am. Stat. Assoc.*, 1988, **83**, 596–610.
44. W. S. Cleveland, E. Grosse and W. M. Shyu, in *Statistical Models in S*, ed. J. M. Chambers and T. J. Hastie, Chapman & Hall, 1992.
45. J. W. Tukey, *Exploratory Data Analysis*, Addison-Wesley, 1977.
46. J. M. Chambers, W. S. Cleveland, B. Kleiner and P. A. Tukey, *Graphical Methods for Data Analysis*, 1983.
47. B. M. Bolstad, R. A. Irizarry, M. Astrand and T. P. Speed, *Bioinformatics*, 2003, **19**, 185–193.
48. R. A. Irizarry, B. Hobbs, F. Collin, Y. D. Beazer-Barclay, K. J. Antonellis, U. Scherf and T. P. Speed, *Biostatistics*, 2003, **4**, 249–264.
49. M. McGee and Z. Chen, *Stat. Appl. Genet. Mol. Biol.*, 2006, **5**, 24.

50. M. E. Ritchie, D. Diyagama, J. Neilson, R. van Laar, A. Dobrovic, A. Holloway and G. K. Smyth, *BMC Bioinf.*, 2006, **7**, 261.
51. L. Gautier, L. Cope, B. M. Bolstad and R. A. Irizarry, *Bioinformatics*, 2004, **20**, 307–315.
52. G. K. Smyth, in *Bioinformatics and Computational Biology Solutions Using R and Bioconductor*, ed. V. C. R. Gentleman, S. Dudoit, R. Irizarry and W. Huber, Springer, New York, 2005, pp. 397–420.
53. S. Beisken, T. Meinl, B. Wiswedel, L. F. de Figueiredo, M. Berthold and C. Steinbeck, *BMC Bioinf.*, 2013, **14**, 257.
54. W. A. Warr, *J. Comput.-Aided Mol. Des.*, 2012, **26**, 801–804.
55. C. Dalvit, *Drug Discovery Today*, 2009, **14**, 1051–1057.
56. T. Maurer, *Methods Enzymol.*, 2011, **493**, 469–485.
57. A. M. Giannetti, N. J. Skelton, W. Wang, B. Bravo, K. W. Bair, T. Baumeister, E. Cheng, L. Crocker, Y. Feng, J. Gunzner-Toste, Y.-C. Ho, R. Hua, B. M. Liederer, Y. Liu, X. Ma, T. O'Brien, J. Oeh, D. Sampath, Y. Shen, C. Wang, L. Wang, H. Wu, Y. Xiao, P.-W. Yuen, M. Zak, G. Zhao, Q. Zhao and P. S. Dragovich, *J. Med. Chem.*, 2014, **57**, 3484–3493.
58. C. L. Baird and D. G. Myszka, *J. Mol. Recognit.*, 2001, **14**, 261–268.
59. D. G. Myszka, *Anal. Biochem.*, 2004, **329**, 316–323.
60. P. Safsten, S. L. Klakamp, A. W. Drake, R. Karlsson and D. G. Myszka, *Anal. Biochem.*, 2006, **353**, 181–190.
61. T. Bravman, V. Bronner, K. Lavie, A. Notcovich, G. A. Papalia and D. G. Myszka, *Anal. Biochem.*, 2006, **358**, 281–288.
62. R. L. Rich, J. G. Quinn, T. Morton, J. D. Stepp and D. G. Myszka, *Anal. Biochem.*, 2010.
63. J. G. Quinn, *Anal. Biochem.*, 2012, **421**, 401–410.
64. J. G. Quinn, *Anal. Biochem.*, 2012, **421**, 391–400.
65. A. M. Giannetti and E. Reese, *Make Decisions Earlier in FBLD with Affinities from Primary Screening The Utility of diSPR™ in Fragment Based Drug Discovery: A Case Study*, 2013. http://www.sensiqtech.com/uploads/file/applications/FBLD_Application_Note.pdf.

CHAPTER 3

Applications of NMR in Fragment-Based Drug Design

ISABELLE KRIMM*[a]

[a]Institut des Sciences Analytiques, UMR CNRS 5280, Université Lyon, 5 rue de la Doua, 69100 Villeurbanne, France
*E-mail: isabelle.krimm@univ-lyon1.fr

3.1 Introduction

Since the publication of structure–activity relationships (SARs) using the NMR method in 1996,[1] NMR spectroscopy has become an even more important technique in drug discovery, in particular for fragment-based drug design (FBDD).[2] NMR detects interactions in a large range of affinities (low μM to mM) and is therefore particularly suitable for fragment screening. NMR can also be used for affinity measurement, competition experiments, and can assist the drug-discovery process at different points, including hit finding, hit validation and hit optimisation.[3]

Structural information is mandatory for fragment-based approaches, not only to drive the fragment evolution, but also to fully understand SARs in a series of fragment analogues. One typically studies fragments that bind the protein target through polar directional interactions involving hydrogen bonds or charge interactions.[4] The resolution of the complex 3D structure is classically done by X-ray crystallography but is not successfully achieved for all fragment hits detected by other biophysical techniques. NMR can play a

RSC Drug Discovery Series No. 47
Fragment-Based Drug Discovery
Edited by Steven Howard and Chris Abell

Published by the Royal Society of Chemistry, www.rsc.org

crucial role in such cases, and offers the possibility to obtain key information regarding the binding site and binding mode of the fragment.

A genuine advantage of NMR is that the states of both the protein and the compounds can be controlled to avoid artefacts. The library can be entirely tested before and after the screening for purity and stability. The protein state can be easily analysed through a simple 1D NMR spectrum. This enables one to check if the protein has been structurally modified *via* oxidation, covalent modification, unfolding, aggregation, or changes in oligomeric states. These advantages enable retention of fragments that would otherwise be eliminated, such as frequent hitters or certain Pan Assay Interference compounds (PAINS).[5] A primary NMR-based fragment screening, when considered as a tool for pharmacophore mapping could thus include even poorly drug-like fragments. The unwanted functions can later be modified into more drug-like compounds.

The use of NMR in FBDD has been reviewed in different books and excellent reviews.[3,6–10] Practical aspects concerning the preparation of the fragment stocks and the NMR tubes have been largely reviewed[11,12] and will be discussed here only briefly. In the first part of this chapter, the applications of the ligand-observed experiments saturation transfer difference (STD),[13] WaterLOGSY[14] and the 2D transferred-NOESY (including ILOE[15] and INPHARMA[16]) will be reviewed. In the second part, the structural information inferred from protein-observed NMR experiments will be considered. While a diverse range of protein-observed experiments is available, only the applications of HSQC-type experiments will be discussed here.

3.2 Ligand-Observed Experiments

The ligand-observed methods discussed here rely on ^{1}H-NMR spectroscopy, but ^{19}F-NMR methods are also applied in the pharmaceutical industry.[17] By contrast with protein-observed experiments, ligand-observed experiments are more sensitive with larger proteins and do not require isotope labelling. Ligand-based methods are mainly known as screening experiments, but can also be used for assessing protein–ligand affinities, and can provide pertinent structural information for the protein–ligand complexes.

When using NMR, one has to know if the exchange between the free and bound state of the molecules is fast or slow (relative to the observed NMR parameter).[18] Upon binding, under conditions of fast exchange (weak K_D [dissociation constant]), the observed NMR signal is a weighted average of the signal in the bound and the free states. In the presence of a high excess of ligand over the receptor, the NMR spectrum is dominated by the chemical shifts of the free ligand, but reflects relaxation properties of the bound state. As a consequence, a simple comparison of the line width in the fragment 1D spectra in the absence or in the presence of the protein target should enable the identification of binders. Unfortunately, the method requires large protein quantities (small ligand-to-protein ratio), except for the TINS approach,[19] where the effect is enhanced by the immobilisation of the protein target.

Consequently methods based on magnetisation transfer, such as the NOESY-based methods, including STD and WaterLOGSY, are preferred.

3.2.1 STD Experiments

3.2.1.1 Principle and Experimental Conditions

The STD experiment consists of applying a radiofrequency irradiation that selectively saturates the protein resonances, without saturating the resonances of the small organic compounds.[13,20] The saturation then spreads within the protein *via* the nuclear Overhauser effect (NOE) and spin diffusion. The experiment sensitivity therefore increases with the molecular weight of the protein. When ligands bind the protein under fast exchange conditions, the saturation will also spread to the ligand. The spectrum recorded with saturation of the receptor (called the *on-resonance* spectrum) is subtracted from the spectrum recorded without saturation of the protein protons (the *off-resonance* spectrum). Therefore, only the NMR signals of binders are observed in the STD spectrum.

STD experiments are performed with a large excess of the ligand. The protein concentration usually ranges from 0.5 μM (protein $M_w > 100$ kDa) to 20 μM (protein $M_w < 20$ kDa), with 200 μM to 1 mM fragment concentration. The saturation frequency of the STD experiment is fixed at the up-field region (1 to −1 ppm) where no ligand signals are observed. To avoid any artefacts (direct saturation of the ligand), it is preferable to saturate at −1 ppm. Nevertheless, for proteins below 50 kDa, the signal-to-noise ratio is greatly improved by saturating at up-field positions (0.5–0.9 ppm), but this should not be used with fragments containing methyl groups. Typical saturation times are 1–2 s. Only one saturation time is used for screening experiments, while more quantitative analysis of the STD results requires several saturation times.[21,22]

3.2.1.2 Recent Pulse Sequence Development

Recently, the clean-STD experiment was proposed to gain sensitivity for proteins with molecular weights below 70 kDa.[23] The strength of the radiofrequency irradiation is increased and ligand excitation artefacts due to the overflow of the saturation are removed, by recording two STD experiments with saturation in down-field and up-field positions. Another STD experiment for observing the ligand signals resonating close to the water signal has been developed by incorporating the WET technique (water suppression enhanced through T_1 effects).[24,25] This method is likely to be useful for 3D fragments that contain more sp^3 carbons.

3.2.1.3 STD as a Tool for Fragment Screening and Fragment Ranking

The STD experiment detects protein–ligand interactions with K_D ranging from 0.1 μM to about 10 mM and is therefore particularly suitable for screening fragment libraries. Due to its capability of detecting mM interactions, the STD experiment detects both specific and unspecific binding ligands. Additionally, STD signals can be due to weak binding at multiple sites. Therefore, to ensure that the STD signal is due to a specific binding event, competition experiments are usually required. For example, the ATP molecule, which binds to kinases with K_D in the 10–100 μM range, was used as a reference molecule to find ATP competitors against protein kinases using STD experiments.[26] A second approach relies on ligand epitope mapping, based on the comparison of the STD signals of the ligand protons (see Section 3.2.1.4).

For a quantitative analysis of the STD signals, Mayer and Meyer proposed the calculation of the STD amplification factor (STD_{AF}),[27] which is the intensity of an STD signal corrected by the excess of ligand:

$$STD_{AF} = \frac{I_{STD}}{I_0} \times \frac{L_T}{P_T} \quad (3.1)$$

where I_{STD} and I_0 are peaks integrals in the STD and 1D experiments recorded with the same parameters, and L_T and P_T are the total concentrations of the ligand and protein, respectively. A saturation curve is obtained by measuring STD_{AF} for a ligand titration series. Clearly, due to the non-specific binding detected in the STD signals, affinities measured by STD are usually underestimated.[28] To suppress the unspecific binding, STD signals can be measured in the absence and the presence of an inhibitor,[29,30] or build-up curves can be used.[21,22] STD experiments can also be used to rank the fragments, by comparing the STD amplification factors (STD_{AF}) of the fragments. Since STD signals are highly dependent on the nature of the protons (aromatic protons have relaxation properties that differ from those of aliphatic protons), it is mandatory to compare STD signals for the same type of protons. The approach is particularly useful for ranking a series of analogous fragments. Recently, a ranking method was proposed that uses competition STD in the search for fragments binding the phenylalanine pocket of RadA.[31] The 5-hydroxyindole was used as a 'reporter' ligand and the STD signals were monitored to see how the STD signals of the reporter were decreased.

3.2.1.4 *STD as a Tool for Binding-Mode Assessment*

Specific binding usually translates into an epitope mapping effect on the STD spectrum.[27,32] The relative intensities of the STD peaks vary within a fragment (each proton has a different STD_{AF}), which provides information regarding ligand orientation: the protons that are in close contact to the protein have a higher STD_{AF} as compared to protons that are not directly in contact with the protein surface. The STD-epitope mapping method has been typically used for large and flexible ligands (such as saccharides) and requires experimental as well as computational efforts for a robust interpretation of the data.[32–35]

In the SOS-NMR[36] approach, protein samples are uniformly deuterated except for specific amino acid types. STD experiments performed on the ligand complexed to a series of these target samples allow definition of the amino acid composition of the ligand-binding site. The STD intensities are converted into distance restraints, enabling the determination of the 3D structure of the protein–ligand complex without the need of the protein resonance assignments.

As illustrated in Figure 3.1, STD spectra can be useful for assessing the binding specificity of fragments, by comparing STD signals of a fragment

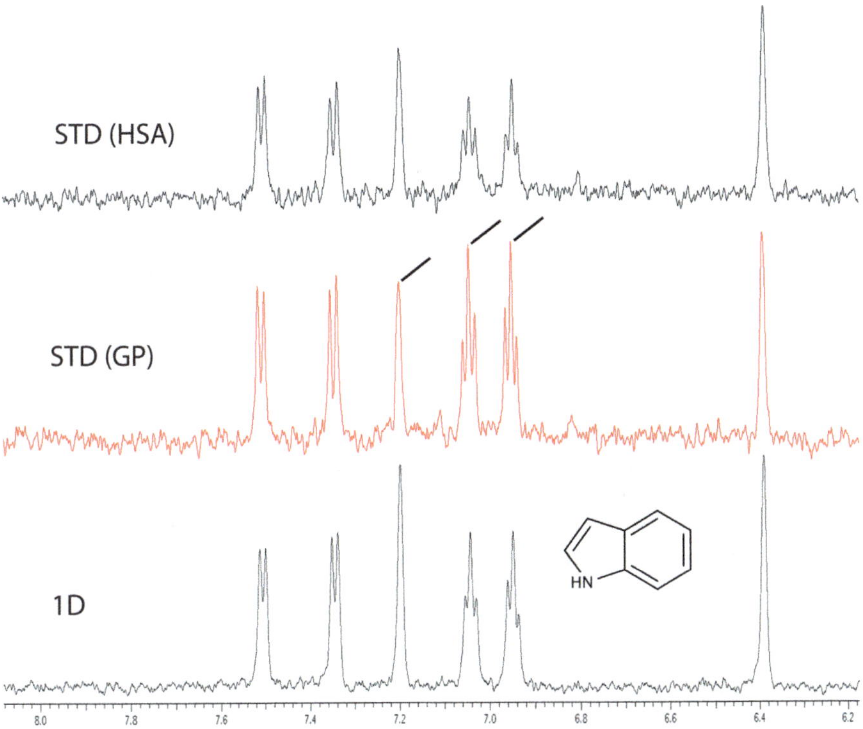

Figure 3.1 STD experiments showing the specific binding of indole to the glycogen phosphorylase protein. 1D and STD experiments for indole (500 μM) were recorded in the presence of two proteins: human serum albumin (HSA, in black) and glycogen phosphorylase (GP, in red) at 1 μM. In the presence of GP, the relative intensities of the STD peaks are different from the peak intensities observed in the 1D spectrum (see the intensities labelled with a black line), due to the epitope mapping effect, therefore demonstrating specific binding in the case of GP. By contrast, in the presence of the HSA protein, no epitope mapping effect is observed: the indole exhibits similar relative intensities in the STD and the 1D spectrum recorded with identical NMR parameters.

binding to various protein targets, or by comparing the STD intensities to those observed in a 1D spectrum recorded with the same NMR parameters (corresponding to the off-resonance STD experiment). This information represents a real benefit from NMR-based screening.

3.2.2 WaterLOGSY Experiments

3.2.2.1 Principle and Experimental Conditions

WaterLOGSY (water–ligand observed *via* gradient spectroscopy),[14,37,38] like the STD experiment, involves magnetisation transfer *via* intermolecular NOE and spin diffusion from the protein to the ligand. Here, the bulk water magnetisation is excited and transferred during the NOESY mixing time to the bound ligand *via* water–ligand–protein or protein–ligand complexes whose rotational correlation times yield negative cross-relaxation rates. Fragments that bind the protein receptor exhibit a negative NOE with water, while non-binders that only interact with bulk water experience much faster tumbling that translates into a positive NOE. Therefore, binders and non-binders are easily discriminated from their opposite sign in a WaterLOGSY spectrum. Since magnetisation can also be transferred by chemical exchange from the water to labile protons of the small organic compound, WaterLOGSY experiments must be recorded in the absence of the protein to track these 'false-positive' signals.

Typically, mixing time for the NOESY ranges from 1 to 3 s. For small proteins ($M_w < 20$ kDa), a concentration of 10 μM is required, while 1 μM is sufficient for larger proteins ($M_w > 60$ kDa). Dalvit reports the application of the method with protein concentrations as low as a few hundred nM.[14,37,38] By comparison with the STD experiment, very large ligand–protein ratios (>1000 : 1) should be avoided, since, in contrast with the STD spectrum, the WaterLOGSY spectrum reflects both the free and bound states of the ligand. We have screened 300 fragments mixed into 5 compound cocktails using the same NMR sample for both STD and WaterLOGSY experiments, against serum albumin as a control. STD and WaterLOGSY gave the same hits, and also the same hit ranking. We nevertheless consider that such results, obtained with proteins with well-defined binding sites, are not necessarily observed with more challenging targets, due to the weak affinities of the fragments. Such an observation has been recently reported by the group of Ciulli in the case of protein–protein interactions.[39] The optimisation of the sample conditions in this case might be crucial.

3.2.2.2 Recent Pulse Sequence Development

Two different pulse sequences were recently reported to improve the sensitivity of the WaterLOGSY experiment by a factor of two. The polarization optimized PO-WaterLOGSY experiment,[40] which is based on the water flip-back strategy, is shown to better conserve the water polarisation, while the aroma-WaterLOGSY[41] was reported for the screening of compounds that contain aromatic groups. The latter experiment should be of interest for avoiding expensive deuterated reagents.

3.2.2.3 WaterLOGSY as a Tool For Fragment Screening and Fragment Ranking

Ideally, since WaterLOGSY spectra can be recorded in much shorter times than STD (typically in a few minutes), a first screen of fragment mixtures can be performed using WaterLOGSY, and then each hit is confirmed separately using both WaterLOGSY and STD. For low-molecular-weight proteins, WaterLOGSY could be preferred to STD as it is less dependent on spin-diffusion efficiency and requires less proton density. For example, among the three methods (STD, WaterLOGSY and NOESY) used to screen 112 compounds against a RNA target, WaterLOGSY was clearly the most sensitive technique.[42]

Theoretically, the estimation of the ligand-binding affinity is feasible, both directly and by competition experiments.[9,38] The free state contribution of the WaterLOGSY signals must be removed from the WaterLOGSY signals of the ligand in the presence of the protein. The WaterLOGSY intensities at different ligand concentration can be fitted to measure the dissociation constant. As for STD, WaterLOGSY is more suitable for fragment ranking by competition experiments.

3.2.2.4 WaterLOGSY as a Tool for Binding-Mode Assessment

The group of Günther recently showed that the WaterLOGSY experiment could be used to distinguish buried protons from solvent-exposed protons. The method called SALMON (solvent accessibility, ligand binding and mapping of ligand orientation by NMR spectroscopy) consists of analysing the WaterLOGSY signals within a ligand molecule.[43] The method requires the minimisation of the spin-diffusion process, by increasing the temperature of the experiment or by reducing the NOESY mixing time (125 ms). The experiment has been applied to ligands of the quinone oxidoreductase 2,[43] and the dehydrogenases AKR1C3 and HSD17β1.[44]

3.2.3 Other Applications of STD and WaterLOGSY Experiments

3.2.3.1 Validation of the Fragment Library

The quality of the fragment library is of major importance for generating robust results and avoiding false-positive hits. Most of the screening artefacts are related to aggregation of the compounds under the experimental conditions. NMR can detect compound aggregation, by the observation of a broadened water resonance or the broadening of ligand resonances. Concentration-dependent signal intensities of the fragments are clear evidence of their solubility. We typically record WaterLOGSY spectra of all the compounds in the absence of the protein in order to test the compound solubility. Compounds exhibiting WaterLOGSY signals typical of aggregates are removed from the library. The mixtures are also checked by WaterLOGSY experiments, since association between molecules can be observed with mixtures. The use of a 1D reference spectrum, a WaterLOGSY spectrum and/or a selective longitudinal relaxation filter spectrum for identified hits dissolved in aqueous solution and in the presence of a water-soluble reference molecule, has been suggested for rapid characterisation of the identity, purity, solubility and aggregation state of the active compound.[45] The approach, called the SPAM filter (for solubility, purity and aggregation of the molecule) enables the accurate measurement of IC_{50} values, and the identification of false negatives.

3.2.3.2 *Probing Synergy for Allosteric Enzymes*

STD and WaterLOGSY experiments can provide information regarding the affinity of the protein–ligand complex. Both can be used to compare the signal of a fragment binding in the presence of a second compound that binds another protein pocket, to provide evidence for synergy between binding in the two pockets. We have performed such experiments using the glycogen phosphorylase (GP), a potential therapeutic target in type-II diabetes.[46] GP is an allosteric enzyme with four distinct binding sites, for which a large number of inhibitors and activators of the protein are known. We have analysed the interaction of GP with 19 fragments resulting from a very simple deconstruction approach of these inhibitors/activators. STD experiments were recorded for each fragment in the presence of others fragments and GP inhibitors. As an example, synergy was clearly observed in the STD and WaterLOGSY experiments between 3-aminobenzonitrile, binding to the 'new-allosteric site', and inhibitors of the active site, and in the presence of caffeine, which binds in a third binding site (Figure 3.2). These observations match the synergy reported for inhibitors of the new-allosteric site and both active site inhibitors and caffeine. This correlation is not always observed. For example, other fragments such as the indole and chloroindole, which both bind at the new-allosteric site do not bind synergistically with active site inhibitors or caffeine.[46]

This study[46] is another example demonstrating how fragments constitute powerful tools to probe protein-binding pockets, not only to identify protein hotspots and key interactions between protein and ligands, but also to reveal synergy relations between various pockets of allosteric enzymes. Nevertheless, one must keep in mind that properties are not necessarily conserved upon fragment evolution.[47]

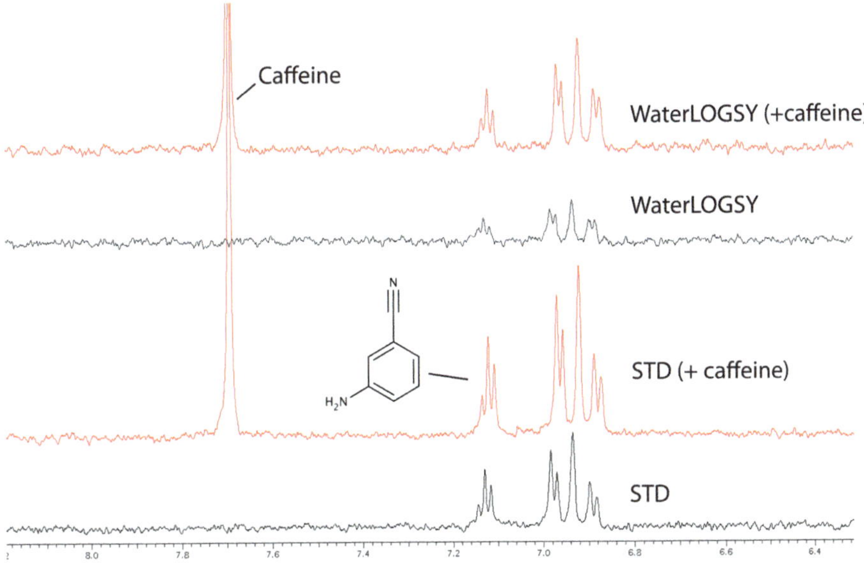

Figure 3.2 STD and WaterLOGSY experiments provide evidence for synergy between the binding of the fragment 3-aminobenzonitrile and caffeine to GP. The STD and WaterLOGSY signals of the fragment increase (+100%) in the presence of caffeine. Protein concentration is 2 μM, fragment concentration is 400 μM and caffeine concentration is 2000 μM.

3.2.4 NOESY Experiments

3.2.4.1 Observation of Transferred-NOE, ILOE and INPHARMA Peaks

The binding of a small-molecular-weight compound to a macromolecular receptor causes a change in the observed NOESY peaks on the 2D NOESY spectrum, leading to the so-called *transferred*-NOEs.[48] For a mixture of fragments in the presence of the protein, the compounds that bind the protein will exhibit strong negative NOEs, whereas non-binders will show weak positive NOEs. If two ligands of the mixture bind *simultaneously* with similar residence times at adjacent sites on the protein surface, forming a ternary complex, negative interligand NOEs can be observed, as reported for the first time in 1997 on the avidin protein using *n*-pentanoic acid and 4-methylimidazolin-2-one, and then later for lactate dehydrogenase and the glucose-6-phosphate dehydrogenase.[15] These intermolecular NOEs are also referred to as ILOEs.[49] Another type of intermolecular NOEs can also be observed, when two ligands bind *competitively* to the same protein-binding pocket (and not *simultaneously* to adjacent hotspots) and exhibit similar residence times. The resulting NOESY cross-peaks are referred as INPHARMA (interligand NOE for pharmacophore mapping) NOEs. These INPHARMA NOEs were reported for the first time in 2005 for the interaction of epothilone A and baccatin with tubulin.[16]

3.2.4.2 Experimental Conditions for Transferred-NOESY Experiments

Transferred-NOESY experiments typically require mixing times of 200–600 ms. To observe the ILOE, one uses longer mixing times (600–800 ms), while INPHARMA peaks have been observed with mixing times of 70–300 ms. The mixing time usually needs to be optimised for each protein, since it is related to the correlation time and the dynamics of the complex. This is particularly crucial for INPHARMA.

The protein concentration used for transferred-NOESY experiments is larger than for 1D experiments, and varies from 5 μM (large proteins) to 50 μM (M_w < 40 kDa) with 0.5–1 mM of fragments. The major problem when looking for ILOEs is the artefacts arising from compound aggregation and non-specific binding. When possible, the experiments should be performed in the presence of a known competitor, to assure that the intermolecular NOEs disappear in the presence of a competitor blocking the binding site. ILOE and INPHARMA experiments both suffer from the fact that cross-peaks are not detected for all ligand pairs which might have been expected to produce such effects. This is possibly due to the kinetics, and could explain why few examples have been reported. For ILOE, it is necessary that two adjacent binding sites exist for the fragments to bind near to each other (see Section 3.2.3.4). For observing INPHARMA NOEs, the affinities of the ligands must be similar (the K_D ratio should be less than 10).[50]

3.2.4.3 NOESY for Fragment Screening

The binding information derived from transferred-NOESY experiments is very similar to that obtained from STD and WaterLOGSY experiments. One advantage of the NOESY experiments reported by the group of Pellecchia is that they can be performed in very large mixtures as compared to 1D-based methods (around 100 compounds). This dramatically increases the throughput.[51] Additionally, the screening of large mixtures theoretically increases the probability of observing ILOEs.

3.2.4.4 ILOE for Ligand-Binding Mode and Inhibitor Design

ILOE experiments can be useful for assessing the binding mode of a fragment hit in the presence of a known compound. For example, ILOE-driven step-by-step reasoning was used for the design of pantothenate synthetase inhibitors using the 3D known crystal structure of the protein in the presence of the pantoate.[52]

ILOE is a powerful method to search for second-site binding ligands, by screening a library in the presence of high concentrations of first-site binders. It can be more appropriate to use first-site binders containing a methyl group, for a better signal-to-noise ratio and an easier assignment procedure.[52] The ILOE approach has been successfully applied by the group of Pellecchia for the design of inhibitors against the Bid, Bcl-xL and Mcl-1 proteins as well as the mitogen-activated protein kinase p38alpha.[51,53–55] More recently, ILOE-based screening led to the identification of ligand pairs that bind in proximal sites within the co-factor binding pocket of human thymidylate synthase,[56] and helped identify adjacent site ligands to fucose binding site norovirus virus-like particles.[57]

3.2.4.5 *INPHARMA Experiments for Assessing Ligand-Binding Site*

When the binding site of a reference fragment is known, INPHARMA experiments rather than competition experiments can be used to identify the binding site of other fragments. This is particularly useful if the competitors are poorly soluble and display low affinity. We have performed such experiments with GP.[58] Caffeine **2** and chloroindole **4** were used as reference ligands for the inhibitor site and new-allosteric site, respectively. NOESY experiments have been performed for six fragments in the presence of the two reference molecules. Intentionally, two frequent hitters (a biphenyl **7** and a naphthalene compound **8**) were included in the experiment. INPHARMA peaks were observed for six pairs, while the seven other pairs showed no INPHARMA peaks. The INPHARMA peaks observed in the presence of the two frequent hitters **7** and **8** are uniform throughout the spectrum, indicating that there is no preferential binding mode for these fragments, suggesting non-specific binding. For the four other fragments, INPHARMA peaks were observed between analogous fragments, and fragments binding to the inhibitor site could be distinguished from fragments binding to the new-allosteric site (Figure 3.3).

3.2.4.6 *INPHARMA Experiments for Assessing Ligand-Binding Mode*

Initially, the INPHARMA experiment was designed to derive the 3D structure of a protein–ligand complex, using the 3D structure of a protein bound to a reference compound.[16] A qualitative version of the approach can be used in FBDD. For example, when comparing two related fragments, one needs to know if the binding mode is conserved, or if the ligand changes induce a modification of the binding mode. To address this question, it is not mandatory to resolve the full 3D structure of the complex. INPHARMA peaks between two compounds in the presence of the protein can yield information regarding the relative binding mode of the two molecules.[58] Remarkably, the analysis does not require the knowledge of the free protein 3D structure.

3.2.4.7 *Distinction of INPHARMA/ILOE Peaks*

A careful analysis of the NOEs should indicate if the peaks correspond to INPHARMA or ILOE peaks. INPHARMA peaks are observed between all pairs of protons of compounds **1** and **2** that are in contact with the same protein

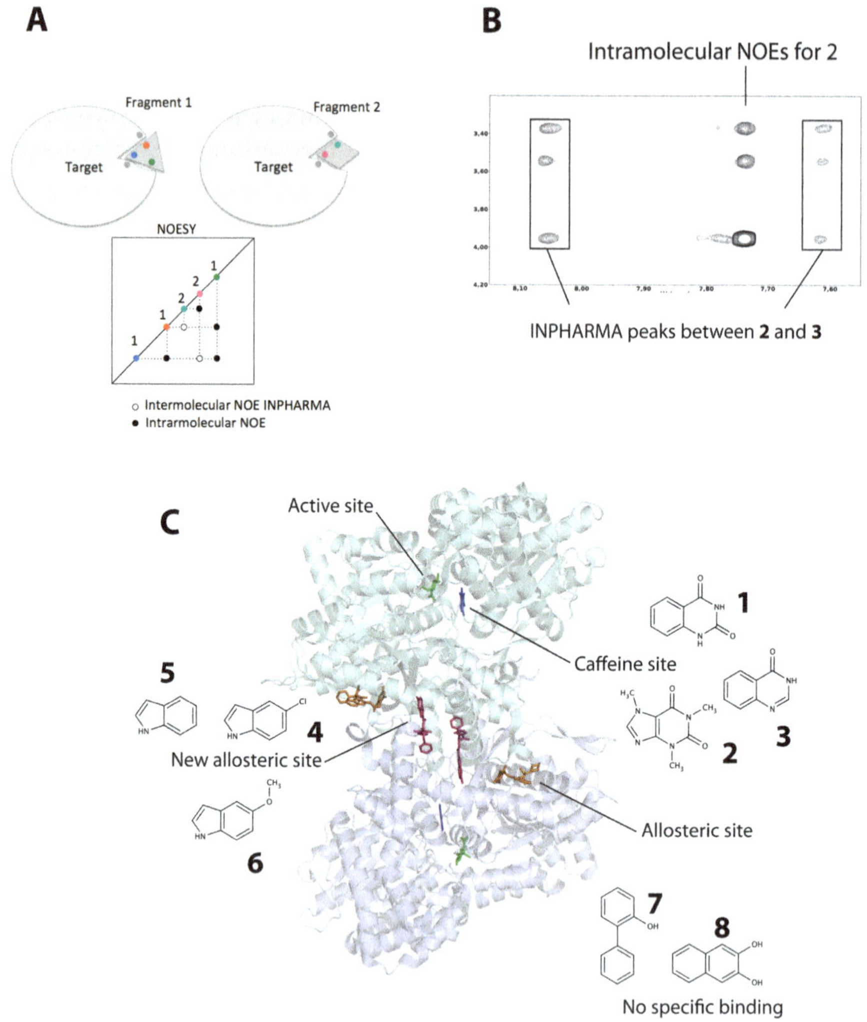

Figure 3.3 INPHARMA experiments for assessing the binding site of fragments. (A) Principle of the INPHARMA experiment. When fragments 1 and 2 bind competitively to the same binding site, NOE peaks are observed between protons of ligand 1 (displayed as blue and red spheres, respectively) and protons of ligand 2 (pink and cyan, respectively) if these protons exchange magnetisation with the same protein proton (in grey). Expected INPHARMA peaks (intermolecular NOEs) are schematically displayed on a theoretical NOESY spectrum. (B) INPHARMA experiment on the GP protein. INPHARMA peaks are observed between methyl groups of caffeine (**2**) and aromatic protons of fragment **3**, showing that fragment **3** binds in the caffeine-binding site. The NOESY experiment is recorded at 600 ms mixing time, with 10 μM GP. (C) 3D structure of the GP protein. Compounds that bind in the four distinct binding sites of the protein are shown in green (active site), blue (inhibitor–caffeine site), orange (allosteric site) and pink (new-allosteric site). Using INPHARMA experiments, compounds **1** and **3** were shown to bind to the caffeine site, compounds **5** and **6** were shown to bind to the new-allosteric site where the chloroindole **4** binds. The frequent hitters **7** and **8** displayed no specific binding.

proton upon binding. By contrast, only protons from compound **1** that are in close proximity to protons from compound **2** in the ternary complex (protein/compound **1**/compound **2**) exhibit NOESY (ILOE) peaks. ILOE peaks can be observed if the protein is deuterated, while INPHARMA requires the transfer *via* the protein protons. The experiment should also be recorded with serum albumin as a control. While it is likely that the fragments also bind HSA, it is unlikely that the ternary complex is formed with HSA, preventing the observation of the ILOE. When feasible, competition experiments between compounds **1** and **2** can be run and should correlate the NOESY data: no competition with ILOE peaks, while a competition should be observed in the case of INPHARMA peaks. Finally, INPHARMA peaks should be observed at mixing times much lower than for ILOE peaks. Consequently it is recommended that NOESY experiments are recorded with mixing times varying from 100 to 600 ms.

3.3 2D Protein-Observed Experiments

One major difference between protein-observed experiments and ligand-observed experiments is the requirement for labelled protein samples for the acquisition of ^{15}N-NMR spectra, ^{13}C-NMR spectra or any other isotope-edited/filtered experiments. Whilst computational tools for the assignment are available,[59] facilitating the task, the protein-observed experiments are still time-demanding for proteins larger than 40 kDa, and require deuteration. Nevertheless, interesting information can be obtained with large proteins by using selective labelling with ^{13}C methyl groups. For these experiments the proteins must be soluble and not aggregate at 50–100 μM and should be stable at 20–30 °C for several days. When protein domains are isolated from the full-length protein, 2D NMR spectra can be used to assess folding, stability and solubility of the different constructs and optimise the domain length.

The nature of protein NMR spectra recorded in the presence of ligands depends on the rate of association/dissociation between the protein and the ligand. When the exchange rate k_{ex} is slow on the chemical shift timescale ($k_{ex} < \Delta\delta$ with $\Delta\delta = \delta_{bound} - \delta_{free}$), the NMR peaks of the free protein disappear and the peaks of the bound protein appear. In this case, the binding of molecules can be observed, but affinity measurement is not feasible and binding-site identification can be rather complicated. When exchange is fast, *i.e.*, when k_{ex} is greater than the chemical shift difference $\Delta\delta$, the NMR peaks move gradually from the δ_{free} chemical shift to the δ_{bound} chemical shift. For ligands with $K_D > 10$ μM, the exchange rate is likely to be fast. Nevertheless, it depends on the k_{on} rate that can sometimes be much slower than diffusion-controlled binding, which is typically 10^9 M^{-1} s^{-1}.

3.3.1 Protein Spectra for Fragment Screening

Protein-observed experiments are more robust than ligand-observed experiments and should be the method of choice for fragment screening against small proteins. One of the most interesting parameters for studying protein–fragment complexes is the protein chemical shift,[60] observed with ^{15}N-HSQC experiments. The ^{15}N-HSQC spectrum, called the *protein fingerprint*, shows one NMR signal for each observable amide group NH in the protein. The spectrum also contains signals for side-chains such as Asn, Gln and Trp residues, while those of Arg and Lys are usually not observed.

Upon fragment binding, the NMR peaks of the binding site residues shift, while the other regions of the protein are not perturbed (unless conformational changes occur). The specific binding of a fragment will generate chemical shift perturbations (CSPs) for a small portion of the protein residues (Figure 3.4). The method does not require knowledge of the protein function or the development of a target-specific assay, and can be done at the very beginning of a project in the absence of the resonance assignment. We have noticed that only about 1/3 of the hits identified by ligand-observed experiments were observed as hits in protein-observed experiments. This can be due to unspecific binding, to very weak binding (>10 mM), or to binding at multiple binding sites with low affinities. These binding events are additive in

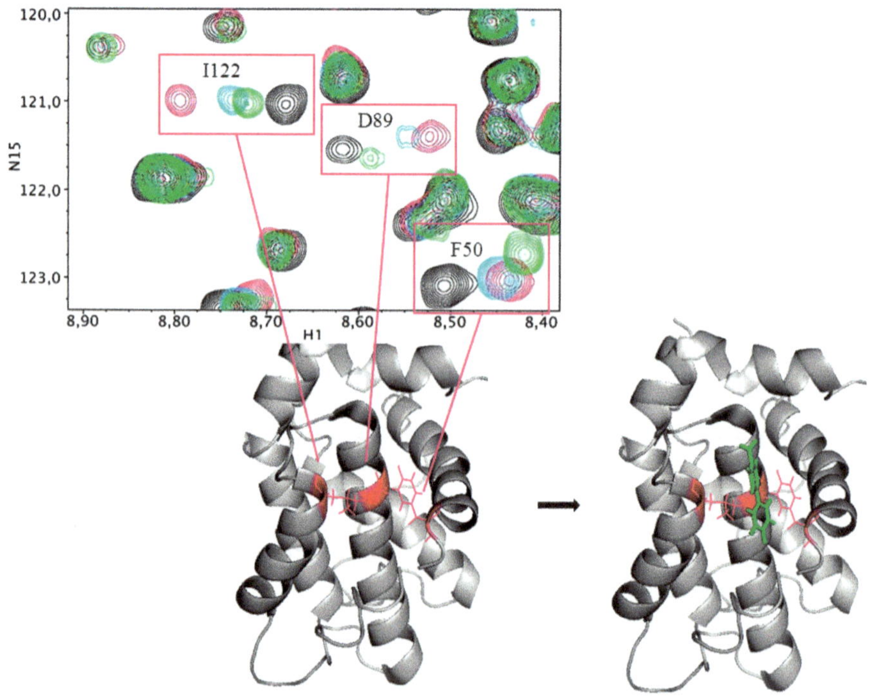

Figure 3.4 Chemical shift perturbations observed on the Bcl-xL protein NMR spectra upon fragment binding. The HSQC spectrum of the free protein (in black) is superimposed on three HSQC spectra of the protein bound to three fragments (pink, green and blue spectra). For the three fragments, significant CSPs are observed for residues F50, D89 and I122. Mapping of the CSPs onto the 3D structure of the protein helps identify the binding site of the fragments.

the STD and WaterLOGSY experiments, leading to a detectable binding signal, while they translate into very small undetectable CSPs in the HSQC spectra.

For protein-observed spectra, the protein concentration required varies from 30 to 200 μM and the ligand concentration from 100 to 1000 μM (DMSO stocks vary from 0.1 to 1 M). For larger proteins (>40 kDa), TROSY-type experiments can be used, in combination with protein deuteration. Using Sofast-HMQC, the group of Fesik reported running 200 spectra in 1 day (15 kDa protein, 40 μM, at 500 MHz with a cryo-probe).[6] Screening by monitoring the methyl groups in 2D ^1H–^{13}C spectra of selectively labelled samples at Leu, Val and Ile residues of proteins is particularly efficient due to the long relaxation time of the methyl resonances. Additionally, methyl groups represent efficient probes for ligand-binding events that involve hydrophobic interactions.

By comparing the CSP induced by a series of ligands, one can ascertain whether the ligands bind in the same binding site or not. When a reference ligand is known and available, it can be used to search for fragments that induce shifts in similar regions of the protein. Against the aspartyl protease, BACE1, peaks corresponding to active site residues were first established by binding known peptide inhibitors. Ten thousand fragments were then screened in clusters of 12 and hits identified which showed a CSP for some of those peaks.[61] It is also possible to search for fragments that induced CSP in other regions, to identify novel protein-binding sites. To avoid artefacts due to solvent interactions, it is mandatory to record protein spectra in the presence of an increasing concentration of DMSO, to remove CSPs due to solvent binding.

3.3.2 Protein Spectra for Ligand Affinity Measurements

NMR spectra of the protein can be used to measure affinities in the low-affinity range (μM to mM), when protein resonances are in fast exchange. The titration does not require the assignment of the protein resonances; therefore affinity can be measured at the beginning of a new drug-design project. If the binding affinity reaches the low μM range, intermediate exchange can be observed. Under these conditions HSQC experiments cannot be used for K_D determination.

Chemical shift changes are quantified for each titration point using a combined chemical shift value (see eqn (3.2)).

$$\Delta = \sqrt{(\Delta\delta_H)^2 + (c\Delta\delta_N)^2} \qquad (3.2)$$

where $\Delta\delta_H$ and $\Delta\delta_N$ are the chemical shift changes of amide proton and nitrogen, respectively. A recent review discusses the different values that have been tested and reported for the scaling factor c, which ranges from 0.1 to 0.25.[60] The observed CSP as a function of ligand concentration is fitted by non-linear regression that accounts for single-site binding, using eqn (3.3):

$$\Delta = \Delta\delta_{max} \times \frac{\left\{[P_T] + [L_T] + K_D - \sqrt{([P_T] + [L_T] + K_D)^2 - 4[P_T][L_T]}\right\}}{2[P_T]} \qquad (3.3)$$

where $\Delta\delta_{max}$ is the maximum chemical shift change at saturation, $[P_T]$ and $[L_T]$ the total protein and ligand concentration, and K_D the dissociation constant. This fitting is possible only if one binding mode is observed for the ligand, whereas multiple binding modes will give rise to additional effects on the chemical shifts, and the progression of the observed Δ upon ligand addition will no longer be linear.

To have an approximate idea of the range of affinities for a series of fragments, a method has been proposed that requires recording 4 HSQC spectra with the ligands at 0, 25, 50 and 800 μM.[62] It is important to keep in mind that the value of $\Delta\delta_{max}$ observed is not related to the affinity and cannot be used to rank the fragments. For an accurate measurement of the K_D, the concentration of the protein and the ligand should be close to K_D.[60] Theoretical simulations show that the protein concentration should ideally be $K_D/2$, while ligand concentrations should span the range from $0.4[P_T]$ to $10[P_T]$ ($K_D/5$ to $5K_D$), with 15 titration points.[60] Additionally, to obtain a precise evaluation of the K_D, it is recommended to fit a large number of titration curves simultaneously.

3.3.3 Protein Spectra for Ligand-Binding Site identification

When the protein resonances (at least the backbone atoms) have been assigned, ^{15}N-HSQC experiments are used to identify the binding site of ligands. Typically, one spectrum is recorded in the free state, and two spectra are recorded in the presence of two ligand concentrations to assign unambiguously the resonances of the bound protein (the peaks shift linearly with the ligand concentration). By contrast, strong binding (sub mM) will make re-assignment mandatory, because only the saturated bound form will be observed in the spectrum.

It is not always straightforward to determine the binding site from ^{15}N-HSQC experiments, because the chemical shift mapping is based on the shifts observed on amine groups, while the protein interaction surface is principally amino acid side-chains. CSPs observed for amine groups do not necessary imply that the corresponding side-chains interact with the ligand. In addition, CSPs are frequently observed far away from the binding site. This can be due to unspecific interactions, or to conformational changes. It is important here to verify that the shifts observed are dependent on the ligand concentration. There is no direct way to separate the chemical shift changes due to the binding, from the chemical shift changes due to conformational change, because in either case the changes are similarly dependent on the bound ligand ratio. Nevertheless, as described in Section 3.3.6, the comparison of experimental fragment-induced CSPs with back-calculated CSPs helps to identify regions of the proteins undergoing re-arrangement upon fragment binding.

McCoy and Wyss have developed a method to identify the binding sites of ligands that contain aromatic rings,[63] in the absence of protein conformational change. They approximate the ring current effect due to aromatic rings by a point-dipole, which is then used to estimate the location of the centre of the electron density of the aromatic ligand. Ligand positions that are allowed by the experimental data are represented by dot densities, and the highest dot density corresponds to the centre of the ligand aromatic ring.

The use of additional nuclei usually provides better discrimination of the binding site, but no consensus has been observed.[60] Therefore, due to the lower requirement in protein sample and shorter time for resonance assignment, the ^{15}N-HSQC experiments remain the best option. There is an exception for large proteins, where the CSPs observed for methyl groups are used. A common way is to specifically label the methyl groups of valine, leucine and isoleucine residues. Very recently, experiments have been recorded with unlabelled proteins, using a bovine serum albumin sample at 1 mM.[64]

3.3.4 Protein Spectra for Finding Second-Site Binders: SAR-by-NMR

A linked-fragment strategy was the first fragment-based approach reported by the Abbott company.[1] The premise of the strategy is the linking of two fragments that can simultaneously bind to proximal sites on a protein would yield high-affinity ligands. 2D NMR can be used to identify the first- and second-site ligands. Structural information of the ternary complex is then required to design linked molecules. The approach is not always feasible. In many cases, only one hotspot is present on a protein surface, and all hits found in a primary screen bind at this site. Additionally, Fesik reports that second-site binders are almost always much weaker in affinity ($K_D > 1$ mM), making their detection difficult.[6] Thus, a second screen must be conducted under conditions in which the first site is occupied. One option is to saturate the first site with a known binder, as small and potent as possible, and to use high fragment concentrations for the second-site screen. Fesik proposes screening a subset of the library containing only fragments with water solubility greater than 5 mM (this may mean re-organisation of the library).[6] This issue is nicely illustrated by the deconstruction of Bcl-xL inhibitors into fragments, showing that the binding site is not systematically conserved. In this case, deconstruction of Bcl-xL inhibitors, which span two hotspots, gave fragments which were shown to bind only to the first hotspot.[47] Second hotspot binding can be achieved only if the first binding pocket is saturated, and the affinities are much weaker.

3.3.5 Protein Spectra for Assessing Ligand-Binding Mode

Computational methods, when combined with NMR experimental restraints, could represent an efficient way to rapidly gather information on ligand–protein complexes without full structure determination. For example, CSPs can be used for docking, to limit the search to the binding site of the ligand. CSPs can also be used as distance restraints, as in the HADDOCK,[65] AutodockFilter,[66] Bigger,[67] and LIGDOCK[68] programs.

Quantitative approaches that rely on the back-calculation of proton CSPs have also been developed, ^{15}N-CSP being much more difficult to interpret. Initially, McCoy and Wyss calculated the ring current shift induced by aromatic rings,[69] using the SHIFTS program from Case's group. This program calculates the chemical shifts of proteins. Thus the ligand effect is calculated using an aromatic amino acid as a probe, and is based on the ring current effect of aromatic rings. These calculations are based on the hypothesis that the ring current effect of aromatic ligands accounts for the largest contribution of the observed ligand-induced CSP on the protein NMR spectra, in the absence of protein conformational change. Other groups have also published examples where the CSP back-calculation is used for post-filtering of protein–ligand structures generated by docking.[70] The ring current effect is calculated with the Haigh–Maillon model:

$$\Delta\delta = fB\sum_{ij} S_{ij}\left(\frac{1}{r_i^{\,3}} + \frac{1}{r_j^{\,3}}\right) \tag{3.4}$$

where f is the ring-specific intensity factor (*e.g.*, 1.00 for benzene-type rings) and B is the target nucleus factor ($B = 7.06 \times 10^{-6}$ Å for amide protons). The sum is over pairs of bonded atoms in the ring with $ij \in \{12, 23, 34, 45, 56, 61\}$ for a six-membered ring, r_i and r_j are the distances from the ring atoms i and j to the amide proton of the protein, S_{ij} is the area of the triangle formed by atom i and j and the amide proton projected onto the plane of the aromatic ring.

In addition to the ring current effect, back-calculation of CSPs can include electric field effects. The group of Gohlke developed a different approach, with the introduction of a hybrid scoring function to guide the docking instead of filtering the docked positions.[71] In this approach, only the ring current effect is back-calculated, using the Haigh–Maillon model.

All these approaches could be extremely valuable in FBDD for assessing the binding modes of the fragments.[72] As shown below, we believe that the CSP back-calculation could also be used for assessing whether the protein undergoes re-arrangement upon fragment binding.

3.3.6 Protein Spectra for Probing Protein Conformational Change Upon Fragment Binding

Recently, Surade and Blundell reviewed various factors that are known to make drug discovery difficult.[73] Amongst those factors, protein structural adaptation for protein–ligand recognition was considered a major problem in rational drug design. Regarding fragments, the same authors recall that while fragments can induce protein re-arrangement, it is less likely, both kinetically and thermodynamically.[73] In a recent review[74] by the Astex company, the comparison of free and fragment-bound crystal protein structures revealed that for 25 biological targets, 12 exhibit re-arrangement larger than 5 Å. For example, in HSP90, some fragments induced the so-called *collapsed helix conformation* around Gly108. As reported,[74] fragments are capable of causing induced fit, providing that there is no substantial energetic cost that cannot be offset by the formation of favourable interactions.

Ligand-induced CSP includes ring current effects (due to aromatic rings), electric fields caused by charged atoms, magnetic anisotropy (due to carbonyl for example), and hydrogen bonds. These direct binding effects, excluding hydrogen bonds, can be calculated using empirical equations. The ring current effects spread around the aromatic ring, perturbing a large number of protons, while hydrogen-bond effects induce CSP contributions for isolated protons. We consider therefore that the back-calculation of the CSP through the ring current effect and the electric field effect are relevant when dealing with aromatic fragments. We have recently proposed that the comparison of experimental CSPs and back-calculated CSPs could be used to assess whether a protein undergoes conformational re-arrangement upon fragment binding.[75] We have performed such an analysis with Bcl-xL, by analysing the complex between Bcl-xL and 4′-fluoro-1,1′-biphenyl-4-carboxylic acid. The biphenyl fragment was identified by the Abbott company as the best hit from 2D NMR screening, and the dissociation constant was calculated to be 300 μM.[76] We have compared the experimental CSP to CSPs calculated for 200 positions docked into the free structure of Bcl-xl (PDB:1R2D).

As illustrated in Figure 3.5, a large number of residues exhibit experimental CSP values much larger than the calculated CSP values. While CSPs are calculated for a saturated binding site, the Bcl-xL binding site is not fully saturated when the experimental CSPs are measured. Therefore, theoretically, calculated and experimental CSP should follow the same pattern, with a scaling factor related to the affinity. Here, large unexplained experimental CSPs are observed, for residues located in and around the binding site, indicating that these are due to a re-arrangement of the protein upon fragment binding (Figure 3.5). To demonstrate that the method is robust, we have performed residual dipolar coupling (RDC) measurements on the free and fragment-bound protein. RDC values depend on the orientation of an NH vector to the alignment tensor of the molecule and are typically used as structural restraints. RDC can also be back-calculated using 3D protein structures and compared to experimental RDC to validate the 3D structures. We have used eight different Bcl-xL 3D structures, in the free, ligand-bound and peptide-bound forms. For the RDC measured in the presence of the fragment, the correlation between measured and back-calculated RDC was clearly better with ligand-bound structures, than with free 3D structures or peptide-bound 3D structures. The RDC experiments confirm that Bcl-xL undergoes conformational changes upon fragment binding and also indicate that the re-arrangement is similar to the one observed in the presence of ligands such as in the structure 2O2M, and not as large as that observed in the presence of peptides such as Bak and Bad. RDC is more informative than CSP, but can be much more difficult to handle. Therefore we suggest using CSP back-calculation to assess whether the protein moves upon binding, and to identify the re-arrangement site. While induced-fit processes can be seen

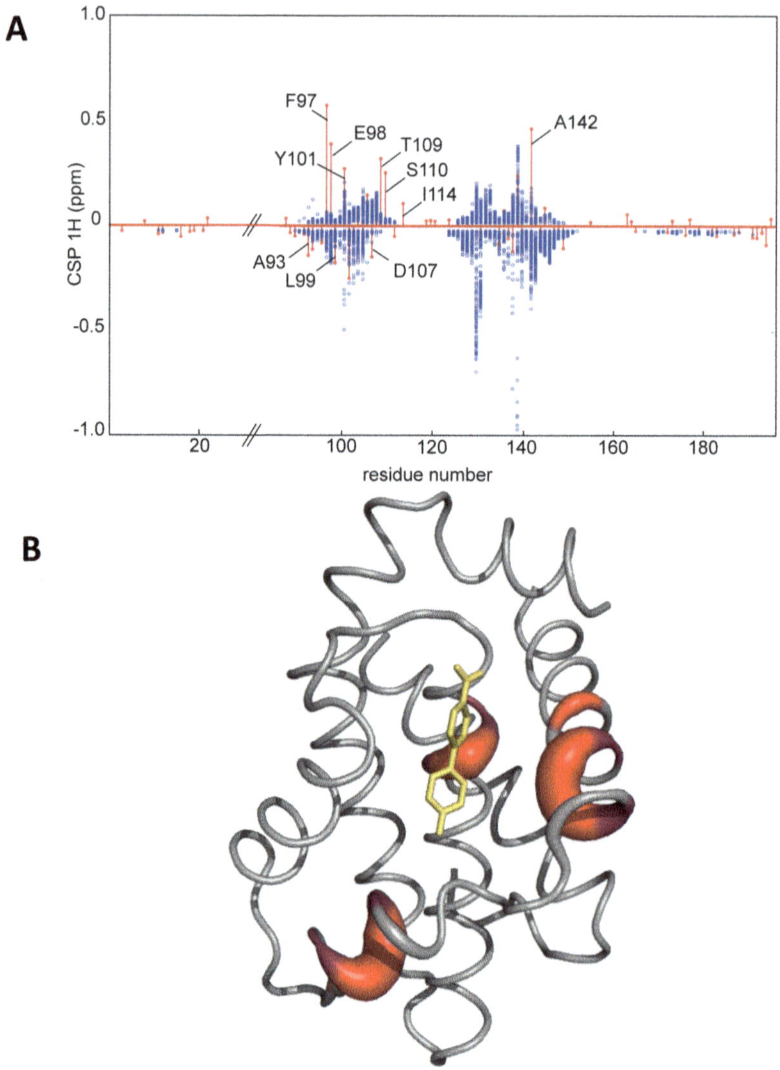

Figure 3.5 Comparison of experimental and back-calculated CSPs for the Bcl-xL protein in the presence of the fragment 4′-fluoro-1,1′-biphenyl-4-carboxylic acid. Adapted from ref. 75. (A) Experimental CSPs are displayed as red lines; CSPs calculated for 200 positions of the fragment 4′-fluoro-1,1′-biphenyl-4-carboxylic acid docked in the Bcl-xL binding site are displayed with blue spheres. The 200 complex structures were obtained using AutoDock. Residues exhibiting large disagreement (>0.06 ppm) between experimental and back-calculated CSPs are labelled. (B) The discrepancies between experimental and back-calculated CSPs are mapped onto the 3D structure of Bcl-xL. The results provide evidence for a conformational re-arrangement of Bcl-xL upon binding to the fragment 4′-fluoro-1,1′-biphenyl-4-carboxylic acid, as confirmed by residual dipolar coupling measurements.

as novel opportunities for interactions, it can also explain why soaking can sometimes fail to generate X-ray structures of protein–fragment complexes. Indeed, the binding pocket has to be accessible *via* a solvent channel, and the protein crystal should be able to tolerate conformational flexibility if an induced-fit process is required. If fragment screening has been carried out by 2D NMR experiments, then we consider that it is valuable to use CSP back-calculation, since no additional experiments are required.

3.4 Conclusion

NMR plays and will continue to play a major role in FBDD. The technique is a robust method for fragment hit identification, enabling the observation of both the protein and the compound. The technique generates few artefacts relative to other biophysical techniques, in particular for protein-observed experiments. Fesik recently claimed that NMR would become an even more valuable technique in FBDD in the future, with the use of shorter acquisition times in NMR experiments, and the enlargement of the fragment libraries.

NMR will also be more and more used as a structural method, with the development of computational tools using only a small number of experimental NMR restraints. NMR, as a reliable method for the detection of binding with very weak affinities in the mM range, will also remain one of the preferred methods for the identification of hits for challenging targets such as protein–protein interactions, or disordered proteins.

References

1. S. B. Shuker, P. J. Hajduk, R. P. Meadows and S. W. Fesik, *Science*, 1996, **274**, 1531.
2. D. Erlanson, in *Fragment-Based Drug Discovery and X-Ray Crystallography*, ed. T. G. Davies and M. Hyvönen, Springer, Berlin, 2012, vol. 317, p. 1.
3. R. Campos-Olivas, *Curr. Top. Med. Chem.*, 2011, **11**, 43.
4. C. Bissantz, B. Kuhn and M. Stahl, *J. Med. Chem.*, 2010, **53**, 5061.
5. J. B. Baell and G. A. Holloway, *J. Med. Chem.*, 2010, **53**, 2719.
6. M. Harner, A. Frank and S. Fesik, *J. Biomol. NMR*, 2013, **56**, 65.
7. H. L. Eaton and D. F. Wyss, in *Methods in Enzymology*, ed. L. C. Kuo, Academic Press, New York, 2011, vol. 493, p. 447.
8. T. Maurer, in *Methods in Enzymology*, ed. L. C. Kuo, Academic Press, 2011, vol. 493, p. 469.
9. C. Dalvit, *Drug Discovery Today*, 2009, **14**, 1051.
10. M. Pellecchia, I. Bertini, D. Cowburn, C. Dalvit, E. Giralt, W. Jahnke, T. L. James, S. W. Homans, H. Kessler, C. Luchinat, B. Meyer, H. Oschkinat, J. Peng, H. Schwalbe and G. Siegal, *Nat. Rev. Drug Discovery*, 2008, **7**, 738.
11. C. A. Lepre, in *Methods in Enzymology*, ed. L. C. Kuo, Academic Press, New York, 2011, vol. 493, p. 219.

12. J. Schultz, in *Fragment-Based Drug Discovery: A Practical Approach*, ed. E. R. Zartker and M. J. Shapiro, 2008.
13. M. Mayer and B. Meyer, *Angew. Chem., Int. Ed.*, 1999, **38**, 1784.
14. C. Dalvit, G. Fogliatto, A. Stewart, M. Veronesi and B. Stockman, *J. Biomol. NMR*, 2001, **21**, 349.
15. D. Li and R. London, *Biotechnol. Lett.*, 2002, **24**, 623.
16. V. M. Sánchez-Pedregal, M. Reese, J. Meiler, M. J. J. Blommers, C. Griesinger and T. Carlomagno, *Angew. Chem., Int. Ed.*, 2005, **44**, 4172.
17. A. Vulpetti and C. Dalvit, *ChemMedChem*, 2013, **8**, 2057.
18. T. Carlomagno, *Annu. Rev. Biophys. Biomol. Struct.*, 2005, **34**, 245.
19. S. Vanwetswinkel, R. J. Heetebrij, J. van Duynhoven, J. G. Hollander, D. V. Filippov, P. J. Hajduk and G. Siegal, *Chem. Biol.*, 2005, **12**, 207.
20. J. L. Wagstaff, S. L. Taylor and M. J. Howard, *Mol. BioSyst.*, 2013, **9**, 571.
21. J. Angulo, P. M. Enríquez-Navas and P. M. Nieto, *Chem.–Eur. J.*, 2010, **16**, 7803.
22. J. Angulo and P. Nieto, *Eur. Biophys. J.*, 2011, **40**, 1357.
23. Y. Xia, Q. Zhu, K.-Y. Jun, J. Wang and X. Gao, *Magn. Reson. Chem.*, 2010, **48**, 918.
24. K. Furihata, S. Shimotakahara, Y. Shibusawa and M. Tashiro, *Magn. Reson. Chem.*, 2009, **47**, 971.
25. K. Urihata, S. Shimotakahara, Y. Shibusawa and M. Tashiro, *Anal. Sci.*, 2010, **26**, 1107.
26. M. A. McCoy, M. M. Senior and D. F. Wyss, *J. Am. Chem. Soc.*, 2005, **127**, 7978.
27. M. Mayer and B. Meyer, *J. Am. Chem. Soc.*, 2001, **123**, 6108.
28. J. Angulo, B. Langpap, A. Blume, T. Biet, B. Meyer, N. R. Krishna, H. Peters, M. M. Palcic and T. Peters, *J. Am. Chem. Soc.*, 2006, **128**, 13529.
29. M. G. Szczepina, R. B. Zheng, G. C. Completo, T. L. Lowary and B. M. Pinto, *ChemBioChem*, 2009, **10**, 2052.
30. Z. Ji, Z. Yao and M. Liu, *Anal. Biochem.*, 2009, **385**, 380.
31. D. E. Scott, M. T. Ehebauer, T. Pukala, M. Marsh, T. L. Blundell, A. R. Venkitaraman, C. Abell and M. Hyvönen, *ChemBioChem*, 2013, **14**, 332.
32. S. Kemper, M. K. Patel, J. C. Errey, B. G. Davis, J. A. Jones and T. D. W. Claridge, *J. Magn. Reson.*, 2010, **203**, 1.
33. V. Jayalakshmi and N. R. Krishna, *J. Magn. Reson.*, 2002, **155**, 106.
34. V. Jayalakshmi and N. Rama Krishna, *J. Magn. Reson.*, 2004, **168**, 36.
35. V. Jayalakshmi, T. Biet, T. Peters and N. R. Krishna, *J. Am. Chem. Soc.*, 2004, **126**, 8610.
36. P. J. Hajduk, J. C. Mack, E. T. Olejniczak, C. Park, P. J. Dandliker and B. A. Beutel, *J. Am. Chem. Soc.*, 2004, **126**, 2390.
37. C. Dalvit, P. Pevarello, M. Tatò, M. Veronesi, A. Vulpetti and M. Sundström, *J. Biomol. NMR*, 2000, **18**, 65.
38. C. Dalvit, M. Fasolini, M. Flocco, S. Knapp, P. Pevarello and M. Veronesi, *J. Med. Chem.*, 2002, **45**, 2610.
39. D. M. Dias, I. Van Molle, M. G. J. Baud, C. Galdeano, C. F. G. C. Geraldes and A. Ciulli, *ACS Med. Chem. Lett.*, 2014, **5**, 23.

40. A. Gossert, C. Henry, M. J. Blommers, W. Jahnke and C. Fernández, *J. Biomol. NMR*, 2009, **43**, 211.
41. J. Hu, P.-O. Eriksson and G. Kern, *Magn. Reson. Chem.*, 2010, **48**, 909.
42. E. C. Johnson, V. A. Feher, J. W. Peng, J. M. Moore and J. R. Williamson, *J. Am. Chem. Soc.*, 2003, **125**, 15724.
43. C. Ludwig, P. J. A. Michiels, X. Wu, K. L. Kavanagh, E. Pilka, A. Jansson, U. Oppermann and U. L. Günther, *J. Med. Chem.*, 2007, **51**, 1.
44. C. Ludwig, P. J. A. Michiels, A. Lodi, J. Ride, C. Bunce and U. L. Günther, *ChemMedChem*, 2008, **3**, 1371.
45. C. Dalvit, D. Caronni, N. Mongelli, M. Veronesi and A. Vulpetti, *Curr. Drug Discovery Technol.*, 2006, **3**, 115.
46. I. Krimm, J.-M. Lancelin and J.-P. Praly, *J. Med. Chem.*, 2012, **55**, 1287.
47. S. Barelier, J. Pons, O. Marcillat, J.-M. Lancelin and I. Krimm, *J. Med. Chem.*, 2010, **53**, 2577.
48. C. B. Post, *Curr. Opin. Struct. Biol.*, 2003, **13**, 581.
49. D. Li, E. DeRose and R. London, *J. Biomol. NMR*, 1999, **15**, 71.
50. J. Orts, C. Griesinger and T. Carlomagno, *J. Magn. Reson.*, 2009, **200**, 64.
51. J. Chen, Z. Zhang, J. L. Stebbins, X. Zhang, R. Hoffman, A. Moore and M. Pellecchia, *ACS Chem. Biol.*, 2007, **2**, 329.
52. P. Sledz, H. L. Silvestre, A. W. Hung, A. Ciulli, T. L. Blundell and C. Abell, *J. Am. Chem. Soc.*, 2010, **132**, 4544.
53. B. Becattini and M. Pellecchia, *Chem.–Eur. J.*, 2006, **12**, 2658.
54. M. F. Rega, B. Wu, J. Wei, Z. Zhang, J. F. Cellitti and M. Pellecchia, *J. Med. Chem.*, 2011, **54**, 6000.
55. B. Becattini, S. Sareth, D. Zhai, K. J. Crowell, M. Leone, J. C. Reed and M. Pellecchia, *Chem. Biol.*, 2004, **11**, 1107.
56. D. W. Begley, S. Zheng and G. Varani, *Chem. Biol. Drug Des.*, 2010, **76**, 218.
57. C. Rademacher, J. Guiard, P. I. Kitov, B. Fiege, K. P. Dalton, F. Parra, D. R. Bundle and T. Peters, *Chem.–Eur. J.*, 2011, **17**, 7442.
58. I. Krimm, *MedChemComm*, 2012, **3**, 605.
59. E. Schmidt and P. Güntert, *J. Am. Chem. Soc.*, 2012, **134**, 12817.
60. M. P. Williamson, *Prog. Nucl. Magn. Reson. Spectrosc.*, 2013, **73**, 1.
61. Y.-S. Wang, C. Strickland, J. H. Voigt, M. E. Kennedy, B. M. Beyer, M. M. Senior, E. M. Smith, T. L. Nechuta, V. S. Madison, M. Czarniecki, B. A. McKittrick, A. W. Stamford, E. M. Parker, J. C. Hunter, W. J. Greenlee and D. F. Wyss, *J. Med. Chem.*, 2009, **53**, 942.
62. L. Fielding, *Curr. Top. Med. Chem.*, 2003, **3**, 39.
63. M. A. McCoy and D. F. Wyss, *J. Am. Chem. Soc.*, 2002, **124**, 11758.
64. M. Quinternet, J.-P. Starck, M.-A. Delsuc and B. Kieffer, *Chem.–Eur. J.*, 2012, **18**, 3969.
65. C. Dominguez, R. Boelens and A. M. J. J. Bonvin, *J. Am. Chem. Soc.*, 2003, **125**, 1731.
66. J. Stark and R. Powers, *J. Am. Chem. Soc.*, 2007, **130**, 535.
67. X. Morelli, A. Dolla, M. Czjzek, P. N. Palma, F. Blasco, L. Krippahl, J. J. G. Moura and F. Guerlesquin, *Biochemistry*, 2000, **39**, 2530.

68. S. N. Rao, M. S. Head, A. Kulkarni and J. M. LaLonde, *J. Chem. Inf. Model.*, 2007, **47**, 2159.
69. M. McCoy and D. Wyss, *J. Biomol. NMR*, 2000, **18**, 189.
70. M. Cioffi, C. Hunter, M. Packer, M. Pandya and M. Williamson, *J. Biomol. NMR*, 2009, **43**, 11.
71. D. González-Ruiz and H. Gohlke, *J. Chem. Inf. Model.*, 2009, **49**, 2260.
72. C. Aguirre, T. Ten Brink, O. Cala, J. F. Guichou and I. Krimm, *J. Biomol. NMR*, 2014, **60**, 147.
73. S. Surade and T. L. Blundell, *Chem. Biol.*, 2012, **19**, 42.
74. C. W. Murray, M. L. Verdonk and D. C. Rees, *Trends Pharmacol. Sci.*, 2012, **33**, 224.
75. C. Aguirre, T. ten Brink, O. Walker, F. Guillière, D. Davesne and I. Krimm, *PLoS One*, 2013, **8**, e64400.
76. A. M. Petros, J. Dinges, D. J. Augeri, S. A. Baumeister, D. A. Betebenner, M. G. Bures, S. W. Elmore, P. J. Hajduk, M. K. Joseph, S. K. Landis, D. G. Nettesheim, S. H. Rosenberg, W. Shen, S. Thomas, X. Wang, I. Zanze, H. Zhang and S. W. Fesik, *J. Med. Chem.*, 2005, **49**, 656.

CHAPTER 4

Current Status and Future Direction of Fragment-Based Drug Discovery: A Computational Chemistry Perspective

IAN D. WALL*[a], MICHAEL M. HANN[a], ANDREW R. LEACH[a], AND STEPHEN D. PICKETT[a]

[a]GlaxoSmithKline, Gunnels Wood Road, Stevenage, Hertfordshire, SG1 2NY, UK
*E-mail: ian.d.wall@gsk.com

4.1 Introduction

In recent years fragment-based drug discovery (FBDD) has evolved from being a niche activity carried out by specialist companies and small groups of enthusiasts within pharmaceutical companies to become an established component of the hit-to-lead strategy of most modern drug-discovery organisations.[1] Indeed, the field has developed to such a level that the first drug identified with FBDD has now reached the market.[2] Best practice is beginning to emerge, such as the desired property space for a fragment library,

RSC Drug Discovery Series No. 47
Fragment-Based Drug Discovery
Edited by Steven Howard and Chris Abell

Published by the Royal Society of Chemistry, www.rsc.org

the importance of biophysical methods in hit identification and characterisation, and the need for extremely close working relationships between medicinal and computational chemists. However, differences in opinion and approach do still exist between various practitioners, such as the importance of biochemical screening and the extent to which X-ray crystallography is needed and used.

Computational chemistry has an important role to play throughout an FBDD programme, from screening-set design to hit expansion and fragment growth and evolution. A vast range of tools and approaches can be and have been applied at these various stages. There are many computational methods that can be transferred from traditional drug-discovery approaches; sometimes their application needs to be modified and in some cases certain tools have been developed specifically to facilitate certain parts of FBDD.

This chapter will touch on experimental aspects, but the major focus will be on the use of computational methods developed to maximise the efficiency of the various stages of FBDD. We will discuss novel tools developed at GSK and the application of existing tools, using case studies to exemplify their utility. The current state-of-the-art will be presented and areas of potential development in the future highlighted. We also compare FBDD with other hit-identification methods and discuss their potential for synergy.

4.2 Fragment Sets

Two key concepts underpin the value of FBDD. The first is the correlation between physico-chemical properties, particularly lipophilicity, and drug attrition.[3,4] Therefore, starting with lower molecular weight and less lipophilic compounds, such as fragments, is likely to be beneficial. Secondly, chemical space is much smaller for fragments (number of heavy atoms (HAs) < 20), than for the molecules that are traditionally screened in methods such as high-throughput screening (HTS) (HAs < 35).[5] Consequently, a well-designed fragment library can cover the available chemical space efficiently with a much smaller screening set. These principles, combined with the throughput of the experimental methods used to test the compounds, have shaped the strategies used to design fragment libraries.

A variety of approaches to fragment library design have been reported.[6–8] These reports, and our personal experiences, suggest a number of qualities a desirable fragment set should exhibit. Compounds in such a set should be:

- Soluble and not aggregate at the (often high) concentrations used in FBDD
- Pure and stable
- Easily followed up (chemically feasible elaboration)
- Diverse and from relevant areas of chemical space
- Attractive to medicinal chemists (avoid undesirable functional groups)

Guidelines have also been proposed for the desired physico-chemical properties of a fragment set such as the *Rule of 3* from Astex.[9] Although other groups have varied their cut-offs from those proposed by Astex,[10–13] the general principles of keeping size, log*P*, *etc.* low still hold. As with many such 'rules' the astute practitioner will see the Rule of 3 as a guideline rather than a rule.

In the design of the GSK set we chose to work back from desired ligand efficiencies, and minimum potencies detectable, to calculate HA and log*P* cut-offs. For example, working on the basis that the minimum reliable level of detection in our assays is dissociation constant (K_D) = 100 μM (pK_D = 4) and a desirable ligand efficiency (LE) for a hit is 0.3, a HA limit of 18 was agreed (Figure 4.1). Another important consideration is avoiding the inclusion of compounds that are known to cause assay interference. Consequently, we chose to remove compounds that had frequently been identified in previous screening campaigns, and those containing structures likely to be associated with interference. Substructures commonly responsible for interference have been reported.[14]

The most appropriate size for fragment sets remains a topic of discussion. Core fragment libraries are diverse sets and hence, the probability of identifying desirable hits can be increased by increasing the size of the set. This needs to be balanced against the throughput of the hit-detection techniques. Our strategy is to have a core fragment set supplemented with a number of additional bespoke sets that can be screened alongside the core set as appropriate for the target. Bespoke sets are designed for a variety of reasons, to target specific protein classes such as kinases[15] or bromodomains,[16] to contain certain chemical moieties such as acids or bases, or for use with

		pK_D					
HAC	~M_w	2	3	4	5	6	7
8	115	0.34	0.51	0.69	0.86	1.03	1.20
9	130	0.30	0.46	0.61	0.76	0.91	1.07
10	144	0.27	0.41	0.55	0.69	0.82	0.96
11	158	0.25	0.37	0.50	0.62	0.75	0.87
12	173	0.23	0.34	0.46	0.57	0.69	0.80
13	187	0.21	0.32	0.42	0.53	0.63	0.74
14	202	0.20	0.29	0.39	0.49	0.59	0.69
15	216	0.18	0.27	0.37	0.46	0.55	0.64
16	230	0.17	0.26	0.34	0.43	0.51	0.60
17	245	0.16	0.24	0.32	0.40	0.48	0.56
18	259	0.15	0.23	0.30	0.38	0.46	0.53
19	274	0.14	0.22	0.29	0.36	0.43	0.50
20	288	0.14	0.21	0.27	0.34	0.41	0.48
21	302	0.13	0.20	0.26	0.33	0.39	0.46
22	317	0.12	0.19	0.25	0.31	0.37	0.44

LE 0.3-0.4 0.4-0.5 0.5-0.7

Figure 4.1 Table of ligand efficiency (LE) as a function of heavy atom count (HAC) and pK_D (−logK_D). This information was used to decide the HA limit for our fragment collection.

particular screening methods including ^{19}F-NMR spectroscopy[17] or X-ray crystallography.[18,19]

Recently there has been a lot of enthusiasm for increasing the three-dimensionality (3D) of fragment libraries. Incorporation of 3D into diversity assessment has been a long-standing challenge within chemoinformatics using concepts such as pharmacophore coverage[20] and shape.[21–23] The challenge is how to interpret the output in the context of multiple conformers. A library diverse in substructure motifs will necessarily encompass the full range of accessible chemotypes (and shapes that they can adopt). Thus in constructing the GSK library we have taken the following approach. The library has a high Fsp3 (fraction of sp^3 atoms) content[24] and good coverage of the 2D substructure frameworks[25] and pharmacophoric coverage as defined by reduced graphs.[26–28] Additionally, we have ensured that the fragments have several accessible vectors for growth, as assessed by substitution patterns of the fragment core in databases of available compounds. Whilst having 3D character in the final clinical candidate is a desirable feature to increase solubility, specificity and ultimately druggability, if the fragment library is well designed this can be added during the optimisation process.

Other groups have taken alternative approaches. Of particular note is the publication, by the 3D consortium, of a library enriched in 3D molecules using principle moments of inertia (PMI) to assess shape.[29] Increasing 3D shape increases the complexity of the compounds and therefore reduces the chance of them being identified as hits, in accordance with our previously published complexity model.[30] One key argument for the development of the library from the 3D consortium is that it will have an increased chance of identifying hits for protein–protein interaction (PPI) targets. This assumption is based on the observation that PPI ligands tend to have more 3D character than ligands of other protein classes, which may be a consequence of having to disrupt large areas of interaction surface. Future screening activities will reveal whether this has been a fruitful strategy.

4.3 Performance of Fragment Sets

We have analysed the performance of the GSK core fragment set against 18 recently screened proteins. The first encouraging finding was that there were very few promiscuous hits with ~0.5% of compounds hitting four or more targets. On the other hand, around 50% of compounds have yet to show activity (in NMR, SPR or biochemical screens) against any of the targets.

Figure 4.2 shows the conversion rate from screening hit to X-ray structure across these 18 proteins. Biochemical hits are more likely to yield a structure than NMR hits: this observation is probably driven, at least in part, by the fact that NMR can generally detect weaker hits than a biochemical assay. SPR has the lowest conversion rate from hit to X-ray, however it should be noted that SPR hits include those compounds that show linear, as opposed to saturable, binding curves and, therefore, will undoubtedly contain some non-specific binders.

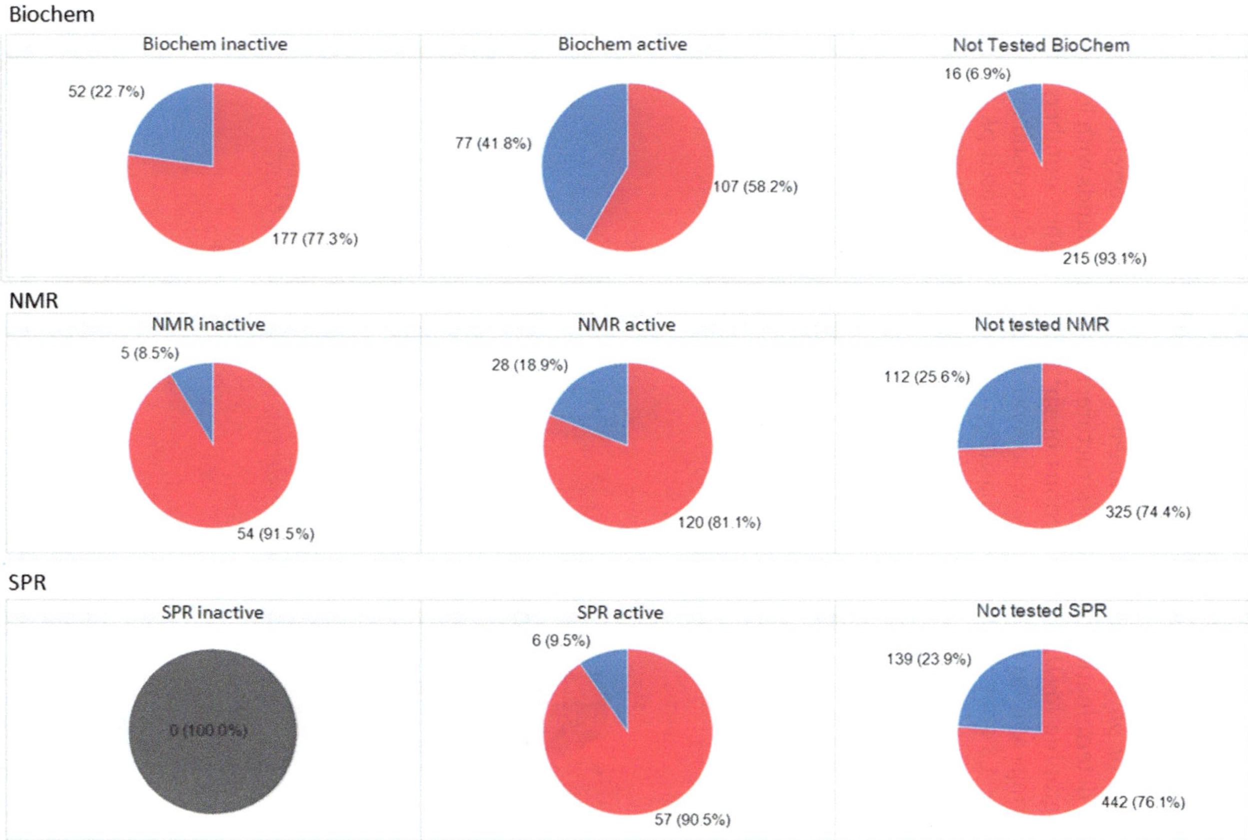

Figure 4.2 Illustration of how fragments perform in X-ray crystallographic screening *vs.* screening method (biochemical assay, NMR, SPR). Pie charts show the number of compounds (percentage in brackets) tried in crystallography that gave a structure (blue) and did not give a structure (pink). Rows represent an assay type and columns whether a compounds was active, inactive, or not tried using a given method. Results are from fragment screens against 18 different proteins.

There are those who believe high-concentration biochemical assays are an inferior method for screening fragment sets when compared to biophysical methods. Although in some cases this is true, in our experience a well-configured biochemical assay can add a lot of useful information. We thus consider the two approaches to be complementary and the use of biophysical and biochemical assays in tandem is, in our view, the most effective approach.

A number of physico-chemical parameters were found to correlate with successful determination of an X-ray structure; those compounds with lower chromatographic log*D*, lower fractions of sp^3 atoms, higher numbers of aromatic rings and no basic centres were more likely to give a crystal structure. Figure 4.3 shows the effect of basicity (Chemaxon calculated $pK_a \geq 8$) on the screening success of the various screening methods. In particular,

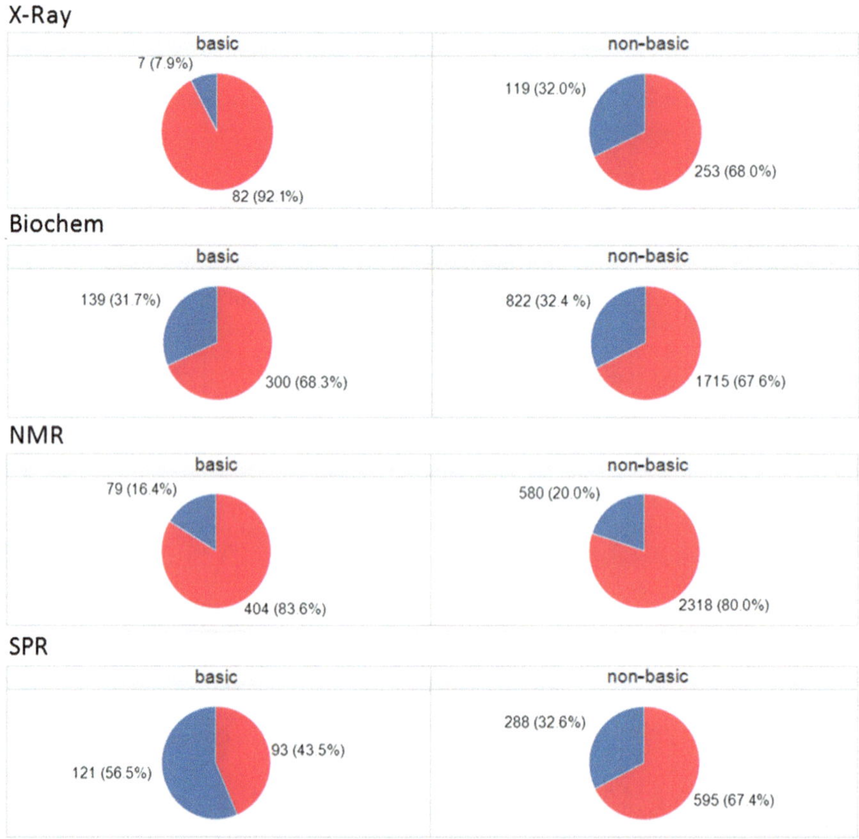

Figure 4.3 Hit rates for basic compounds *vs.* non-basic compounds according to experimental method. Pie charts show the percentage of compounds that appear as active (blue) and inactive (pink). The rows are the four methods (X-ray, biochemical assay, NMR and SPR) and the columns are the presence or absence of a basic centre.

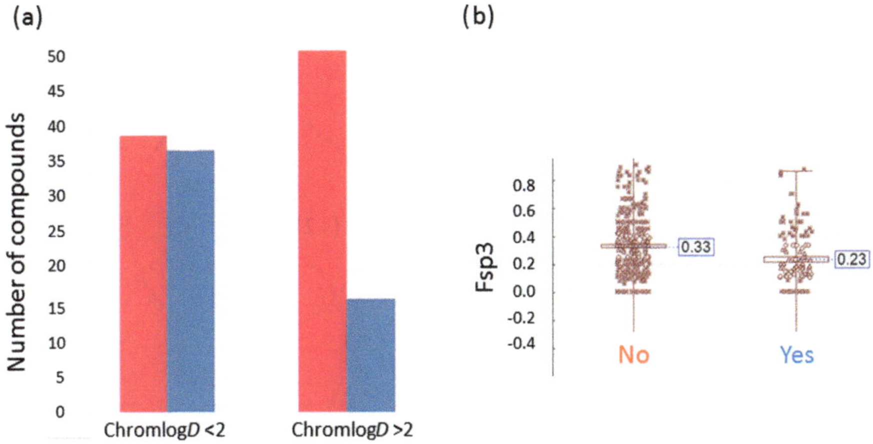

Figure 4.4 Relationship between physico-chemical properties and crystallographic success. (a) Compounds tried in crystallography that gave a structure (blue) and did not give a structure (pink). (b) Fraction of sp^3 (Fsp3) centres as a function of compounds tried in crystallography that gave a structure (Yes) and did not give a structure (No).

it is interesting that SPR success increases when a basic centre is present, whereas X-ray success is reduced. It is possible this is due to basic centres correlating with false positives in SPR. Figure 4.4(a) shows that compounds with a chromatographic $\log D < 2$ are much more likely to give a crystal structure than those with chromatographic $\log D > 2$. Figure 4.4(b) shows that those compounds which give a crystal structure have a lower fraction of sp^3 centres than those where no structure can be obtained.

Further analysis of the success in obtaining crystal structures for hits from various screening methods is given in Figure 4.5, this time for a single illustrative project, comparing STD (saturation transfer difference) NMR to SPR. The figure clearly reveals fragments that are independently identified with more than one method, are more likely to give structures – a commonly observed trend. Hits that are active in NMR and SPR are more likely to yield a crystal structure than those that are active in just one method. A particularly eye-catching result, in this case, was that all the compounds that showed both saturable binding curves in SPR and activity in an NMR assay gave a structure.

4.4 Screening Hit to X-Ray

The majority of FBDD programmes use X-ray crystallography as a hit-validation step. The reasons for this are two-fold. Firstly, the observation of a crystal structure gives added confidence that weakly binding hits obtained from screening are real and acting through an appropriate mechanism (*e.g.*, binding to a site of interest and in a one-to-one stoichiometry). Secondly, optimisation of the hit series can often be directed more easily with structural information to guide the design. Consequently, generating as many structures as possible for

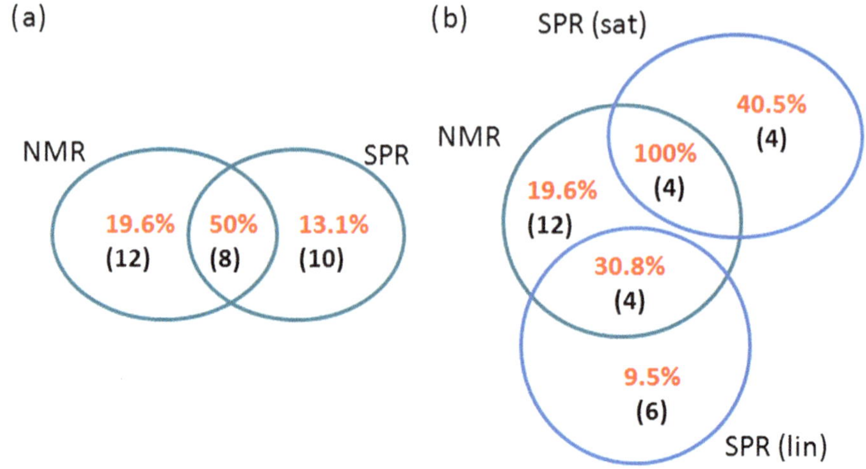

Figure 4.5 The percentage (in red) of screening hits tried in protein X-ray crystallography that resulted in a crystal structure. Absolute number of successful structures is shown in black. Hits are grouped according to the screening method (NMR and/or SPR) in which the hit was active. (a) NMR *vs.* SPR and (b) NMR *vs.* SPR; where SPR is split into linear binding curves (lin) and saturable binding curves (sat).

genuine binders is essential to maximise the number of validated hits and ultimately the success of the programme. Given that crystallographic throughput can be limited, analyses such as those presented above can help prioritise those hits most likely to be successful in crystallography.

Another important consideration is how to pursue those screening hits that do not initially yield a crystal structure. For a number of programmes we have observed that using substructure and similarity searching around these hits can identify close analogues that do produce structures. Figure 4.6 shows an example from one of our earliest fragment programmes, which initially highlighted the importance of this approach.

To build on this observation, we have established a comprehensive workflow, called FindAnalogues, for searching both GSK and external compound collections for analogues of initial hits (Figure 4.7). FindAnalogues uses a combination of substructure searching, Tversky similarity[31,32] and reduced graph similarity.[26] The methods were chosen and combined to provide an efficient way of identifying compounds of a similar size to the original hit. Molecules are pre-processed prior to substructure searching to facilitate identification of simple changes such as carboxylic acid to amide. The use of reduced graph descriptors allows for variation in heterocyclic rings whilst maintaining the overall pharmacophoric nature of the molecule. This type of modification is sometimes referred to as *fragment optimisation*, that is to say the interactions made by the initial fragment hit are optimised before efforts to grow and introduce new interactions are attempted. In FindAnalogues, filters are generally set to allow the addition of a single small moiety to the hit molecule, typically permitting an increase in size of up to 5–6 HAs.

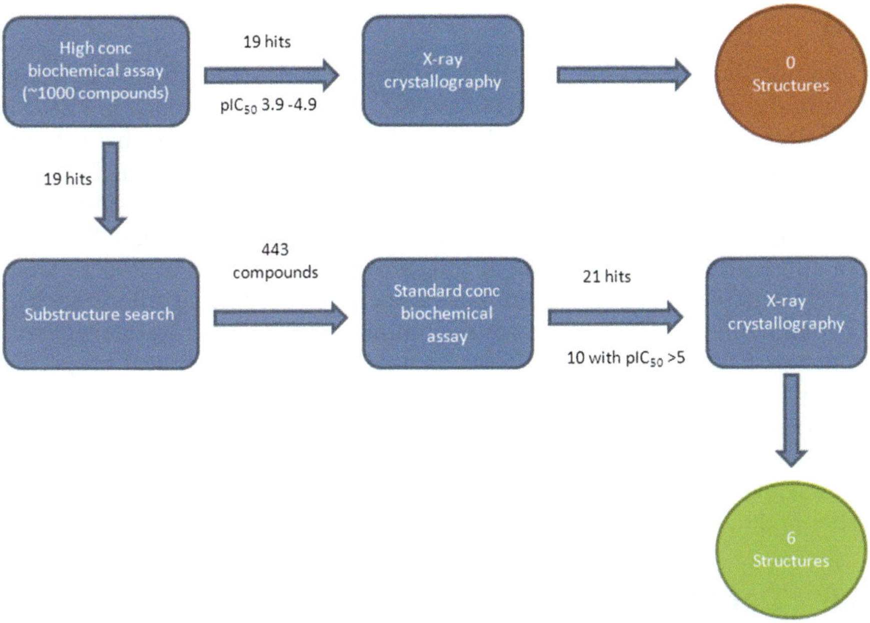

Figure 4.6 Flow chart showing progression of hits to X-ray crystallography directly and after analogue searching.

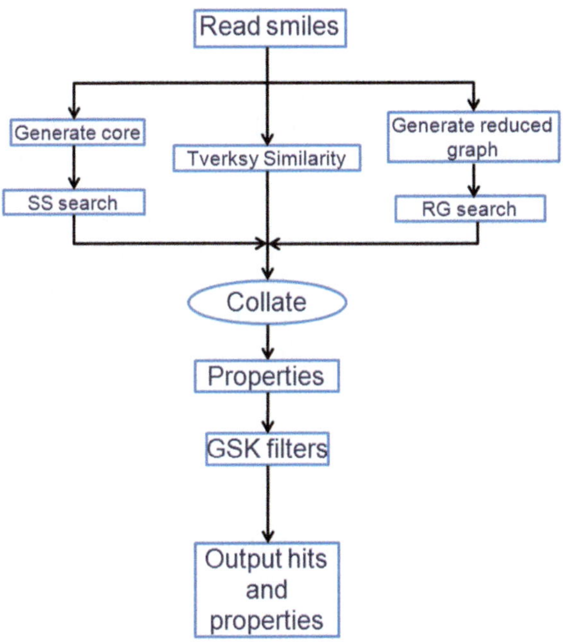

Figure 4.7 FindAnalogues workflow. The FindAnalogues programme uses a selection of substructure and similarity searching methods to search for analogues of fragment hits in compound databases.

4.5 Hit Clustering

Once hits have been identified they are generally clustered to aid identification of the most appropriate hits to pursue. A common approach in FBDD is to cluster the compounds by binding mode and growth vectors, that is to say compounds that make similar pharmacophoric interactions and can be grown to access similar regions of the pocket are clustered together. This process is normally done manually, although methods such as protein–ligand interaction fingerprints[33] could be used to add some automation to the process. Once the clusters have been identified they can be prioritised using criteria such as LE, potential to target key areas of the binding pocket, structure–activity relationships (SARs) from any available analogues, physico-chemical properties and synthetic tractability. To facilitate the exercise, we have developed a simple Pipeline Pilot™ workflow to summarise the key information relating to a given cluster, an example of which is given in Figure 4.8. This summary sheet allows quick and easy visualisation of a representative structure, average properties and the strengths and weaknesses of a particular cluster.

4.6 Data Handling and Communication

Success in FBDD, and more generally in structure-based drug design, is widely accepted to depend on close interaction between medicinal chemist, computational chemist and crystallographer. At GSK, we are firm believers in this philosophy and in the need to make chemistry and design decisions based on data-driven hypotheses, not on ease of synthesis or poorly considered diversity. To facilitate this interaction, two important tools have been developed. The first tool, known as Cornershop,† combines key project data from primary assays, selectivity assays, liability assays and measured or calculated physico-chemical properties in a single visualisation environment. The Cornershop queries databases of experimental data and accesses models of molecular properties, pulling all the data for compounds relevant to the project into a CSV (comma separated values) file. The resulting file can then be viewed in a preferred visualisation tool, typically at GSK this is Spotfire.[34] A screenshot of the web interface to Cornershop is shown in Figure 4.9. The Cornershop updates automatically, or at user request, and emails alerts when new data have been added. Spotfire templates can be incorporated to control the default view a user accesses. In addition the tool tracks how the average properties, such as molecular weight, log*P* and LE, of a series change over the course of time; this is valuable both for monitoring the progress of a project as it evolves and for reviewing the project strategy after its completion. Accessing these data through Spotfire also makes interfacing with our standard (Q)SAR and Matched Molecular Pair tools[35] straightforward, allowing data summaries to be produced easily, rather than relying on less reliable alternatives such as Excel spreadsheets, pieces of paper, or team members'

---

†So named by analogy with the typically larger "data mart".

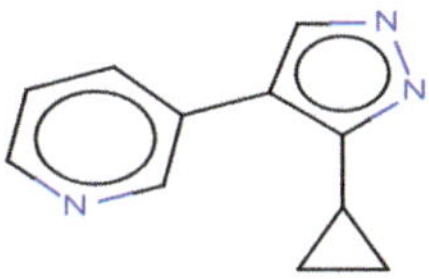

Calculated Properties	
Regno	
mw	185.23
clogP	1.2
tPSA	41.57
NumHeavy	14
Acceptors	2
Donors	1
Rotbonds	2

Measured Properties	
Regno	
Target pIC50	4.8
LigEff	0.47
LigEffAT	0.46
CLND_sol	>=391 uM
CHROM_LOGD_PH74	1.6
HSA_PCT_BIND	84.2
AGP_PCT_BIND	53.0

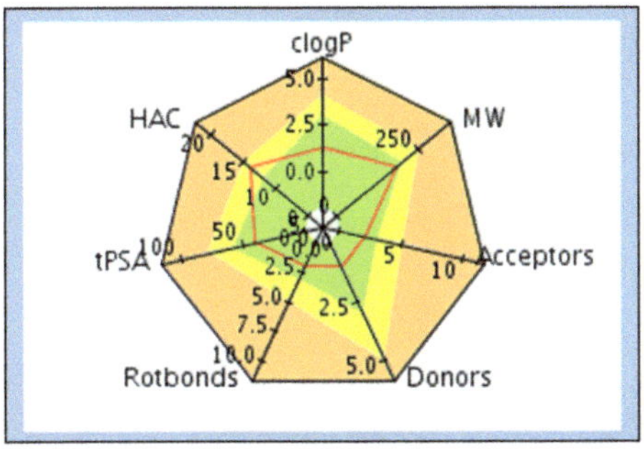

Summary Comments					
Properties of Validated Hit Series	Current data pIC50	LE	LLEAT	Does the series hit the criteria	Comments
LE between 0.3 & 4.3					
Best in cluster					
Clear X-ray binding mode					
Astex rule of 3					
Synthetic tractability					
Expectation that elaborated molecules will be patentable					
Off target effects					
EXP					
Stability data					

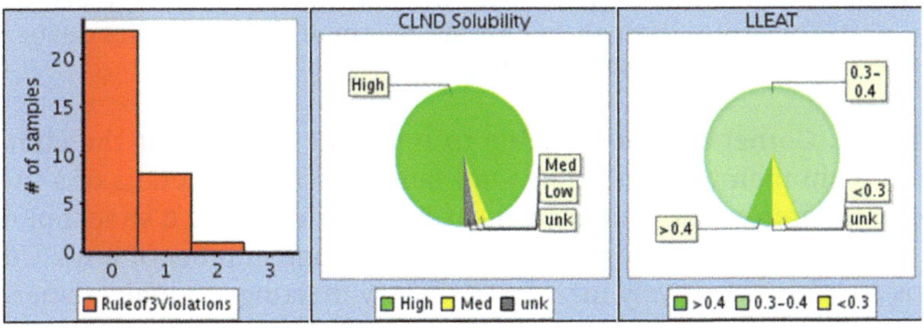

Figure 4.8 Automatically generated template which summarises the key information relating to a given cluster of fragment hits. Includes both properties for a representative compound and also average properties for the cluster. Comments describing the strengths and weaknesses of a particular cluster are also captured. Generated with Pipeline Pilot (http://accelrys.com/products/pipeline-pilot/).

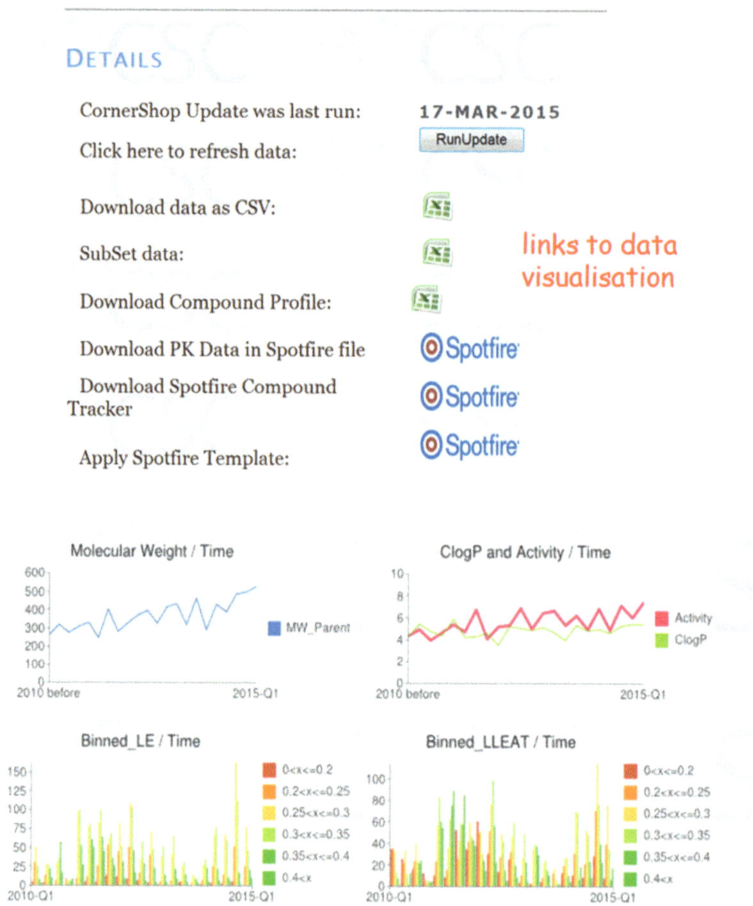

Figure 4.9 Cornershop: a computational tool for managing project data. The tool is automatically refreshed with new data. Data can be visualised and presented through Spotfire. Average molecular properties can be monitored *via* 'property tracking'.

memories. Cornershop is available to scientists throughout the company. Project team members, managers not familiar with the details of a project, or other interested individuals can easily access data, see a snapshot of the current status and quickly query the data in a huge number of ways. Cornershops are also extensively used in chemistry meetings to make data-driven decisions. Calculated and measured properties (biochemical, DMPK and physico-chemical) can be easily accessed and SAR questions can be answered on the fly. This information facilitates group discussion, aids compound design, and allows scientists to prioritise which compounds should be progressed through the screening cascade.

The second tool, Structure Viewer, is a web interface for visualising 3D protein–ligand structures, including electron density where available,

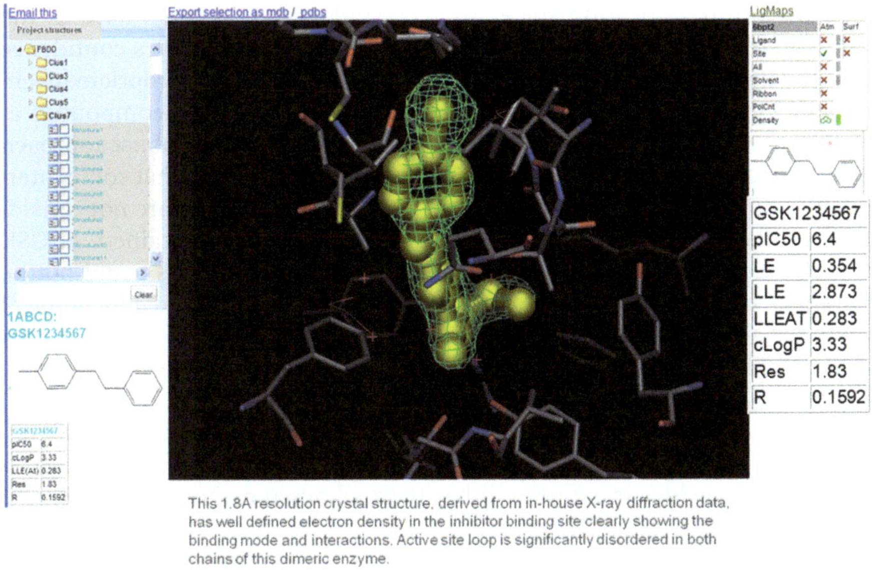

Figure 4.10 Protein Structure Viewer allowing access to multiple aligned crystal structures alongside basic biological and physico-chemical data.

alongside simple biological and physico-chemical data. The Structure Viewer pulls crystallographic information from the corporate database of X-ray structures and superposes them in a common reference frame, before storing them. Dockings or other information contained in PDB files can be imported manually and stored. Basic biological and physico-chemical data are extracted from the project Cornershop and incorporated in the view. The crystallographer's comments are also included in the interface so any key information contained is not overlooked. Such an easily accessible web tool for visualising protein structures makes structure-based design possible for all interested team members, as individuals or as small groups. The Structure Viewer tool is also widely used in chemistry brainstorming meetings to stimulate ideas for interactions that could be targeted, strained conformations that could be stabilised and to rule out ideas that clearly are not compatible with the protein structure. This also makes it quick to identify more complex considerations that will need detailed modelling outside the meeting. The Structure Viewer interface is shown in Figure 4.10.

4.7 X-Ray Crystallography

One of the central tenets of FBDD has been that a high-throughput crystallography system is essential to the success of a programme. Structural data are particularly important at two distinct stages – the hit-confirmation stage and the hit-optimisation stage.

At the hit-identification stage, X-ray crystallography is often used as the ultimate test of the validity of a hit. Obtaining a structure gives confidence that the ligand is interacting with the protein through an appropriate mechanism of action, an important consideration when screening compounds at high concentration and detecting low levels of potency. This robust approach has the drawback of being prone to false negatives, meaning that some interesting compounds with an appropriate mechanism of action are not considered validated hits because an X-ray structure cannot be obtained. At GSK we have examples of compounds for which crystal structures could not be obtained that were validated by orthogonal biophysical and biochemical methods and successfully optimised. Conversely, it should not be forgotten that it is also possible for electron density data to be misinterpreted, effectively leading to X-ray false positives.

One should also be cautious about inferring a causal link between crystallographically observed binding modes and activity in a bioassay, particularly if the ligand resides in a pocket other than a known orthosteric (substrate) binding site. For example, activity in a biochemical assay may be due to non-specific assay interference (*i.e.*, a false positive) rather than a mechanism associated with a crystallographically observed binding mode. In other cases, activity in a biochemical assay may be associated with a protein conformation which is not accessible within the environment of a protein crystal, perhaps due to occlusion by symmetry related protein molecules. Again, any crystallographically observed binding modes may not be biologically relevant.

In other instances it has been possible to use crystallography to provide insight into the binding of weakly active compounds, which cannot be detected by other methods. For compounds whose K_D values cannot be determined by traditional assays, crystallography can be used as a binary (binder, non-binder) assay to assess the success of early rounds of chemistry. Such an approach does require discipline in limiting the amount of chemistry that is carried out before potency is increased.

In the hit-optimisation stage, structural information is the primary data used in the design process. It is often proposed that increasing the amount of structural data available increases the chance of success. Indeed, some groups advocate solving crystal structures of almost every compound synthesised. Structural information is undoubtedly a very valuable asset, but the added value of obtaining numerous structures of closely related compounds with similar activities is questionable. We advocate the approach of generating structural data that is likely to provide key insight, rather than routinely generating structures without questioning the value. A typical strategy would involve initially generating structures that are representative of the molecules of interest and then focussing on compounds that raise structural questions. These include unexpectedly weak or potent compounds, those targeting specific interactions, or those predicted to have close contacts that could cause conformational changes in the ligand or protein.

A frequently asked question is whether a successful FBDD programme can be carried out with little or no structural information. At GSK, fragment series have been successfully optimised without any X-ray structures for the given series. However, this typically still requires crystal structures of the protein with other chemotypes bound, and potent ligands in other series, allowing a combination of docking and ligand-based design to drive the compound optimisation.

It is also possible to supplement a single or very small number of X-ray structures by using NMR to derive constraints to guide the docking of ligands.[36] The question of whether an FBDD programme could be carried out totally in the absence of structural information, simply using ligand-based design and traditional medicinal chemistry approaches, remains unanswered. There is no inherent reason why such an approach should not be viable, but we are not aware of any cases where it has successfully been carried out.

4.8 Structure-Based Design

FBDD projects normally generate large numbers of crystal structures, both in the hit-identification and hit-to-lead stages for the reasons outlined above. Clearly with such an emphasis on generating structural information, structure-based design is the pre-dominant modelling approach in FBDD.

At the simplest level, structure-based design is carried out by visualisation of binding modes and interactions, looking at satisfaction of hydrogen bonds, surface complementarity, unfilled subpockets and other manual design approaches. Such analysis is now commonplace amongst experienced computational and medicinal chemists and is facilitated by tools such as the Structure Viewer, mentioned above. This availability of structural data and the close link between modelling, crystallography and medicinal chemistry has put design at the heart of projects and has been key to the success of the fragment approach.

Undoubtedly the most widely used modelling tool in structure-based design is protein–ligand docking. This will not be discussed in detail in this chapter as it has been extensively reviewed elsewhere, both in terms of the underlying methodology[37] and specifically the docking of fragments.[38] In particular there has been discussion around the accuracy of predicting binding modes for fragments compared to larger molecules. However, data from Astex[38] suggests that the accuracy of binding mode prediction is not correlated with molecular size or weight, but with the LE of the compound. Docking methods are used within the FBDD community both as a lead optimisation tool and as a virtual screening tool to identify additional ligands to supplement standard screening libraries in the hit-identification stage. Suffice to say that docking is central to FBDD projects, still requires a lot of careful and balanced interpretation, but can be a very useful tool in hypothesis generation, especially when combined with some of the other techniques discussed below.

One of the key early objectives in an FBDD programme is to optimise the LE of the initial hit. The thorough investigation of a range of closely related structures at this point is important. In particular, optimising the surface contact between ligand and receptor can give good boosts in potency. In Figure 4.11(a) it can be seen that at first glance the shape match between the ligand and the pocket looks good. However, replacing a hydrogen atom with a methyl group (Figure 4.11(b)) gives an exquisite fit and a boost in potency of greater than 1 log(the fluorine analogue gives an intermediate boost in potency).

A very clear example of how the potency of a fragment hit can be optimised with a minimal amount of change to the structure can be found in the HSP90 example from Murray *et al.*[39] They took a 790 μM hit, and by adding six HAs were able to obtain a boost in potency of >1 000 000 fold.

Once the LE of the initial hit has been optimised, the focus turns to evolving the molecule towards a lead and ultimately a drug. Since fragment hits are weak binders, this will inevitably involve increasing the target potency of the compound. To give the best chance of success it is important to do this in conjunction with monitoring and optimising the physico-chemical properties, secondary pharmacology, toxicity risks and other factors that contribute

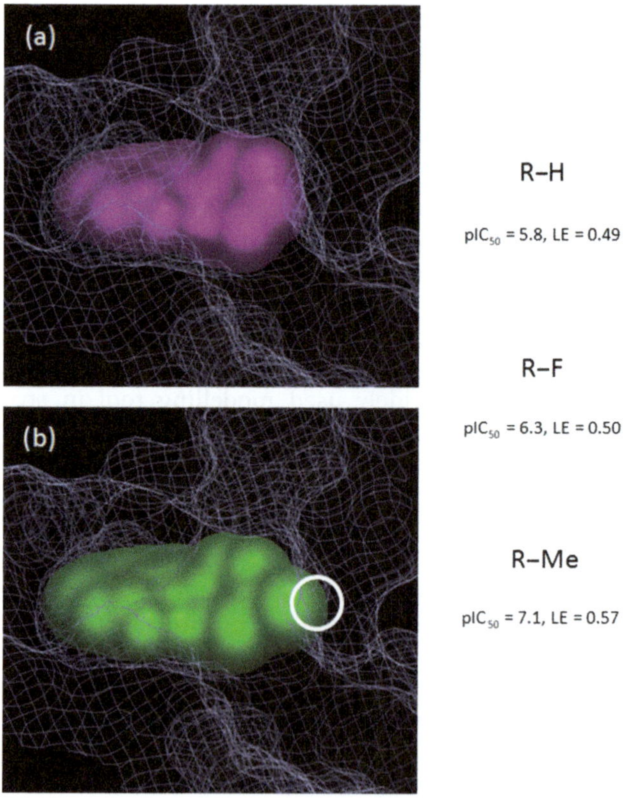

Figure 4.11 (a) Structure of fragment (magenta) showing a small unfilled space within the binding site. (b) Addition of a methyl group circled in the green image gives an exquisite fit to the binding site and a large increase in potency. An intermediate-sized fluoro substituent gives an intermediate increase in potency.

to producing a drug. Two primary approaches can be taken to improving the potency of hits; fragment linking or targeted fragment growth. In our experience the fragment-linking method can improve potency, but generally leads to larger, less ligand efficient compounds – a trend that seems to hold true in literature examples.[40–42] A recent poll on the Practical Fragments blog (http://practicalfragments.blogspot.co.uk) suggests the preference for fragment growth is shared by the wider community, with a much larger percentage of respondents having found fragment growth to work well than those who have found fragment linking to work well. It is, however, difficult to get a good assessment of the tractability of the linking approach due to the limited number of real-life examples where compounds bind in two sites close enough to be linked. When adopting the fragment-growing approach, the growth can be directed either by targeting unsatisfied protein interactions, or targeting interactions made by other fragments without specifically linking them. A more in-depth summary of the various approaches to fragment optimisation, summarising literature examples, has been produced by Congreve *et al.*[43]

Condensing the information from a large number of crystal structures into a concise summary of interaction hotspots within an active site can be a valuable way of exploiting extensive structural data. One simple approach to this is the Hotspots tool, within the Goldmine suite[44] from the Cambridge Crystallographic Data Centre (CCDC). This tool simply calculates the propensity for a given pharmacophoric feature at a given point in the active site across a set of pre-aligned crystal structures. Figure 4.12 shows an example of this applied to 60 liganded crystal structures of a phospholipid-binding protein. It

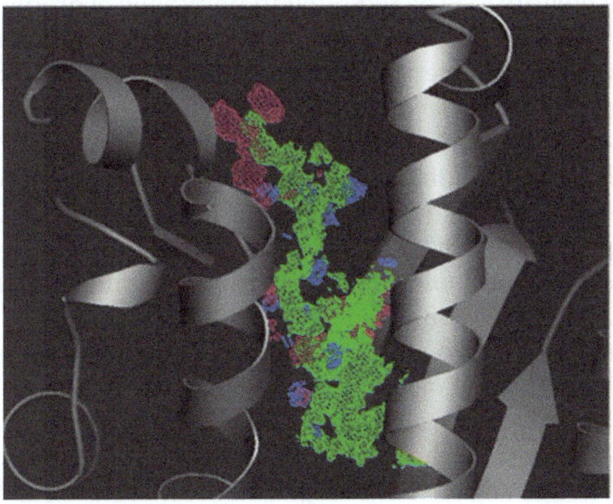

Figure 4.12 Use of the Hotspots tool from Goldmine to summarise pharmacophoric hotspots for a phospholipid-binding protein; based on 60 liganded crystal structures. The analysis reflects the propensity for hydrogen-bond acceptors (pink), hydrogen-bond donors (blue) and lipophilic motifs (green).

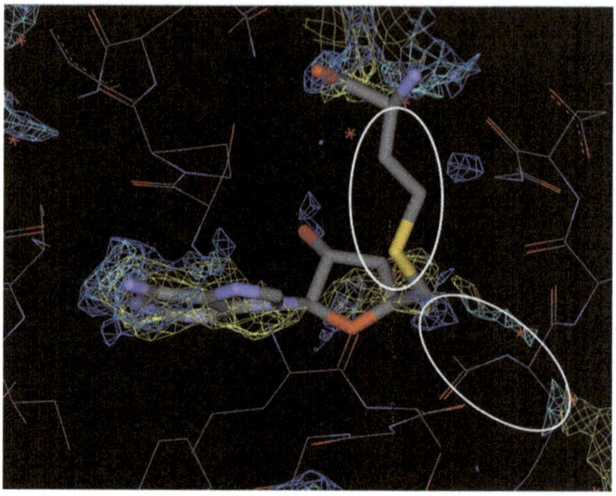

Figure 4.13 GRID maps for a SAM binding protein. Contours show regions where binding of a methyl probe (yellow), water probe (light blue) and positively charged nitrogen (dark blue) are preferred. The adenine binding region (left), where all fragment hits bound, is favourable for all probes. Extension of compounds into the circled regions, where there is very little contouring, gave no increase in potency.

shows a clear preference for hydrogen-bond acceptors in the top (phosphate binding) region of the active site and a clear preference for hydrophobic features in the middle and bottom (lipid binding) regions. Interestingly it also highlights some areas in the lipid-binding region where hydrogen-bonding groups may be tolerated. Such information can be very useful in identifying regions where polarity may be tolerated in nominally hydrophobic regions.

Another way of mapping potentially interesting information onto the active site of a protein is using Superstar from the CCDC.[45] This tool looks at the propensity of pharmacophoric features in an active site by reference to their frequency of observation across known crystal structures (*i.e.*, the PDB or CCDC).

Alternatively, computational tools can be used to predict binding hotspots in active sites. One tool used extensively at GSK is GRID.[46] GRID uses a range of computational probes that are systematically placed at grid points across the active site, and the molecular interaction potential is calculated for a given probe at each grid point. A molecular interaction field is built up over the site for each probe, and can be contoured at a given energy level. Figure 4.13 shows the application of GRID to an *S*-adenosyl methionine (SAM) binding enzyme. In this case, strong contours can be seen in the region where the adenine of SAM binds, on the left of the pocket as shown. When a fragment screen was carried out against this target, all the hits clustered in this region. As the compounds were grown and extended towards the right-hand side of the pocket, potency was increased to around 100 nM (from hits of around 10 μM). However, attempts to extend into the circled regions yielded compounds that made additional interactions but did not result in an increase in potency. In this case, GRID contours are observed in regions where

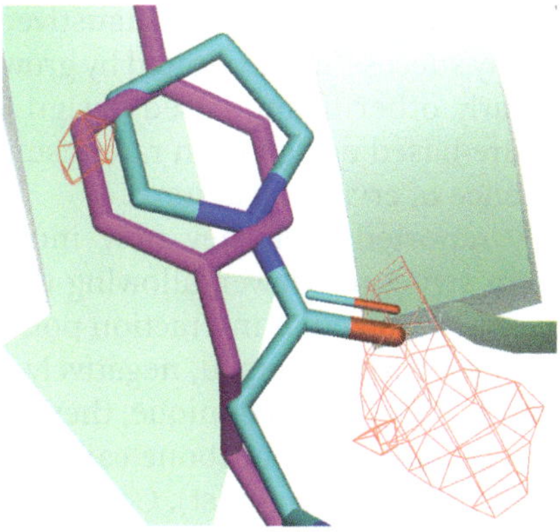

Figure 4.14 Application of a GRID for compound optimisation. Identification of a carbonyl GRID feature (red mesh) suggested replacement of the lipophilic phenethyl motif (magenta) with a more polar acyl pyrrolidine motif (cyan). Solubility was increased whilst maintaining potency.

productive interactions could be made, and are absent in regions where it was not possible to achieve potency gains.

Figure 4.14 shows a second example of the application of GRID. In this case the magenta compound with a *para*-substituted phenyl ring was the lead molecule. Adjacent to the linker, GRID predicts a region of the pocket with a high propensity for carbonyl moieties (red mesh). By changing the linker and the ring from the carbon-linked phenyl to an acyl pyrrolidine, it was possible to maintain potency whilst introducing polar atoms and removing an aromatic ring, resulting in an improvement in solubility.

The output from the aforementioned active site mapping tools has been incorporated into Structure Viewer, making it accessible to a broad audience and facilitating its use in project meetings, brainstorming sessions and so forth.

4.9 Ligand-Based Design

The wealth of protein structural information in FBDD has made structure-based design the dominant modelling technique reported for FBDD projects. However, it is important not to be blinkered by this information alone, and to recognise that there are many powerful ligand-based modelling tools that can be used to complement the structural information. Methods for summarising and understanding SARs, assessing ligand strain and overlaying different classes of ligands have been shown to impact drug-discovery programmes and their utility should be exploited on FBDD targets.

SAR analysis tools can often add important understanding to the design process. For example we have found that the use of systematic R-group analysis can highlight important nuances in data that are not easily picked up by less exhaustive methods. In a recent programme, such an analysis revealed that occupying a given subpocket gave an important boost to potency, a fact that had been overlooked in previous, less exhaustive, analyses. This information was subsequently successfully exploited by growing a fragment series into this region. Similarly other ligand-based design tools have been used to complement structure-based design or in rare instances to optimise fragment series in the absence of crystal structures.

Astex have recently reported that they now incorporate Cresset field points[47] in their protein structure viewer, allowing the user ready access to this information. These are molecular interaction points, representing areas where interactions with positively charged, negatively charged or hydrophobic probes are favourable. Using this technique, they were able to identify an unfavourable interaction between a backbone carbonyl and a phenyl group of the ligand concerned (an IAP antagonist). Changing the phenyl ring for a more electron-deficient pyridine ring led to an increase in activity.[48]

Understanding the internal strain in a ligand in its bioactive conformation is an important part of rational design. Any strain that is introduced in a ligand, by binding in a conformation other than the global energy minimum in solution, represents an energetic cost paid on binding. Therefore any modifications that can be made to the ligand to energetically favour the bioactive conformation and maintain the existing protein–ligand interactions will result in a boost in potency. In fact, it can be argued that this is one of the most reliable ways to boost potency since there are no other competing energetic considerations. Conversely, the energetic benefit from forming a hydrogen bond is harder to predict because the gain from forming the interaction is competing with the de-solvation cost of the two partners.

Assessment of the strain in the bioactive conformation can be carried out in two ways. A knowledge-based approach can be used to look at the frequency of occurrence of the bound conformation across small-molecule crystal structures of related compounds; tools from the CCDC such as Conquest and Mogul are most commonly used. Alternatively, the energetic cost can be estimated by comparing the internal energy of the ligand in the bound conformation with that of a minimised form; for best results this should be done with a quantum mechanical method. Such an assessment is based on purely enthalpic considerations. Recently Openeye have announced they are developing a method that incorporates an entropic component to give the strain free energy.[49]

One of the most important concepts in FBDD is putting design at the heart of the project. We wholeheartedly endorse this approach, and believe this should be the case for all projects, not just FBDD projects. However, it is important to understand how to use design appropriately. At times this means designing specific compounds to answer specific questions, but at other times it involves choosing a range of compounds to probe a particular interaction. To this end, we believe that library design has an important role to play at some stages of an FBDD project. At times this may be driven by protein structural information, but approaches such as using experimental design to cover SARs and property space as efficiently as possible also have a role to play.

4.10 Integrated Hit Identification

FBDD has often been seen as an alternative to, or even in competition with, more traditional hit-identification methods such as HTS, focussed set screening (based on existing knowledge of a target to focus the selection of compounds to screen) and encoded libraries technology (ELT). ELT uses DNA encoding of chemical entities and enables the affinity-based screening of in the order of 10^9 compounds.[50] At GSK we have been in a position to compare a number of these different hit-discovery techniques and we have found them all to have strengths and weaknesses as summarised in the information in Table 4.1.

However, and most importantly, we have found these different methods to be highly synergistic. Thus while ELT has a tendency to find larger and more lipophilic molecules but with good affinities, fragment screening delivers

Table 4.1 Summary of strengths and weaknesses of a variety of hit-identification approaches.

Potential strengths	Potential weaknesses
Fragment screening	
Utilises the reduced complexity approach to increasing hit rate	Very sensitive biophysical methods (SPR, NMR, X-ray *etc.*) needed to detect weak binding.
Focus is on ligand-efficient starting points	Cost of chemistry follow up required to establish a lead-quality molecule.
Efficient sampling of chemical diversity	Primarily limited to structure-enabled targets
Fragments play to the strengths of structure-based design and biophysics which are enabled at the outset	Reductionist approach may oversimplify complexity of interactions – *i.e.*, cooperativity is lost
Aims to build only the interactions required	Without continued attention to optimisation indices like LE and LLE it is very easy to waste a good starting point
Focussed screening	
In silico selections possible from the widest diversity of tangible compounds using 2D or 3D selection methods	Prior knowledge of target may be wrong or limiting!
Acoustic dispensing makes cherry picking from in-house collections viable	Even state-of-the-art virtual screening methods suffer from false positives and negatives. Docking and scoring algorithms still poor
Good availability of diverse compounds from suppliers	
Encoded libraries technology	
Huge numbers of compounds can be screened $>10^9$	Chemistry must tolerate water and oxygen. Reactions can be done with 70–80% organic co-solvents (*e.g.*, CH_3CN, DMF, DMA, *etc.*)

(*continued*)

Table 4.1 (*continued*)

Potential strengths	Potential weaknesses
Affinity selection, and thus tends to identify highly selective, very potent compounds. Frequently with unique mechanisms of action	Complexity, size and lipophilicity of molecules tends to be high. This is inherent in split-and-pool strategies, increasing the number of quality molecules in the library also comes with the downside of incorporating others with high M_W and lipophilicity
Efficient screening process with minimal infrastructure compared to HTS	Encoded libraries cover pockets of chemical space in significant depth, but have not yet reached the diversity of chemical space covered by HTS collections
Linker attachment point is an advantage for use of the molecule as a probe or in bifunctional molecules	Cost of chemistry to confirm hits off DNA
High- throughput screening	
Diversity and breadth of chemotypes considered is very high with proven track record of delivering most diverse leads	Compound collection costs are high due to replacing compounds and adding new diversity. Cost of capital equipment for collection storage and screening
Complex molecules display intramolecular cooperativity which may be absent in fragments	The need to miniaturise assays can cost time and impact quality, leading to high false-positive rate. Combined with scale, creates need for orthogonal assay development and triage for follow up
Robustness based on automation and miniaturisation	Perception of dated approach although success rate suggests this is erroneous in terms of impact
Broadly applicable to both biochemical and cellular assays	Seen as expensive and slow, but not so once infrastructure is in place

weak hits with good LE and lipophilic LE (LLE). These fragments can synergistically assist the interpretation of other screening outputs by suggesting active substructures found within ELT (or HTS) hits. This allows rapid de-convolution of more complex hits into the essential core activity and also suggests where mass can be added without abrogating binding.

This realisation has led to FBDD being used in parallel with other methods, with data from the various approaches being used to complement each other and philosophies being transferred between methodologies. Hence, concepts such as prioritising hits based on LE and cutting back hits to various fragments to understand the key contributions to binding, have become common practice in HTS follow up. Similarly, we have had a number of successes evolving our fragment hits either by directly hybridising with functionality from HTS or ELT hits, or targeting interactions made by HTS or ELT compounds in fragment series.

4.11 Opportunities and Future Developments

A number of computational tools for FBDD projects have been outlined along with successful applications. However, there are also areas where we believe new tools could be usefully developed to enhance the ability of computational chemistry to improve the efficiency and effectiveness of fragment optimisation. Those areas are summarised below along with suggestions for addressing these gaps.

As highlighted above, one of the first tasks when progressing a fragment hit is to optimise the LE of the starting point by trying to increasing the potency without substantially increasing the molecular weight. This is generally done by using protein structural information to suggest compounds to target any obvious unexploited interactions. However, the process is often incomplete and there is a temptation to drive quickly for potency by adding functionality without due regard for LE. For example, imagine the trivial case of identifying an indole as a hit, then consider the various regioisomers of azaindole and indazole, and for each, the possibility of methylating or halogenating at the various available positions – the number of possible combinations rapidly becomes enormous. There appears to be the opportunity here for a more systematic approach of identifying an exhaustive list of possible analogues, within a given set of constraints, and assessing them computationally. Unfortunately, most computational techniques, such as protein–ligand docking, do not have the sensitivity to distinguish between such closely related analogues. It would be interesting to consider the possibility that quantum mechanics based methods such as hybrid quantum mechanics/molecular mechanics (QM/MM) might be able to impact in this area.

The unprecedented amount of structural data obtained for many fragment projects represents both a challenge and an opportunity for the modelling community. Whilst having large numbers of structures and utilising them in traditional ways certainly adds to the information a project has available, there seems to be an opportunity for more work to be done on summarising all the structural data in a systematic way. In particular, it would be interesting to see methods developed that go beyond the hotspots-type approaches described above, which for example would have a systematic way of incorporating such information in compound design or provide a summary of any protein flexibility observed across a range of structures.

The analysis of the increasing numbers of crystal structures of each protein has also revealed that induced pockets are quite commonly observed, mostly through side-chain rotations but occasionally through backbone movements. We have a number of examples of the successful targeting of induced pockets to boost potency, typically observed in crystallography and then exploited. The lack of robust tools for predicting induced pockets seems to be a gap in modelling capabilities. Docking tools such as induced fit docking (IDF) from Schrodinger[51] do exist, but they do not systematically look for induced pockets, but rather investigate whether side-chain rotations are possible in the presence of a prescribed list of ligands.

One of the most active areas of current research in structure-based design tools is water modelling. The commercially available tools in this area are evolving rapidly and an increasing number of examples are emerging of their successful application.[52–54] Although there is a long way to go to fully validate these tools and understand the best ways of applying them, it is clear they have the potential to be of value in FBDD. The ability to identify water molecules to target for displacement, those to interact with, and binding hotspots where clusters of higher energy water exist, would add valuable information to the systematic fragment-growth approach advocated in FBDD.

4.12 Druggability

An interesting additional use of fragment screening is to determine the druggability of a protein target.[55] *Druggability* refers to the probability of a particular protein being able to bind small, drug-like molecules with high affinity and specificity and is a key component of understanding the overall tractability of a proposed target for drug discovery. Overall, target tractability is clearly related to both biological tractability (*i.e.*, does the target link to a disease) and druggability (*i.e.*, does the target bind compounds in an efficacious manner). Both are necessary for success. The most highly validated target from a biological perspective is of no value for small-molecule drug discovery if it is not able to bind a small molecule – as is often the case with PPI targets. Finding ways to effectively judge the druggability of a new target is thus an essential component of drug discovery. If this can be done early in a programme without committing the time and money required to develop an HTS, then this would be advantageous. Following the early suggestions of Hajduk *et al.*[55] and other work by Chen and Hubbard[56] and Edfeldt *et al.*[57] we have found 2D NMR using N^{15}-labelled protein to be an effective way of judging druggability. This is relatively easy to set up and involves screening a well-chosen and diverse fragment library against the protein target of interest. Observation of specific shifts in protein peaks is an indication that small molecules can bind in a specific manner and provides some encouragement that the protein is druggable. If the sequence assignment is available (or achievable) then additional information can be derived to indicate the physical location of ligand binding. If protein crystal structures are available then it is possible to follow up with X-ray crystallography to gain further insights into fragment-binding modes. Of course, fragment screening gives no information on the functional consequences of the molecules that bind and thus, such druggability assessments are not the same as lead identification although the latter may follow. Absence of binding by any members of a well-chosen fragment library is a good indication of the lack of druggability of a target.

4.13 Conclusions

FBDD has matured to a point where it is now a commonly used and valued technique in biopharmaceutical companies of all sizes. Computational chemists have been amongst the leading advocates of the approach and good computational chemistry is considered central to the success of an FBDD programme. A large number of tools has been developed to design screening sets, maximise the value derived from the resulting output and to direct the optimisation of these hits into leads and clinical candidates.

We have discussed how existing tools have been used or modified for use in FBDD and also the development of new tools to help with specific tasks. The tool GRID, one of the first structure-based design tools developed, and in many ways the *in silico* inspiration for the FBDD approach, continues to provide important design insight to our FBDD programmes. We have adapted some fundamental molecular similarity tools for the requirements of following up fragment hits, and shown how their use can maximise the number of validated hits (those giving crystal structures) identified. New tools have been developed to summarise key structural and biological data making it straightforward to have data-based discussions during the design process.

We have also highlighted the importance of using a broad range of experimental and computational tools in FBDD. Although a large amount of structural information is available, ligand-based design methods can give important additional insight and the value of appropriately designed arrays should not be overlooked in conjunction with specific individual compounds. FBDD is increasingly being used alongside other hit-identification methods, and synergies have been identified both in terms of cross-fertilisation of approaches and philosophies, and hits from one method being advanced using information from another.

Finally, a number of areas have been summarised where gaps in computational methodology exist. There is a great opportunity for the computational chemistry community to fill the gaps and further increase the impact of computational chemistry on FBDD. The large amounts of structural data generated on typical FBDD programmes provide an excellent dataset to develop these methods, and thus push forward our capabilities not just in FBDD, but in structure-based design more generally.

References

1. M. Baker, *Nat. Rev. Drug Discovery*, 2013, **12**, 5.
2. G. Bollag, J. Tsai, J. Zhang, C. Zhang, P. Ibrahim, K. Nolop and P. Hirth, *Nat. Rev. Drug Discovery*, 2012, **11**, 873.
3. M. P. Gleeson, *J. Med. Chem.*, 2008, **51**, 817.
4. P. D. Leeson and B. Springthorpe, *Nat. Rev. Drug Discovery*, 2007, **6**, 881.
5. G. M. Makara, *J. Med. Chem.*, 2007, **50**, 3214.
6. R. E. Hubbard, I. Chen and B. Davis, *Curr. Opin. Drug Discovery Dev.*, 2007, **10**, 289.
7. A. Schuffenhauer, S. Ruedisser, A. L. Marzinzik, W. Jahnke, M. Blommers, P. Selzer and E. Jacoby, *Curr. Top. Med. Chem.*, 2005, **5**, 751.
8. N. Blomberg, D. A. Cosgrove, P. W. Kenny and K. Kolmodin, *J. Comput.-Aided Mol. Des.*, 2009, **23**, 513.
9. M. Congreve, R. Carr, C. Murray and H. Jhoti, *Drug Discovery Today*, 2003, **8**, 876.

10. H. Jhoti, G. Williams, D. C. Rees and C. W. Murray, *Nat. Rev. Drug Discovery*, 2013, **12**, 644.
11. I. J. Chen and R. E. Hubbard, *J. Comput.-Aided Mol. Des.*, 2009, **23**, 603.
12. G. Siegal, E. Ab and J. Schultz, *Drug Discovery Today*, 2007, **12**, 1032.
13. H. Koster, T. Craan, S. Brass, C. Herhaus, M. Zentgraf, L. Neumann, A. Heine and G. Klebe, *J. Med. Chem.*, 2011, **54**, 7784.
14. J. B. Baell and G. A. Holloway, *J. Med. Chem.*, 2010, **53**, 2719.
15. P. Bamborough, M. J. Brown, J. A. Christopher, C. W. Chung and G. W. Mellor, *J. Med. Chem.*, 2011, **54**, 5131.
16. C. W. Chung, A. W. Dean, J. M. Woolven and P. Bamborough, *J. Med. Chem.*, 2012, **55**, 576.
17. A. Vulpetti and C. Dalvit, *ChemMedChem*, 2013, **8**, 2057.
18. M. J. Hartshorn, C. W. Murray, A. Cleasby, M. Frederickson, I. J. Tickle and H. Jhoti, *J. Med. Chem.*, 2005, **48**, 403.
19. S. S. Antonysamy, B. Aubol, J. Blaney, M. F. Browner, A. M. Giannetti, S. F. Harris, N. Hebert, J. Hendle, S. Hopkins, E. Jefferson, C. Kissinger, V. Leveque, D. Marciano, E. McGee, I. Najera, B. Nolan, M. Tomimoto, E. Torres and T. Wright, *Bioorg. Med. Chem. Lett.*, 2008, **18**, 2990.
20. S. D. Pickett, J. S. Mason and I. M. McLay, *J. Chem. Inf. Comput. Sci.*, 1996, **36**, 1214.
21. J. A. Grant, M. A. Gallardo and B. T. Pickup, *J. Comput. Chem.*, 1996, **17**, 1653.
22. *ROCS 3.2.0.4*, OpenEye Scientific Software, Santa Fe, NM. http://www.eyesopen.com
23. P. C. Hawkins, A. G. Skillman and A. Nicholls, *J. Med. Chem.*, 2007, **50**, 74.
24. F. Lovering, J. Bikker and C. Humblet, *J. Med. Chem.*, 2009, **52**, 6752.
25. G. W. Bemis and M. A. Murcko, *J. Med. Chem.*, 1996, **39**, 2887.
26. G. Harper, G. S. Bravi, S. D. Pickett, J. Hussain and D. V. Green, *J. Chem. Inf. Comput. Sci.*, 2004, **44**, 2145.
27. K. Birchall and V. J. Gillet, *Methods Mol. Biol.*, 2011, **672**, 197.
28. K. Birchall, V. J. Gillet, P. Willett, P. Ducrot and C. Luttmann, *J. Chem. Inf. Model.*, 2009, **49**, 1330.
29. A. D. Morley, A. Pugliese, K. Birchall, J. Bower, P. Brennan, N. Brown, T. Chapman, M. Drysdale, I. H. Gilbert, S. Hoelder, A. Jordan, S. V. Ley, A. Merritt, D. Miller, M. E. Swarbrick and P. G. Wyatt, *Drug Discovery Today*, 2013, **18**, 1221.
30. M. M. Hann, A. R. Leach and G. Harper, *J. Chem. Inf. Comput. Sci.*, 2001, **41**, 856.
31. A. Tversky, *Psychol. Rev.*, 1977, **84**, 327.
32. A. R. Leach and V. J. Gillet, *An Introduction to Cheminformatics*, Kluwer Academic Publishers, Dordrecht, 2003.
33. Z. Deng, C. Chuaqui and J. Singh, *J. Med. Chem.*, 2004, **47**, 337.
34. *TIBCO Spotfire*, TIBCO Software Inc., http://spotfire.tibco.com/
35. J. Hussain and C. Rea, *J. Chem. Inf. Model.*, 2010, **50**, 339.
36. U. Schieborr, M. Vogtherr, B. Elshorst, M. Betz, S. Grimme, B. Pescatore, T. Langer, K. Saxena and H. Schwalbe, *ChemBioChem*, 2005, **6**, 1891.

37. G. L. Warren, C. W. Andrews, A. M. Capelli, B. Clarke, J. LaLonde, M. H. Lambert, M. Lindvall, N. Nevins, S. F. Semus, S. Senger, G. Tedesco, I. D. Wall, J. M. Woolven, C. E. Peishoff and M. S. Head, *J. Med. Chem.*, 2006, **49**, 5912.
38. M. L. Verdonk, I. Giangreco, R. J. Hall, O. Korb, P. N. Mortenson and C. W. Murray, *J. Med. Chem.*, 2011, **54**, 5422.
39. C. W. Murray, M. G. Carr, O. Callaghan, G. Chessari, M. Congreve, S. Cowan, J. E. Coyle, R. Downham, E. Figueroa, M. Frederickson, B. Graham, R. McMenamin, M. A. O'Brien, S. Patel, T. R. Phillips, G. Williams, A. J. Woodhead and A. J. Woolford, *J. Med. Chem.*, 2010, **53**, 5942.
40. A. O. Frank, M. D. Feldkamp, J. P. Kennedy, A. G. Waterson, N. F. Pelz, J. D. Patrone, B. Vangamudi, D. V. Camper, O. W. Rossanese, W. J. Chazin and S. W. Fesik, *J. Med. Chem.*, 2013, **56**, 9242.
41. T. Oltersdorf, S. W. Elmore, A. R. Shoemaker, R. C. Armstrong, D. J. Augeri, B. A. Belli, M. Bruncko, T. L. Deckwerth, J. Dinges, P. J. Hajduk, M. K. Joseph, S. Kitada, S. J. Korsmeyer, A. R. Kunzer, A. Letai, C. Li, M. J. Mitten, D. G. Nettesheim, S. Ng, P. M. Nimmer, J. M. O'Connor, A. Oleksijew, A. M. Petros, J. C. Reed, W. Shen, S. K. Tahir, C. B. Thompson, K. J. Tomaselli, B. Wang, M. D. Wendt, H. Zhang, S. W. Fesik and S. H. Rosenberg, *Nature*, 2005, **435**, 677.
42. A. W. Hung, H. L. Silvestre, S. Wen, A. Ciulli, T. L. Blundell and C. Abell, *Angew. Chem., Int. Ed.*, 2009, **48**, 8452.
43. M. Congreve, G. Chessari, D. Tisi and A. J. Woodhead, *J. Med. Chem.*, 2008, **51**, 3661.
44. Goldmine, Cambridge Crystallographic Data Centre, http://www.ccdc.cam.ac.uk/
45. M. L. Verdonk, J. C. Cole and R. Taylor, *J. Mol. Biol.*, 1999, **289**, 1093.
46. P. J. Goodford, *J. Med. Chem.*, 1985, **28**, 849.
47. T. J. Cheeseright, M. D. Mackey, J. L. Melville and J. G. Vinter, *J. Chem. Inf. Model.*, 2008, **48**, 2108.
48. C. Johnson, Fragment based approaches to potent antagonists of protein–protein interactions, in *Frontiers in Medicinal Chemistry*, San Francisco, 2013.
49. Freeform, Openeye Scientific Software, Santa Fe, NM, http://www.eyesopen.com
50. M. A. Clark, R. A. Acharya, C. C. Arico-Muendel, S. L. Belyanskaya, D. R. Benjamin, N. R. Carlson, P. A. Centrella, C. H. Chiu, S. P. Creaser, J. W. Cuozzo, C. P. Davie, Y. Ding, G. J. Franklin, K. D. Franzen, M. L. Gefter, S. P. Hale, N. J. Hansen, D. I. Israel, J. Jiang, M. J. Kavarana, M. S. Kelley, C. S. Kollmann, F. Li, K. Lind, S. Mataruse, P. F. Medeiros, J. A. Messer, P. Myers, H. O'Keefe, M. C. Oliff, C. E. Rise, A. L. Satz, S. R. Skinner, J. L. Svendsen, L. Tang, K. van Vloten, R. W. Wagner, G. Yao, B. Zhao and B. A. Morgan, *Nat. Chem. Biol.*, 2009, **5**, 647.
51. W. Sherman, T. Day, M. P. Jacobson, R. A. Friesner and R. Farid, *J. Med. Chem.*, 2006, **49**, 534.
52. C. Barillari, J. Taylor, R. Viner and J. W. Essex, *J. Am. Chem. Soc.*, 2007, **129**, 2577.

53. A. Bortolato, B. G. Tehan, M. S. Bodnarchuk, J. W. Essex and J. S. Mason, *J. Chem. Inf. Model.*, 2013, **53**, 1700.
54. Y. Yang, F. C. Lightstone and S. E. Wong, *Expert Opin. Drug Discovery*, 2013, **8**, 277.
55. P. J. Hajduk, J. R. Huth and C. Tse, *Drug Discovery Today*, 2005, **10**, 1675.
56. I. J. Chen and R. E. Hubbard, *J. Comput.-Aided Mol. Des.*, 2009, **23**, 603.
57. F. N. Edfeldt, R. H. Folmer and A. L. Breeze, *Drug Discovery Today*, 2011, **16**, 284.

CHAPTER 5

Fragment Screening of G Protein-Coupled Receptors

MILES CONGREVE*[a] AND ROBERT COOKE[a]

[a]Heptares Therapeutics Ltd, BioPark, Broadwater Road, Welwyn Garden City, Hertfordshire, AL7 3AX, UK
*E-mail: miles.congreve@heptares.com

5.1 Introduction to G Protein-Coupled Receptors

G Protein-coupled receptors (GPCRs) are membrane-associated proteins that transmit a range of signals from outside to inside the cell. There are thought to be 390 members in the human genome (excluding olfactory receptors), making this the largest family of membrane-bound receptors.[1] Binding of a specific agonist ligand induces a conformational change within a GPCR from its ground-state conformation to its activated state. In this activated state, binding of an intracellular G protein or of β-arrestin may occur which then leads to cell signaling.[2] GPCRs have a common fold in which an extracellular N-terminus and an intracellular C-terminus are connected by seven transmembrane α-helices (7TM). In most receptors the agonist binding site is located in the centre of the 7TM bundle towards the extracellular face. The receptors are divided into four major families based on sequence similarity[1–4] (Class A, Class B, Class C and Frizzled). Class A is the Rhodopsin family and is the largest with approximately 300 members. This family can be further subdivided depending on the types of agonist ligands the receptors recognise; including the aminergics, chemokines, glycoprotein hormones

RSC Drug Discovery Series No. 47
Fragment-Based Drug Discovery
Edited by Steven Howard and Chris Abell

Published by the Royal Society of Chemistry, www.rsc.org

and neuropeptides. The next largest family is Class B which is subdivided into the secretin and adhesion receptors of which there are 15 and 33 members respectively. The secretin receptors recognise large peptides and peptide hormones such as calcitonin gene-related peptide (CGRP), corticotropin-releasing factor (CRF), glucagon, glucagon-like peptide (GLP), parathyroid peptide hormone (PTH) and secretin. Class B receptors have a large N-terminal extracellular domain which is thought to be involved in agonist binding; the large peptidic agonist ligands bind both to this domain and to the 7TM binding site.[5] Class C GPCRs include the family of metabotropic glutamate (mGlu) receptors. Class C receptors also have a large extracellular N-terminal domain. In contrast to the other receptor families this extracellular domain, known as the *Venus fly trap* domain, is the site of the endogenous agonist binding site. Upon binding of the small-molecule agonist (such as glutamate or GABA; γ-aminobutyric acid) a conformational change in the venus fly trap is believed to transmit the signal by inducing activation of the 7TM receptor.[6] Finally, the Frizzled family of GPCRs includes 10 Frizzled (FZD) receptors and the smoothened receptor (SMO). FZD receptors bind Wnt glycoproteins whereas SMO is activated by formation of a complex with another membrane protein called patched.[7]

Even though GPCRs have been the targets for many drug-discovery programmes for several decades, it was only in 2000 that the first crystal structure of a GPCR, rhodopsin, was reported.[8] After this major accomplishment further structures were not resolved until relatively recently, but in the period 2008–2013, structures of 21 different GPCR family members were published. In most cases the receptors are complexed in the 7TM-binding site with either a small molecule or a peptide ligand. In addition to rhodopsin, structures have been reported for 18 Class A receptors: β_2-adrenergic receptor (β_2AR),[9,10] β_1-adrenergic (β_1AR),[11–13] adenosine A_{2A} receptor (A_{2A}),[14,15] sphingosine 1-phosphate 1 receptor (S_1P_1),[16] chemokine receptor CXCR4,[17] dopamine D_3 receptor (D_3),[18] histamine H_1 receptor (H_1),[19] muscarinic acetylcholine M_2 and M_3 receptors,[20,21] neurotensin receptor,[22] opioid receptors δ,[23] κ,[24] μ,[25] and nociceptin,[26] protease-activated receptor 1 (PAR1),[27] serotonin receptor 1B (5-HT_{1B}),[28] serotonin receptor 2B (5-HT_{2B}),[29] and chemokine receptor CCR5.[30] Structures have also been reported for the smoothened receptor (SMO),[31] and more recently from Class B the corticotropin releasing factor 1 receptor (CRF_1)[32] and the glucagon receptor.[33]

This dramatic progress has been driven by the development of two complementary technologies to overcome the instability of GPCRs when removed from a membrane. The first method involves introduction of a fusion protein or monoclonal antibody fragments into the crystallisation construct sequence to improve crystal contacts and help promote crystal growth, coupled with the use of highly potent stabilising ligands and lipidic cubic phase (LCP) crystallisation conditions.[34,35] The second method is the introduction of several point mutations into the receptor sequence that significantly increase the thermostability of the protein construct and allow purification in short-chain detergents, compatible with crystallisation by vapour diffusion or *via* LCP.[36,37] Receptors stabilised in this way are known as *StaR®* *proteins*.[38] This method biases the GPCR towards a specific conformation, derived from the pharmacology of the ligand used during the stabilisation process, enabling structures of the same GPCR to be determined in either the inactive or active forms. The StaR technology does not depend on the use of highly stabilising ligands for crystallisation and co-complexes can be solved with relatively weak antagonists or agonists.[39,40]

GPCRs have been an important target family for drug discovery historically and remain an active area of research today, with many of the World's top selling drugs acting at these membrane receptors.[41] Many of the GPCR targets of current interest for drug discovery are challenging – particularly those which lack small endogenous ligands from which empirical drug discovery can be leveraged.[42] However, a recent computational analysis of GPCR X-ray structures, focusing on factors that can potentially influence ligand binding, such as the topology of binding pockets and the energetics of water molecules within them, revealed that these membrane receptors typically have druggable sites, comparable with those of soluble enzyme targets.[43] Interestingly, even challenging drug targets such as the CXCR4 chemokine receptor have regions showing encouraging features, consistent with the existence of low molecular weight ligands. In each case, GPCR binding sites were found to be relatively deep and ligands were partially or entirely encapsulated by the protein. In contrast, enzyme-binding sites are often solvent-exposed grooves in the protein surface.[43]

In this chapter we will present approaches to and validation of the use of fragment-based drug discovery (FBDD) for GPCRs. The druggability analysis discussed above suggested that GPCRs should be suitable for FBDD screening approaches, presenting buried binding site 'hotspots' that should deliver productive interactions.[43] In addition, the recent availability of protein–ligand structural information for modelling and the growing potential to solve fragment–receptor co-complex structures, is starting to open up the possibility of using structure-based drug design (SBDD) for this class of membrane receptors. The evidence presented here supports the proposition that fragment hit identification and evolution of fragments to useful lead series is a promising strategy for GPCR targets.

5.2 Fragment Screening of Wild-Type GPCRs

Over the last 15 years, fragment screening has developed as a useful alternative to high-throughput screening (HTS) to identify hits for therapeutic targets.[44–46] Elsewhere in this book FBDD is discussed in detail in the context of soluble protein targets. Fragment libraries are usually a subset of larger, diverse HTS screening collections that contain only very small molecules, usually defined as 100–300 Da in size. Fragment libraries are often further filtered to obey the *Rule of 3* (molecular weight must be less than 300 Da, lipophilicity ($\log P$) must be less than 3, the number of hydrogen-bond donors must be no higher than 3 and the number of hydrogen-bond acceptors must be no higher than 3).[47] Due to their small size and simplicity, the chance of a binding event occurring is higher for a fragment than for a larger molecule where mismatches of shape and functionality are very likely to occur that act to preclude binding.[48] Usually fragments are screened at higher than normal HTS concentrations to increase the chance of detecting hits. Their small size means they tend to have lower affinity for protein targets than drug-sized molecules and because of this, a range of biophysical approaches has been developed and applied to fragment-based screening.[49] An important consideration is that, despite their often low affinity, the binding affinity of fragment hits per non-hydrogen atom (termed the ligand efficiency (LE))[50] is generally higher than for larger hit molecules. This often makes fragment hits good starting points for medicinal chemistry campaigns. FBDD is usually driven by the use of protein–ligand X-ray data to support SBDD beginning from the fragment hits identified.[51] The advent of X-ray structural information for GPCR targets over the last five years is therefore an opportunity to utilise FBDD for this class of membrane proteins. This, coupled with advances in protein engineering and biophysics, has also started to allow sensitive fragment screening methods to be employed with these difficult-to-handle proteins.

In perhaps the first example of FBDD applied to GPCRs, before structural information was available, AstraZeneca developed a workflow to screen routinely a 20 000-member fragment library against all of its new targets. High-concentration screening (HCS) in a biochemical assay identified hits for several GPCR targets that were progressed using analogue screening to hit series with demonstrable structure–activity relationships (SARs).[52] Relatively little else has been reported in the literature in which fragment screening using HCS approaches has been applied to GPCR targets. Much more recently, de Esch *et al.* have described the identification of fragment hits for the histamine H_4 receptor using a radioligand-binding assay and screening at 10 μM concentration.[53] Fifty-six hits from a library of 1010 molecules were identified, giving a promising 6% hit rate. However, counter-screening of the library against a ligand-gated ion channel target, serotonin 5-HT_{3A}, in a fluorescence-based functional assay, showed a disappointing degree of overlap between hits for the two targets. Perhaps unexpectedly, the authors observed that the dual active fragments had a higher molecular complexity than those active at only 1 of the targets, as exemplified by **1**, H_4 $pK_i = 7.0$, inactive at 5-HT_{3A} and **2** H_4 $pK_i = 8.2$, 5-HT_{3A} $pK_i = 5.9$ (Figure 5.1).

An elegant approach to fragment screening of GPCRs has been reported by Stoddart *et al.*, in which a fluorescence-based binding assay was developed for use in living cells.[54] An interesting feature of this technique was that the integrity of the membrane environment of the receptor was maintained under physiological conditions. Potentially, this might confer advantages for ligand discovery because intracellular signalling proteins, which would not be present in assays using isolated receptors, might allosterically influence the binding of molecules to GPCRs.[55] A fluorescently labelled adenosine A_3 receptor antagonist was developed from a standard adenosine receptor ligand called XAC (xanthine amine congener; *N*-(2-aminoethyl)-2-[4-(2,6-dioxo-1,3-dipropyl-2,3,6,7-tetrahydro-1*H*-purin-8-yl)phenoxy]acetamide).

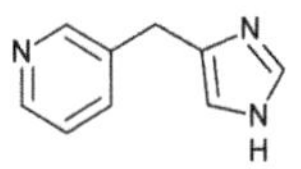

1 H_4 pK_i = 7.0

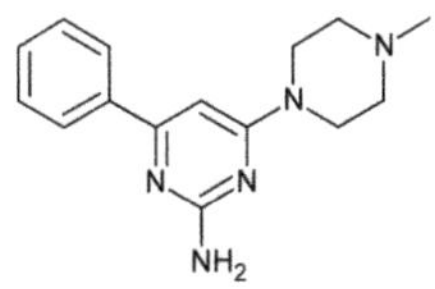

2 H_4 pK_i = 8.2

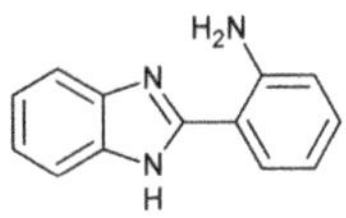

3 A_3 pK_i = 6.4

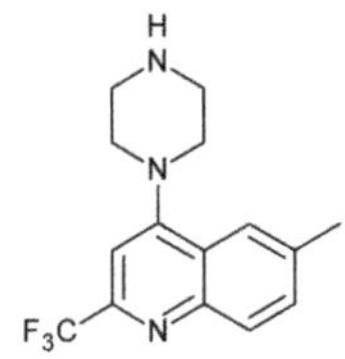

4 β_2AR pK_D = 7.8

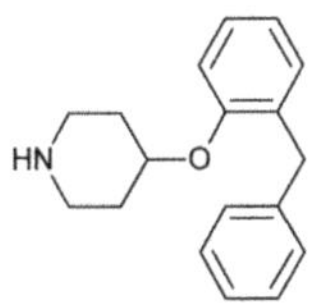

5 H_1 pK_i = 8.2

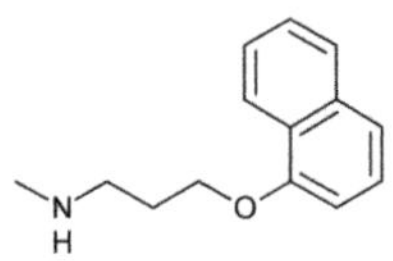

6 H_1 pK_i = 6.1

Figure 5.1 Histamine H_4 receptor, adenosine A_3 receptor, β_2 adrenergic receptor, and histamine H_1 receptor fragment screening hits.

XAC has high affinity and a slow off-rate at the receptor. The authors successfully developed a high-content screening approach using a confocal imaging plate reader to detect binding of literature antagonists. Affinities of the antagonists in competition-binding format were in good agreement with the potencies of the molecules in a functional assay readout; the technique was then extended to screen 248 fragments, identifying 38 hits at a single concentration of 1 mM. In follow up concentration response experiments, DP 01095 (**3**, Figure 5.1) had the highest affinity, pK_i = 6.4. For other hits, affinity values as low as pK_i = 4.0 were measurable, demonstrating the utility of the method to detect weakly binding fragments.

Surface plasmon resonance (SPR) is a highly sensitive biophysical technique developed to study binding interactions using label-free optics that are capable of measuring very small changes in refractive index at the surface of a *biosensor* chip.[56–58] Over the past decade SPR has developed into the most widely applied generic approach to characterise ligand binding, including fragment screening for soluble targets. Typically, experiments are conducted with the protein target of interest immobilised on the biosensor surface with test compounds in solution flowing over the chip. Upon ligand binding, the increase in molecular weight of the complex changes the refractive index resulting in a detectable signal. An obvious limitation of this biophysical approach is that the protein construct needs to remain stable and correctly folded in solution in order to be successfully immobilised onto the SPR chip and to be suitable for the subsequent screening process. It is thus not surprising that reports of applications of SPR for GPCRs are rare. In one example, the wild-type chemokine receptor CCR5 was successfully trapped on a biosensor chip and probe molecules successfully detected binding to the protein.[59] Subsequently, the approach was extended to the screening of a set of 656 fragments against the immobilised wild-type β_2 adrenergic receptor.[60] Five fragment hits with binding affinities in the range pK_D 4.7–7.8 were confirmed in dose–response experiments following primary screening at a single concentration; the most active was compound **4** (Figure 5.1). This study shows that biophysical screening approaches can be used for membrane proteins that have been solubilised and then trapped on chips, although perhaps limited to cases where the protein is well expressed and has a reasonably high inherent stability. In our own experience most wild-type GPCRs are not sufficiently stable when solubilised for use with SPR, even when using mild detergents and various lipid additives. Instead, engineering of the GPCR to introduce stability (described below) is usually required in order to utilise SPR and other biophysical approaches for fragment screening.

The availability of protein–ligand X-ray data for Class A GPCRs has greatly facilitated virtual screening approaches to hit identification and this has become an area of active interest. Interestingly, virtual fragment screening has been attempted for several targets and one example is given here. The histamine H_1 receptor structure has been used for virtual screening of a large library of 757 728 fragment-like compounds, containing fewer than 22 non-hydrogen atoms.[61] First, the library was processed to retain only molecules carrying a formal charge of +1, reducing the test set to 108 790 fragments. This ensured that the fragments had the potential to form ionic interactions with $Asp107^{3.32}$ (Asp107 at position 32 in TM3, the superscripted numbers are the Ballesteros–Weinstein numbering system used for GPCRs).[62] The assumption was that this salt bridge would be an absolute requirement for binding to the receptor. The docking program PLANTS was then used to carry out the virtual screen. Next, fragments forming an interaction with $Asp107^{3.32}$ in the predicted binding pose were selected, leaving 95 147 molecules. Further post-processing was then conducted, using interaction fingerprints derived from the contacts made by the ligand bound in the X-ray complex, doxepin (PDB:3RZE).[19] The most promising 354 fragments were next categorised using Tanimoto-based similarity analysis with known H_1 antagonists. Finally, the virtual hits were visually inspected and 30 fragments were selected for experimental screening. Of the 26 compounds available for purchase, 19 fragments exhibited H_1 affinity, giving an impressive 73% experimental hit rate with hits in a potency range of 10 μM to 6 nM, with compounds **5** and **6** shown as representative hit molecules in Figure 5.1. Overall, this successful screen demonstrates that, at least for relatively tractable Class A GPCRs, computational selection of fragments is a valuable approach to hit generation.

5.3 Fragment Screening of Stabilised GPCR Constructs

5.3.1 Advantages of StaR Constructs over Wild-Type Receptors

In addition to enabling the structural studies of GPCRs as described above, the StaR proteins generated by thermostabilisation have properties which lend themselves to biophysical investigations of ligand binding, overcoming some limitations inherent in wild-type, unstabilised receptors. Most importantly the StaR proteins are stable when removed from the cell membrane and solubilised in a suitable detergent, so can be handled as a protein in solution. Consequently techniques which are compatible with low concentrations of detergents (*e.g.*, 0.1% dodecylmaltoside, DDM) can be used to investigate stabilised receptors. The detergent is required to form a micelle around the lipophilic protein, allowing it to remain folded in solution. Furthermore, biasing the receptor into the desired pharmacological state (by use of a suitable ligand during the selection of mutations) has the advantage of presenting the required conformation of the ligand-binding site as a screening target. The ability of StaR proteins to recapitulate the pharmacological profiles anticipated from *pure agonist* and *pure inverse agonist* forms has been described previously.[63] Lastly, the stabilisation process makes the receptors more resistant to harsh conditions. For example, StaR proteins can maintain structure and binding ability in 10% DMSO solution after solubilisation from the cell membrane, which enables test compounds with a wide range of solubilities to be used, and in the case of fragment screening, fragments to be tested at higher concentrations than is otherwise possible. As discussed earlier, the ability to screen fragments at a high concentration is often considered important in FBDD in order to allow detection of low-affinity hits.

Use of the conformational thermostabilisation approach has enabled StaR constructs to be studied in a range of biophysical investigations of compound binding, with actual or potential application to fragment screening and/or hit confirmation. This includes SPR, target-immobilised NMR spectroscopy (TINS), capillary electrophoresis (CE), HCS by radioligand binding, and thermal melting (T_m) assays. Each of these approaches to screening or characterising low-affinity fragment binding is outlined briefly below.

5.3.2 SPR Screening

Investigation of a broad range of GPCR StaR proteins using SPR instrumentation and methodology has been carried out in our laboratories. The SPR methodology is used in two modes. In the first approach, a panel of proteins with individual mutations introduced in the ligand-binding site are generated and then probed with a selection of compounds. The purpose of this is to generate a body of site-directed mutagenesis (SDM) data to help guide efforts to dock the test compounds, thereby understanding their binding modes to enable SBDD.[64] In addition to monitoring changes in affinities, these Biophysical Mapping™ (BPM) experiments can highlight changes in association and dissociation rate constants.[64] Although not specifically a method for FBDD, BPM can be very useful for confirming that a fragment is binding specifically to the target receptor and, if so, which binding site it engages with. BPM has advantages over the usual SDM methodology because it is a direct binding method, *i.e.*, it does not require competition with a probe ligand, such as a radiochemical. This makes interpretation of the results more straightforward as it is possible to assess directly the differences in binding induced by each mutation for each compound studied. It is also a relatively high-throughput approach (compared to SDM) because for each mutant StaR construct a library of compounds can be screened, although in practice 10–30 molecules are more typically used.

In the second, more classical, approach libraries of fragments can be tested at relatively high concentrations (up to 500 μM) against the target StaR protein.[65] Currently, a library of 3000–4000 compounds can be screened routinely in 2–3 weeks on a Biacore T200 instrument. The SPR experiment is configured with the protein on the chip and compounds injected as discrete samples. Typically the StaR protein is tagged with poly-histidine and immobilised to a nickel nitrilotriacetic acid chip; this also serves to orientate the receptors in an appropriate conformation. The increased stability means the same sample can be continually washed and re-used during the SPR screen, meaning the protein requirements are very low. As little as 2–4 μg of protein when immobilised can give a good signal (greater than 2000 resonance units on a Biacore T200 machine), and can last for at least 3000–4000 samples. The viability of the immobilised protein can be checked periodically with a standard, and if leaching is an issue the protein can be covalently coupled to the resin through a well-established and robust methodology. To ensure maximum sensitivity so that fragment-sized compounds can be detected reliably, the accurate matching of solvent in sample and control channels is essential. Unlike competition approaches, SPR is a direct binding method, hence fragments which give a positive signal could be binding anywhere on the protein, not just in the same site as the labelled probe. Depending on the system, this could be a positive or negative feature, but it does mean that fragment screening hits from SPR need to be further characterised by methods that can give structural insights. It should also be noted that non-specific and non-saturable binding can occur with SPR and hits from a fragment screen at a single concentration (or even two concentrations) need to be rigorously checked for selectivity against control proteins and for dose–response. In our laboratories, experience of screening *vs.* multiple StaR proteins with essentially the same library of fragments has generated a rich body of screening information, helping significantly in interpretation of the screening results for each subsequent screen carried out.

From the extensive fragment screening conducted by SPR with StaR proteins, results from both an adenosine A_{2A} receptor StaR and a β_1 adrenergic receptor StaR have so far been published.[66] In both cases, fragment hits were readily identified that had confirmed activity in orthogonal assays. The method depended on first isolating the purified stabilised receptor and then attaching it to the SPR chip, as described above. The construct A_{2A} StaR1 was captured onto nickel nitrilotriacetic acid-functionalised chips *via* C-terminal His-10 affinity tags, and the activity of the protein was then validated with a set of standard ligands. These ligands were chosen to have a range of affinities, which were found by SPR to be in good agreement with the values measured using wild-type A_{2A} in cellular assays.[66] A library of fragments was then screened at 200 μM, with the inclusion of periodic injections of a positive control ligand DPCPX (1,3-dipropyl-8-cyclopentyl-xanthine). Importantly, DPCPX was found to perform consistently and the receptor therefore remained active throughout the study. Of the library samples screened, approximately 10% of the fragments were found to bind specifically above the background (*i.e.*, a channel on the chip with no receptor bound to the surface) and were therefore followed up in a concentration–response format. These initial hits had binding affinities (K_D) between 5 and 10 μM by SPR. As an additional characterisation the hits were screened *vs.* the β_1 adrenergic receptor StaR as part of the primary screen, a feature of the SPR instrument being that it has multiple channels allowing more than one protein to be screened in parallel and also allowing a negative control to be included in the screening protocol. This meant that the A_{2A} SPR hits could be categorised as selective or non-selective binders for the A_{2A} StaR *vs.* the β_1 adrenergic receptor StaR. Finally, as part of the design of this study, simple xanthines were included in the library, including caffeine and theophylline, which are known to bind to the adenosine A_{2A} receptor with relatively low affinity. Importantly for this proof-of-concept study, a number of these simple heterocyclic fragments were readily identified as selective binders to A_{2A} StaR.[66] These fragments generally bound in a well-behaved manner showing essentially 1:1 stoichiometry to the receptor. In a subsequent structural biology study, a co-complex of caffeine was solved in the A_{2A} StaR2 X-ray system.[39] This small molecule binds in a similar place to the heterocyclic portion of larger ligands co-crystallised to this receptor and the complex is illustrated in Figure 5.2, indicating the binding mode and the key hydrogen-bonding interaction with the receptor. This structure is an example of a fragment–StaR co-complex, demonstrating that very small and weakly active ligands can, in the best cases, be crystallised in the target GPCR of interest.

5.3.3 TINS Screening

TINS is an NMR method developed to detect weak interactions between proteins and small molecules and has typically been used with soluble proteins. Due to the inherent instability of membrane proteins in solution, the TINS method cannot be readily applied to GPCRs. However, combining the TINS approach with the enhanced stability of StaR proteins has allowed for fragment screening and hit identification. In an initial example of TINS fragment

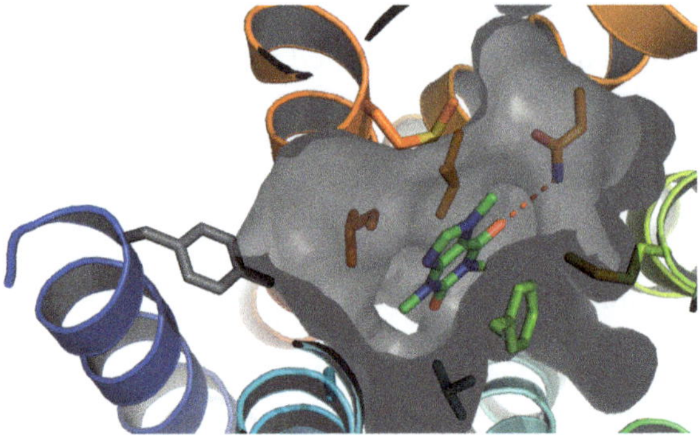

Figure 5.2 A portion of the X-ray structure of caffeine bound to the adenosine A_{2A} StaR (PDB:3RFM).[39] Protein helices are represented as ribbons, with the side-chains of amino acids forming the binding pocket displayed with solid bonds. The surface of the ligand-binding pocket is depicted in grey, within which caffeine is displayed in atom colours. A hydrogen bond between caffeine and the side-chain of $Asn253^{6.55}$ is represented by a red dashed line.

screening for GPCRs, the β_1 adrenergic receptor was thermostabilised, solubilised in detergent (decyl maltoside; DM) and immobilised onto a resin.[66] Screening of a fragment library of approximately 600 molecules confirmed that combination of the TINS approach with a StaR protein could be used to detect weak fragment hits. In follow up experiments a number of these hits were confirmed to bind to the receptor in an orthogonal radioligand-binding assay. In another report, the TINS method was applied to screening of an A_{2A} StaR, which was immobilised on sepharose resin and screened at 500 μM with a library of approximately 500 fragments.[67] OmpA was used in a reference cell to determine specificity of hits; OmpA is a rare example of a stable membrane protein and is known to exhibit minimal specific binding to small molecules. Using a cut-off ratio between the StaR and OmpA of ≤0.7 for the well-resolved ^{1}H-NMR signals, 94 fragments bound specifically to A_{2A} StaR2 and 6 fragments specifically to OmpA. The 94 selective A_{2A} hits were then screened in a radioligand-binding assays with the wild-type receptor, using both [^{3}H]-ZM241385 (an antagonist; 4-(2-{[7-amino-2-(furan-2-yl)[1,2,4]triazolo[1,5-α][1,3,5]triazin-5-yl]amino}ethyl)phenol) and [^{3}H]-NECA (an agonist; *N*-ethyl-5′-carboxamido adenosine).[67] Competitive binding and dissociation rates of the radioligands were examined with the fragments in order to characterise further the hits and establish if they were acting as orthosteric or allosteric ligands. Overall, competitive orthosteric ligands with diverse chemical structures were identified, with pIC_{50} values in the range 2.7–4.2 in concentration–response studies with [^{3}H]-ZM241385. The hits were also able to displace [^{3}H]-NECA, the agonist ligand, as would be expected for orthosteric binders. Furthermore, in a radioligand-binding assay with the A_1 receptor they could displace [^{3}H]-DPCPX (a reasonably selective antagonist ligand for the A_1 receptor). The lack of receptor subtype selectivity of the fragment hits is perhaps not surprising owing to the sizes of the molecules and because of the high homology between the adenosine A_{2A} and A_1 receptors. Further work is warranted to determine whether it will be possible to generate more potent lead-like molecules from these hits. As well as the orthosteric hits, several non-competitive ligands were also detected using the TINS technique. Measurement of the effects of this set of binders on the dissociation rates of radioligands at both A_1 and A_{2A} receptors indicated that modulators with several different profiles had been identified. Some ligands decreased the off-rate of at least one of the radioligands and others increased the off-rate, suggesting that both positive and negative allosteric modulators had been identified.[67]

5.3.4 CE Screening

CE is a sensitive screening technique which has been demonstrated to be able to identify low molecular weight, low-affinity, ligands for heat shock protein 90 in mobility shift competition assays.[68] The technique requires a competitor probe ligand which binds to the biological target of interest and uses a capillary and electrodes with a high-voltage power supply to separate charged and uncharged particles.[69] In a proof-of-concept study, capillary electrophoresis was utilised to detect interactions between a GPCR and a library of fragments. Again using A_{2A} StaR1 (described above) and an appropriate probe ligand it was possible to detect the binding of ligands with a range of potencies, including the fragment-sized molecule, caffeine.[70]

5.3.5 T_m Screening

The screening of fragments by monitoring their ability to increase the temperature at which a protein unfolds (T_m) has been widely used for soluble proteins,[71] but has not been reported for membrane proteins to our knowledge. In our laboratory we have been able to use T_m screening with GPCRs as either the wild-type receptors (if a T_m can be measured on the solubilised receptor) or on partially stabilised receptors (containing small numbers of stabilising mutations). However, this is not an approach so far developed for primary screening, but rather as confirmation of hits detected by other methods. In particular the ability of a fragment to stabilise a StaR or wild-type GPCR further when solubilised from the cell membrane is a good proof of a specific and direct interaction between the fragment and the protein. This approach holds promise for future development as a screening paradigm for GPCRs, but more research will be required to facilitate this and efforts are underway in our laboratories to this end.

5.3.6 Radioligand Competition Binding

Displacement of a radioligand from its binding site is a robust and extremely well-characterised approach for HTS screening and has been successfully applied to fragment screening of both GPCRs and soluble proteins.[44–46,52,53] In our own laboratories we have extensively used this method for GPCR targets for which ligands are known and a radioligand-binding assay can therefore be configured. To facilitate the approach, a number of new GPCR radiochemicals have been developed, allowing HCS screening to be carried out using a library of 3000–4000 fragments. As mentioned earlier, use of the StaR constructs as reagents for screening has the advantage of allowing higher concentrations of DMSO to be used in the assay buffer, enabling screening of libraries at high concentrations (such as 300 μM) whilst ensuring the potential ligands remain in solution.

As part of the conformational thermostabilisation methodology to produce the StaR construct itself, a specific potent ligand is added which is either an agonist or antagonist to the GPCR, biasing the StaR being generated to this same pharmacology. This can have an important consequence during any subsequent screening of the StaR. If the pharmacological form trapped in the StaR in question is poorly populated in the wild type (such as in the agonist conformation), the hits bind more tightly to the StaR than to the wild type, facilitating their detection.[63] This can be an important advantage of using the StaR *vs.* the wild-type receptor for fragment screening. When screening an agonist StaR in a binding assay, it is sometimes possible to detect fragments that also then show some functional agonist behaviour in cell-based assays (unpublished results). This tendency of the StaR to prefer fragments with the same pharmacological bias is not limited to HCS screening, but has been more obvious than when using other screening techniques, perhaps because the radiochemical being used will also be binding to one pharmacological state (*i.e.*, it is either an agonist or antagonist and the assay is a competition with the ligand, rather than a direct binding approach).

5.4 β_1 Adrenergic Receptor Case Study

A detailed case study of the application of FBDD to a representative GPCR is now described and has also been published elsewhere.[72] As previously described, the fragment screen of the β_1 adrenergic StaR was carried out using SPR with a small fragment library (approximately 650 fragments) and was screened in tandem with an adenosine A_{2A} receptor StaR.[66] By screening two StaR constructs together and comparing the results, the hits were more easily distinguished from non-specific binders. During the screening, positive control compounds for each receptor were screened periodically, confirming that the immobilised proteins maintained their ability to bind small molecules selectively. As expected most of the test fragments showed no significant binding to either receptor, however, a number bound more or less equally to both receptors and were considered to be non-selective binders of no further interest. It was observed that subsets of fragments could be detected which bound specifically to either the A_{2A} StaR (as described earlier) or to the β_1 adrenergic StaR. The selective β_1 adrenergic StaR hits were then retested in concentration series to obtain affinity information. Amongst the selective binders to the β_1 adrenergic StaR were 1-[3-(trifluoromethyl)phenyl]piperazine and

2-(piperazin-1-yl)quinoline (**7** and **8**, Figure 5.3), which were estimated to have encouraging binding affinities (K_D 16 μM (pK_D 4.80) and 5.6 μM (pK_D 5.25), respectively) and good LEs (0.41 and 0.48, respectively).[50]

Interestingly, these hits are similar to those identified by Aristotelous *et al.*, described earlier, in an SPR screen of the β_2 adrenergic receptor (compound **4**, Figure 5.1).[60] However, despite piperazines being very well-known in GPCR research as a class of ligands, to our knowledge, related molecules had not been identified for β adrenergic receptors before these two fragment-screening campaigns, the vast majority of ligands known instead containing an ethanolamine functionality (*e.g.*, the ligands in Figure 5.4).

In order to validate hit compounds **7** and **8**, an orthogonal assay format was required to demonstrate unambiguously specific binding to the receptor. Therefore, the hits from the screen and a set of commercially available analogues of similar size and complexity to **7** and **8** were assembled and then screened in a radioligand membrane binding assay with human wild-type β_1 adrenergic receptor and [^{3}H]-dihydroalprenolol. The affinity data (together with LE,[50] ligand lipophilicity efficiency (LLE),[73] and clog$P$[74] values) for a selection of these analogues is shown in Table 5.1. Gratifyingly, all of the commercially available analogues tested had high LEs and bound with moderate-to-high affinity. Compared to the hit **8**, the simplified phenyl piperazine **9** had slightly improved affinity, and the dichloro analogue

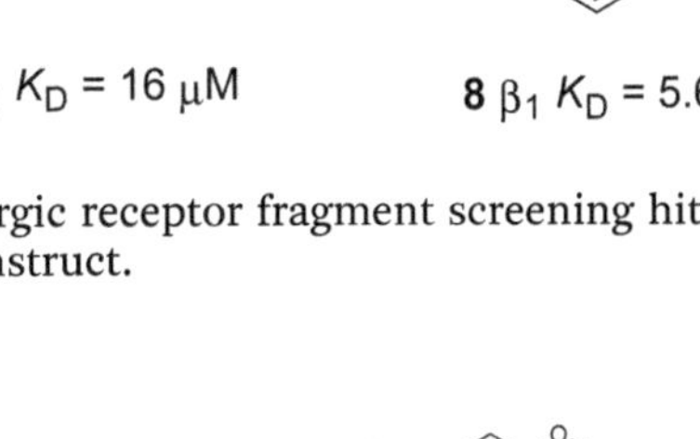

7 β_1 K_D = 16 μM **8** β_1 K_D = 5.6 μM

Figure 5.3 β_1 Adrenergic receptor fragment screening hits from SPR screening of a StaR construct.

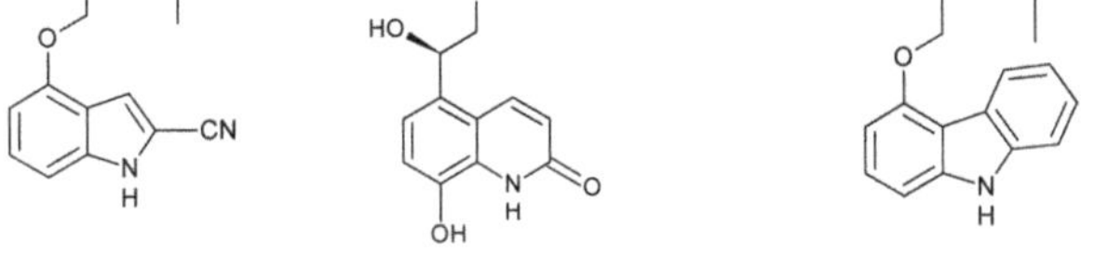

15 cyanopindolol **16** carmoterol **17** carazolol

Figure 5.4 Chemical structures of cyanopindolol **15**, carmoterol **16** and carazolol **17**.

10 was even more potent in the binding assay. However, more polar analogues tested in this series were somewhat weaker and less ligand efficient.[72] Other disubstituted phenyl groups also gave good affinities, with chloro- (including **10**) and methyl- variants yielding higher affinities than tri-fluoromethyl or methoxy derivatives and where 2,3- or 3,5-regiochemistry was found to be preferred over the 3,4-isomers.[72] Quinoline derivatives

Table 5.1 β_1 Adrenergic receptor binding affinities, LE, LLE and clog*P* values of **8–14**.

Compound	Chemical structure	β_1 pK_i[a]	LE[50]	clog$P$[74]	LLE[73]
8	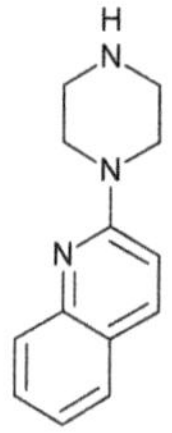	5.20	0.44	1.59	3.61
9	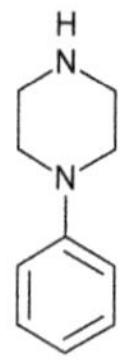	5.87	0.67	1.11	4.76
10	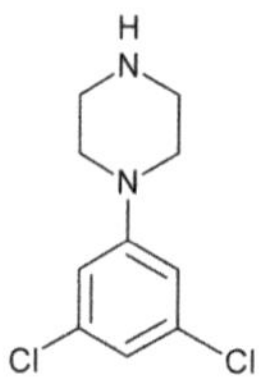	7.07	0.69	3.03	4.04
11	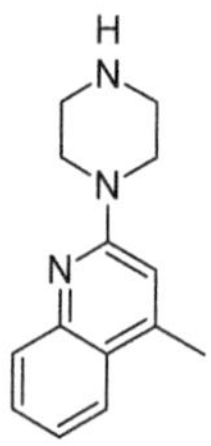	6.65	0.53	2.05	4.60
12	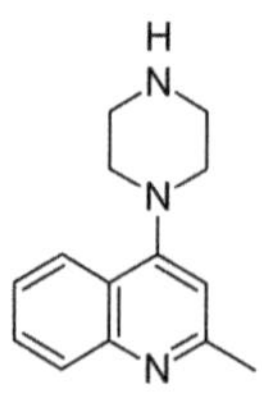	5.80	0.47	1.44	4.36

13	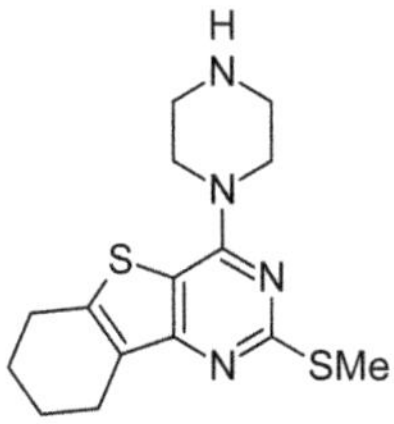	6.70	0.43	3.69	3.01
14	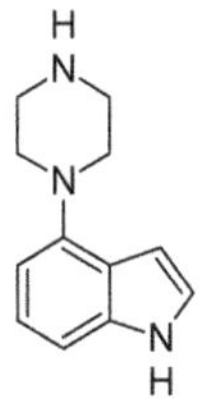	7.17	0.65	1.03	6.14

[a]$pK_i = -\log_{10} K_i$.

were also selected and screened specifically to explore the initial SPR hit **8**; **8** was included as a control and returned highly comparable affinity in the radioligand-binding assay to that from the SPR screen (pK_i 5.20 *vs.* pK_D 5.25 respectively). Compounds **11–13** are included in the table and are representative, demonstrating good scope in terms of SAR and the potential to increase binding affinity. Finally, the indole-containing analogue **14** was included to explore a modelling hypothesis and was found to have excellent affinity and ligand efficiency. This hypothesis and its validation by co-crystallisation of compound **14** in the β_1 adrenergic StaR is described below.

The selection of the compounds described above was made by consideration of the likely binding mode of each of the analogues and the original hits **7** and **8**. The β_1 adrenergic crystal structure of a number of inverse agonists, antagonists and partial and full agonists had already been solved with the turkey β_1 adrenergic StaR.[11–13] The human and turkey receptors are 82% identical in the transmembrane and loop regions (with the exclusion of ICL3) and the residues that comprise the ligand-binding pocket are 100% identical. In addition, comparison with published crystal structure information of the human β_2 adrenergic receptor stabilised by a fusion protein or by an antibody indicates that the binding sites of the two crystal systems are very similar.[75] The turkey receptor therefore provides an excellent model for studying ligand–receptor interactions for this target. Careful consideration was given to the interactions observed in the β_1 adrenergic receptor X-ray structures with cyanopindolol **15**, carmoterol **16** and carazolol **17** (Figure 5.4).[11–13] The hits **7** and **8** and other molecules being considered for screening were docked into the X-ray structure of the β_1 adrenergic receptor and examined in relation to these ligands. The small size of fragments means it is often difficult to dock these ligands with any great confidence due to the multitude of ways they can potentially fit into a receptor-binding site. However, by careful examination of the site in question and how other ligands bind within it, credible binding poses can often be rationalised. Indeed, a complete druggability analysis of the binding site was carried out to identify the region most likely to be occupied by the fragment (Figure 5.5).[43] Water and CH aromatic probes showed the hotspots most favourably occupied by polar or by hydrophobic portions of each ligand, an approach we have shown previously to be predictive for highly ligand-efficient molecules.[43,50] As a specific example, the selection of indole-containing fragment **14** was driven by comparison of its docking pose with the X-ray structures of cyanopindolol **15**, carmoterol **16** and carazolol **17**. Figure 5.5 shows overlays of the subsequently solved X-ray complex of **14** (described below), with those of **15** and **16**. The NH of the indole ring in **14** overlays clearly with the NH of the indole of **15** and the heterocycle of **16**.

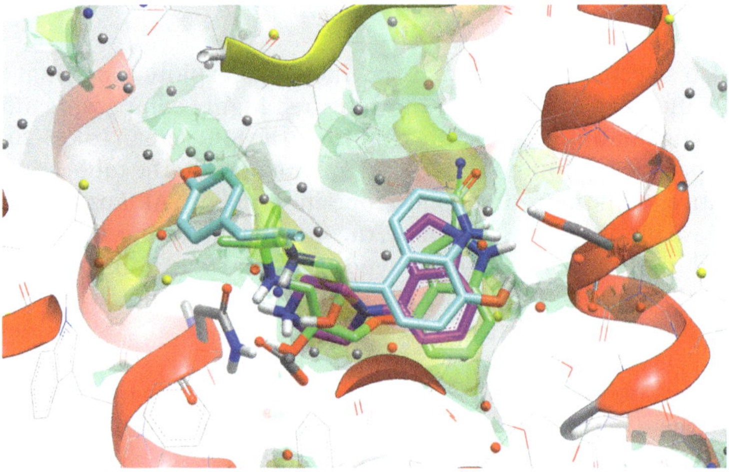

Figure 5.5 Overlay of the β_1 adrenergic ligand cyanopindolol **15** (ligand with green carbons), X-ray structure (PDB:2VT4), carmoterol **16** (cyan carbon atoms) (PDB:2Y02) and the indole fragment **14** (lilac carbon atoms) (PDB:3ZPQ), showing the common interaction of the NH groups with Ser211$^{5.42}$. The druggability of the binding site is shown with the GRID maps (adjacent lipophilic and H-bonding hotspots) and 'unhappy' red and yellow coded waters from a WaterMap analysis. GRID maps are contoured (transparent solid) and coloured in the following manner: aromatic CH probe (lipophilic) in yellow at −2.8 kcal mol^{-1}, water (OH_2) probe in green at −6.0 kcal mol^{-1}, CH_3 methyl group probe in grey at 1 kcal mol^{-1} which defines the pocket surface in terms of how close a ligand carbon atom can go. WaterMap (Schrodinger) water clusters (shown as large spheres) have been colour-coded in red if predicted to have a free energy (ΔG) > 3.5 kcal mol^{-1}, in yellow if their predicted ΔG is between 2.0 and 3.5 kcal mol^{-1}, in grey if ΔG is −1.0 to 2.0 kcal mol^{-1} and in blue if ΔG < −1.0 kcal mol^{-1}. All WaterMap free-energy estimations are relative to bulk solvent.

The NH of the heterocycles in each case was proposed to form a hydrogen bond to Ser211$^{5.42}$. Compound **14** was found to have significantly increased affinity for the receptor (Table 5.1) and represents an interesting lead compound for future optimisation.

The identification of the piperazine-based fragments shown in Table 5.1, having relatively high affinity and high solubility, provided an opportunity to obtain co-structures of this interesting class of molecules with the β_1 adrenergic StaR. As described above, the X-ray structural information can be used to confirm the modes of binding of the fragments proposed in docking experiments (Figure 5.5). In addition, these data can facilitate further binding-mode comparisons with existing liganded X-ray structures and allow for SBDD to be used to develop a series further. Using a consideration of affinity and LE, indole **14** and the structurally dissimilar quinoline **11** were selected for crystallography. The β_1 adrenergic StaR was co-crystallised with fragments **11** and **14** as previously described[11,12] and readily gave structures, determined to resolutions of 2.7 Å and 2.8 Å respectively. An examination of the binding modes indicated that the piperazine rings in the 2 molecules were located between Asp121$^{3.32}$ on helix 3 (H3) and Asn329$^{7.39}$ on H7, and their aromatic substituents were positioned adjacent to H5 (Figure 5.6C and D). It appears that the piperazine rings in these molecules substitute for the ethanolamine core present in other adrenergic receptor ligands; in addition, the structural roles of the phenyl substituents are similar to those of the indole-carbonitrile and carbazole head-groups of cyanopindolol **15** and carazolol **17** (Figure 5.6A and B). As might be expected for fragments, the number of ligand–receptor interactions observed in the β_1 StaR–**11** and β_1 StaR–**14** structures is lower than those in the crystal structures with cyanopindolol **15** and carazolol **17**.[11–13] In **15** and **17** the ethanolamine secondary amine and β-hydroxyl groups both form hydrogen bonds to the Asp121$^{3.32}$ and Asn329$^{7.39}$ side-chains. However, fragments **11** and **14** form just one interaction from the piperazine ring with Asp121$^{3.32}$ and Asn329$^{7.39}$. Perhaps as a consequence of this, there appears to be greater variation in the orientation of the piperazine rings than is shown by the ethanolamine groups in carazolol **17** and cyanopindolol **15** (Figure 5.6). Another interesting consideration is the differences in pharmacological activity between cyanopindolol **15** (a partial agonist with sympathomimetic activity) and carazolol **17** (an inverse agonist) that may in part result from the differences in the conformation of Ser211$^{5.42}$ observed in the structures with these ligands[11–13] (Figure 5.6A and B). Fragment **11** does not form an interaction with Ser211$^{5.42}$ whilst the indole NH of **14** does form a hydrogen-bonding interaction with this residue. This suggests that using the detailed structural information now available, the targeted addition of further substituents to these fragments could serve not only to increase their affinities further, but also to modulate their pharmacological activities.

The case study described above illustrates that the now well-understood FBDD path to lead identification can be used successfully for GPCRs. Biophysical hits have been developed into lead molecules with good SAR and

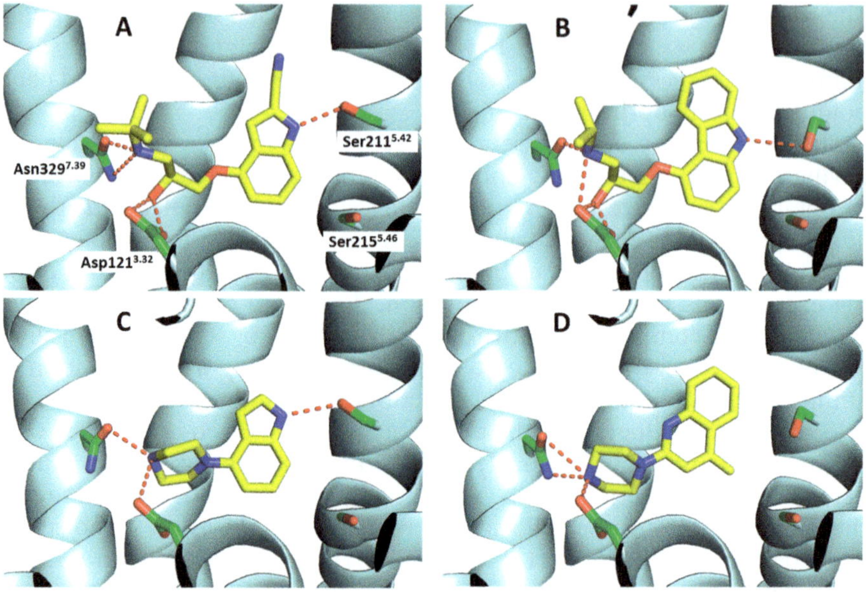

Figure 5.6 Sections of β_1 adrenergic receptor structures, viewed approximately in the membrane plane with the extracellular side uppermost. (A) Complex with cyanopindolol **15** (PDB:2VT4). (B) Complex with carazolol **17** (PDB:2YCW). (C) Complex with **14** (PDB:3ZPR). (D) Complex with **11** (PDB:3ZPR). Atoms are coloured accordingly: ligands/fragments: C, yellow; O, red; N, blue; selected receptor side-chains: C, green; O, red; N, blue. Potential hydrogen bonds and polar contacts are shown as red dashes. Global alignment of the receptor structures was performed using PYMOL (align) and the resulting positions of the ligands are depicted.

crystallised to reveal their ligand–receptor binding modes and interactions. The approach described here validates the use of fragment-based screening and SBDD approaches to membrane proteins and we believe has much promise for future drug-discovery efforts on a broad range of GPCR targets.

5.5 The Future of Fragment Screening with Membrane Proteins

In this chapter we have outlined the early development, validation and implementation of fragment screening and FBDD for GPCRs. The area has been reviewed, giving examples of studies using wild-type receptors and also our own approaches using stabilised constructs, engineered to allow these membrane-associated proteins to be interrogated in solution. A case study from our own work has been described in some detail, in which all the important components of FBDD were applied; namely – screening of a library of small fragments using a biophysical screening method, hit validation using an orthogonal method, demonstration of fragment SAR, evolution of the original fragment hits to more potent and ligand-efficient leads using SBDD methods and the successful structure determination of representative molecules in the hit series using X-ray crystallography, providing a starting point for further optimisation. Overall, the examples presented above support the notion that FBDD can be successfully applied to GPCRs. Biophysical methods, which have been well validated for soluble proteins, can now be used for membrane proteins provided that they are prepared in a suitable form. It should also be noted that new biophysical techniques for detecting ligand binding to GPCRs, such as thermophoresis are being reported,[76] which may develop into further means of primary or secondary screening for FBDD. The development of biophysical assays for fragment screening of other membrane proteins, such as ion channels, has also been described in recent years.[77,78]

A criticism of the FBDD exemplar studies to date is that in each case the GPCRs that have been screened were well-understood targets, in which many small-molecule ligands have already been described. Indeed, in the case of histamine and the β adrenergic systems the endogenous ligands are themselves fragment sized and many fragment-sized ligands pre-dated the fragment screening approaches described earlier. In the case of the adenosine A_{2A} system, small xanthine ligands such as caffeine and theophylline were already known to bind to this GPCR. As such, one might argue that FBDD was highly likely to be successful and sensitive biophysical screening methods may not have been required. However, it is important to see these studies as early proof-of-concept experiments analogous to the early days of FBDD on soluble targets approximately 10 years ago. At that time publications focused on the screening methodology and used relatively tractable targets such as PDE4, Urokinase, FKBP12, DNA gyrase and CDK2.[79–83] The targets themselves were not actually so important, but instead the establishment of the methods and the FBDD concepts themselves. It is our belief that the published literature for FBDD with GPCRs should be seen as being at the same stage of development as FBDD on soluble targets was 10 years ago; the data presented in this chapter should be seen in this context.

Over the last 4 or 5 years we have investigated fragment-based screening on more than a dozen GPCR targets, using StaR constructs, by at least one of the approaches described above and using libraries of approximately 500 to 4000 fragments. In general terms the observation has been that GPCR targets have been at least as tractable to fragment screening as soluble targets, in terms of the number of confirmed hit series identified per target that can be progressed. The current limitation is of course that GPCRs are still challenging systems to crystallise, in order to generate protein–ligand binding data to facilitate SBDD. However, it has been our experience that credible binding poses can be generated for fragment hits, as in the case study presented earlier, using molecular modelling combined with SDM data or BPM.[64] As a hit series is developed we have typically been able to generate X-ray crystal structures with lead compounds and utilise 'true' SBDD during the lead optimisation phase of our discovery projects.

A recent success story for our FBDD platform is the identification of a pre-clinical candidate molecule as a negative allosteric modulator (NAM) of the mGlu5 receptor (metabotropic glutamate receptor).[84] Here a high-concentration radioligand-binding assay was used to screen a large library of fragments. Multiple fragment hits were identified, and several were evaluated by simple enumeration synthetically and by purchasing of commercial analogues. The initial hit validation exercise was driven by molecular modelling and docking into a homology model of the mGlu5 receptor, which is a Class C GPCR. In one series, very rapid progress was made in the medicinal chemistry: large increases in affinity were observed with very small numbers of compounds being synthesised. In this series, a candidate molecule was subsequently identified by synthesis of fewer than 100 compounds and with less than one year's effort. This FBDD story will be the subject of future publications and is included in the discussion here to underline that FBDD is being used effectively in our laboratory for important GPCR drug targets.[84]

5.6 Conclusions

In conclusion, the field of FBDD for GPCRs is still relatively in its infancy but its promise is enormous. Indeed, the growing number of available GPCR X-ray structures together with new biology linking GPCRs to disease is stimulating a renewed interest in GPCRs as drug targets across a wide range of therapeutic areas. A number of highly validated targets previously considered intractable or 'undruggable' are now being revisited. As new X-ray structures emerge, most recently for Class B receptors, the research directed at GPCRs is sure to increase further. If the field of SBDD for enzymes serves as a precedent, then many future medicines targeted at GPCRs are likely to be derived using fragment- and structure-based drug design. In particular, access to StaR constructs, facilitating HCS and direct binding biophysical experiments such as SPR screening and X-ray crystallography, should provide a valuable platform for GPCR FBDD in the future.

Note added in proof: After the submission of this chapter the structures of two class C GPCRs have been reported, including the X-ray structure of the mGlu5 receptor in complex with a NAM.[85,86] These data supported the FBDD project briefly described above during the latter stages of the lead optimisation process.

Acknowledgements

The authors would like to thank Fiona Marshall, John Christopher, Julie Cansfield, Susanna Wood and Janie Anka-Lufford for their help in proof reading and finalising this chapter, and to Benjamin Tehan and Jonathon Mason for preparing figures. Figures were produced using PYMOL (Schrodinger) and VIDA (OpenEye). The WaterMap software is from Schrodinger.

References

1. M. C. Lagerström and H. B. Schiöth, *Nat. Rev. Drug Discovery*, 2008, **7**, 339.
2. V. Katritch, V. Cherezov and R. C. Stevens, *Annu. Rev. Pharmacol. Toxicol.*, 2013, **53**, 531.
3. A. J. Venkatakrishnan, X. Deupi, G. Lebon, C. G. Tate, G. F. Schertler and M. M. Babu, *Nature*, 2013, **494**, 185.
4. D. M. Rosenbaum, S. G. Rasmussen and B. K. Kobilka, *Nature*, 2009, **459**, 356.
5. A. J. Harmar, *Genome Biol.*, 2001, **2**, 3013.1.
6. S. Urwyler, *Pharmacol. Rev.*, 2011, **63**, 59.
7. G. Schulte and V. Bryja, *Trends Pharmacol. Sci.*, 2007, **28**, 518.
8. K. Palczewski, T. Kumasaka, T. Hori, C. A. Behnke, H. Motoshima, B. A. Fox, I. Le Trong, D. C. Teller, T. Okada, R. E. Stenkamp, M. Yamamoto and M. Miyano, *Science*, 2000, **289**, 739.
9. V. Cherezov, D. M. Rosenbaum, M. A. Hanson, S. G. F. Rasmussen, F. S. Thian, T. S. Kobilka, H.-J. Choi, P. Kuhn, W. I. Weis, B. K. Kobilka and R. C. Stevens, *Science*, 2007, **318**, 1258.
10. S. G. F. Rasmussen, H.-J. Choi, D. M. Rosenbaum, T. S. Kobilka, F. S. Thian, P. C. Edwards, M. Burghammer, V. R. P. Ratnala, R. Sanishvili, R. F. Fischetti, G. F. X. Schertler, W. I. Weis and B. K. Kobilka, *Nature*, 2007, **450**, 383.
11. T. Warne, M. J. Serrano-Vega, J. G. Baker, R. Moukhametzianov, P. C. Edwards, R. Henderson, A. G. W. Leslie, C. G. Tate and G. F. X. Schertler, *Nature*, 2008, **454**, 486.
12. T. Warne, R. Moukhametzianov, J. G. Baker, R. Nehmé, P. C. Edwards, A. G. W. Leslie, G. F. X. Schertler and C. G. Tate, *Nature*, 2011, **469**, 241.
13. R. Moukhametzianov, T. Warne, P. C. Edwards, M. J. Serrano-Vega, A. G. W. Leslie, C. G. Tate and G. F. X. Schertler, *Proc. Natl. Acad. Sci. U. S. A.*, 2011, **108**, 8228.
14. V.-P. Jaakola, M. T. Griffith, M. A. Hanson, V. Cherezov, E. Y. T. Chien, J. R. Lane, A. P. IJzerman and R. C. Stevens, *Science*, 2008, **322**, 1211.
15. F. Xu, H. Wu, V. Katritch, G. W. Han, K. A. Jacobson, Z.-G. Gao, V. Cherezov and R. C. Stevens, *Science*, 2011, **332**, 322.
16. M. A. Hanson, C. B. Roth, E. Jo, M. T. Griffith, F. L. Scott, G. Reinhart, H. Desale, B. Clemons, S. M. Cahalan, S. C. Schuerer, M. G. Sanna, G. W. Han, P. Kuhn, H. Rosen and R. C. Stevens, *Science*, 2012, **335**, 851.
17. B. Wu, E. Y. T. Chien, C. D. Mol, G. Fenalti, W. Liu, V. Katritch, R. Abagyan, A. Brooun, P. Wells, F. C. Bi, D. J. Hamel, P. Kuhn, T. M. Handel, V. Cherezov and R. C. Stevens, *Science*, 2010, **330**, 1066.
18. E. Y. T. Chien, W. Liu, Q. Zhao, V. Katritch, G. W. Han, M. A. Hanson, L. Shi, A. H. Newman, J. A. Javitch, V. Cherezov and R. C. Stevens, *Science*, 2010, **330**, 1091.
19. T. Shimamura, M. Shiroishi, S. Weyand, H. Tsujimoto, G. Winter, V. Katritch, R. Abagyan, V. Cherezov, W. Liu, G. W. Han, T. Kobayashi, R. C. Stevens and S. Iwata, *Nature*, 2011, **475**, 65.

20. K. Haga, A. C. Kruse, H. Asada, T. Yurugi-Kobayashi, M. Shiroishi, C. Zhang, W. I. Weis, T. Okada, B. K. Kobilka, T. Haga and T. Kobayashi, *Nature*, 2012, **482**, 547.
21. A. C. Kruse, J. Hu, A. C. Pan, D. H. Arlow, D. M. Rosenbaum, E. Rosemond, H. F. Green, T. Liu, P. S. Chae, R. O. Dror, D. E. Shaw, W. I. Weis, J. Wess and B. K. Kobilka, *Nature*, 2012, **482**, 552.
22. J. F. White, N. Noinaj, Y. Shibata, J. Love, B. Kloss, F. Xu, J. Gvozdenovic-Jeremic, P. Shah, J. Shiloach, C. G. Tate and R. Grisshammer, *Nature*, 2012, **490**, 508.
23. S. Granier, A. Manglik, A. C. Kruse, T. S. Kobilka, F. S. Thian, W. I. Weis and B. K. Kobilka, *Nature*, 2012, **485**, 400.
24. H. Wu, D. Wacker, M. Mileni, V. Katritch, G. W. Han, E. Vardy, W. Liu, A. A. Thompson, X.-P. Huang, F. I. Carroll, S. W. Mascarella, R. B. Westkaemper, P. D. Mosier, B. L. Roth, V. Cherezov and R. C. Stevens, *Nature*, 2012, **485**, 327.
25. A. Manglik, A. C. Kruse, T. S. Kobilka, F. S. Thian, J. M. Mathiesen, R. K. Sunahara, L. Pardo, W. I. Weis, B. K. Kobilka and S. Granier, *Nature*, 2012, **485**, 321.
26. A. A. Thompson, W. Liu, E. Chun, V. Katritch, H. Wu, E. Vardy, X.-P. Huang, C. Trapella, R. Guerrini, G. Calo, B. L. Roth, V. Cherezov and R. C. Stevens, *Nature*, 2012, **485**, 395.
27. C. Zhang, Y. Srinivasan, D. H. Arlow, J. J. Fung, D. Palmer, Y. Zheng, H. F. Green, A. Pandey, R. O. Dror, D. E. Shaw, W. I. Weis, S. R. Coughlin and B. K. Kobilka, *Nature*, 2012, **492**, 387.
28. C. Wang, Y. Jiang, J. Ma, H. Wu, D. Wacker, V. Katritch, G. W. Han, W. Liu, X.-P. Huang, E. Vardy, J. D. McCorvy, X. Gao, E. X. Zhou, K. Melcher, C. Zhang, F. Bai, H. Yang, L. Yang, H. Jiang, B. L. Roth, V. Cherezov, R. C. Stevens and H. E. Xu, *Science*, 2013, **340**, 610.
29. D. Wacker, C. Wang, V. Katritch, G. W. Han, X.-P. Huang, E. Vardy, J. D. McCorvy, Y. Jiang, M. Chu, F. Y. Siu, W. Liu, H. E. Xu, V. Cherezov, B. L. Roth and R. C. Stevens, *Science*, 2013, **340**, 615.
30. Q. Tan, Y. Zhu, J. Li, Z. Chen, G. W. Han, I. Kufareva, T. Li, L. Ma, G. Fenalti, J. Li, W. Zhang, X. Xie, H. Yang, H. Jiang, V. Cherezov, H. Liu, R. C. Stevens, Q. Zhao and B. Wu, *Science*, 2013, **341**, 1387.
31. C. Wang, H. Wu, V. Katritch, G. W. Han, X.-P. Huang, W. Liu, F. Y. Siu, B. L. Roth, V. Cherezov and R. C. Stevens, *Nature*, 2013, **497**, 338.
32. K. Hollenstein, J. Kean, A. Bortolato, R. K. Y. Cheng, A. S. Doré, A. Jazayeri, R. M. Cooke, M. Weir and F. H. Marshall, *Nature*, 2013, **499**, 438.
33. F. Y. Siu, M. He, C. de Graaf, G. W. Han, D. Yang, Z. Zhang, C. Zhou, Q. Xu, D. Wacker, J. S. Joseph, W. Liu, J. Lau, V. Cherezov, V. Katritch, M.-W. Wang and R. C. Stevens, *Nature*, 2013, **499**, 444.
34. M. Caffrey and V. Cherezov, *Nat. Protoc.*, 2009, **4**, 706.
35. M. Caffrey, D. Li and A. Dukkipati, *Biochemistry*, 2012, **51**, 6266.
36. M. J. Serrano-Vega, F. Magnani, Y. Shibata and C. G. Tate, *Proc. Natl. Acad. Sci. U. S. A.*, 2008, **105**, 877.
37. F. Magnani, Y. Shibata, M. J. Serrano-Vega and C. G. Tate, *Proc. Natl. Acad. Sci. U. S. A.*, 2008, **105**, 10750.

38. Y. Shibata, J. F. White, M. J. Serrano-Vega, F. Magnani, A. L. Aloia, R. Grisshammer and C. G. Tate, *J. Mol. Biol.*, 2009, **390**, 262.
39. A. S. Doré, N. Robertson, J. C. Errey, I. Ng, K. Hollenstein, B. Tehan, E. Hurrell, K. Bennett, M. Congreve, F. Magnani, C. G. Tate, M. Weir and F. H. Marshall, *Structure*, 2011, **19**, 1283.
40. G. Lebon, T. Warne, P. C. Edwards, K. Bennett, C. J. Langmead, A. G. Leslie and C. G. Tate, *Nature*, 2011, **474**, 521.
41. M. Congreve, C. J. Langmead, J. S. Mason and F. H. Marshall, *J. Med. Chem.*, 2011, **54**, 4283.
42. M. Congreve, C. Langmead and F. H. Marshall, *Adv. Pharmacol.*, 2011, **62**, 1.
43. J. S. Mason, A. Bortolato, M. Congreve and F. H. Marshall, *Trends Pharmacol. Sci.*, 2012, **33**, 249.
44. D. A. Erlanson, *Top. Curr. Chem.*, 2012, **317**, 1.
45. G. Chessari and A. J. Woodhead, *Drug Discovery Today*, 2009, **14**, 668.
46. C. W. Murray, M. L. Verdonk and D. C. Rees, *Trends Pharmacol. Sci.*, 2012, **33**, 224.
47. M. Congreve, R. Carr, C. Murray and H. Jhoti, *Drug Discovery Today*, 2003, **8**, 876.
48. M. M. Hann, A. R. Leach and G. Harper, *J. Chem. Inf. Comput. Sci.*, 2001, **41**, 856.
49. D. A. Erlanson, R. S. McDowell and T. O'Brien, *J. Med. Chem.*, 2004, **47**, 3463.
50. A. L. Hopkins, C. R. Groom and A. Alex, *Drug Discovery Today*, 2004, **9**, 430.
51. M. Congreve, C. Langmead, J. Mason and F. Marshall, *J. Med. Chem.*, 2011, **54**, 4283.
52. J. S. Albert, N. Blomberg, A. L. Breeze, A. J. Brown, J. N. Burrows, P. D. Edwards, R. H. Folmer, S. Geschwindner, E. J. Griffen, P. W. Kenny, T. Nowak, L. L. Olsson, H. Sanganee and A. B. Shapiro, *Curr. Top. Med. Chem.*, 2007, **7**, 1600.
53. M. H. P. Verheij, C. de Graaf, G. E. de Kloe, S. Nijmeijer, H. F. Vischer, R. A. Smits, O. P. Zuiderveld, S. Hulscher, L. Silvestri, A. J. Thompson, J. E. van Muijlwijk-Koezen, S. C. R. Lummis, R. Leurs and I. J. P. de Esch, *Bioorg. Med. Chem. Lett.*, 2011, **21**, 5460.
54. L. A. Stoddart, A. J. Vernall, J. L. Denman, S. J. Briddon, B. Kellam and S. J. Hill, *Chem. Biol.*, 2012, **19**, 1105.
55. T. P. Kenakin, *Br. J. Pharmacol.*, 2012, **165**, 1659.
56. T. Neumann, H. D. Junker, K. Schmidt and R. Sekul, *Curr. Top. Med. Chem.*, 2007, **7**, 1630.
57. S. Perspicace, D. Banner, J. Benz, F. Muller, D. Schlatter and W. Huber, *J. Biomol. Screening*, 2009, **14**, 337.
58. I. Navratilova and A. L. Hopkins, *ACS Med. Chem. Lett.*, 2010, **1**, 44.
59. I. Navratilova, J. Besnard and A. L. Hopkins, *ACS Med. Chem. Lett.*, 2011, **2**, 549.
60. T. Aristotelous, S. Ahn, A. K. Shukla, S. Gawron, M. F. Sassano, A. W. Kahsai, L. M. Wingler, X. Zhu, P. Tripathi-Shukla, X.-P. Huang, J. Riley, J. Besnard, K. D. Read, B. L. Roth, I. H. Gilbert, A. L. Hopkins, R. J. Lefkowitz and I. Navratilova, *ACS Med. Chem. Lett.*, 2013, **4**, 1005.

61. C. De Graaf, A. J. Kooistra, H. F. Vischer, V. Katritch, M. Kuijer, M. Shiroishi, S. Iwata, T. Shimamura, R. C. Stevens, I. J. de Esch and R. Leurs, *J. Med. Chem.*, 2010, **54**, 8195.
62. J. A. Ballesteros and H. Weinstein, *Methods Neurosci.*, 1995, **25**, 366.
63. K. A. Bennett, B. Tehan, G. Lebon, C. G. Tate, M. Weir, F. H. Marshall and C. J. Langmead, *Mol. Pharmacol.*, 2013, **83**, 949.
64. A. Zhukov, S. P. Andrews, J. C. Errey, N. Robertson, B. Tehan, J. S. Mason, F. H. Marshall, M. Weir and M. Congreve, *J. Med. Chem.*, 2011, **54**, 4312.
65. R. L. Rich, J. Errey, F. Marshall and D. G. Myszka, *Anal. Biochem.*, 2011, **409**, 267.
66. M. Congreve, R. L. Rich, D. G. Myszka, F. Figaroa, G. Siegal and F. H. Marshall, *Methods Enzymol.*, 2011, **493**, 115.
67. D. Chen, J. C. Errey, L. H. Heitman, F. H. Marshall, A. P. Ijzerman and G. Siegal, *ACS Chem. Biol.*, 2012, **7**, 2064.
68. C. Austin, S. N. Pettit, S. K. Magnolo, J. Sanvoisin, W. Chen, S. P. Wood, L. D. Freeman, R. J. Pengelly and D. E. Hughes, *J. Biomol. Screening*, 2012, **17**, 868.
69. D. Hughes, WO 2011036476, Selcia Ltd, 2011.
70. S. P. Andrews and B. Tehan, *Med. Chem. Commun.*, 2013, **4**, 52.
71. J. K. Kranz and C. Schalk-Hihi, *Methods Enzymol.*, 2011, **493**, 277.
72. J. A. Christopher, J. Brown, A. S. Doré, J. C. Errey, M. Koglin, F. H. Marshall, D. G. Myszka, R. L. Rich, C. G. Tate, B. Tehan, T. Warne and M. Congreve, *J. Med. Chem.*, 2013, **56**, 3446.
73. P. D. Leeson and B. Springthorpe, *Nat. Rev. Drug Discovery*, 2007, **6**, 881.
74. ACD clog*P* calculator v12.02 within ACD/ChemSketch release 12.00 was used to calculate clog*P* values, http://www.acdlabs.com
75. T. Warne, J. Chirnside and G. F. X. Schertler, *Biochim. Biophys. Acta*, 2003, **1610**, 133.
76. S. A. I. Seidel, P. M. Dijkman, W. A. Lea, G. van den Bogaart, M. Jerabek-Willemsen, A. Lazic, J. S. Joseph, P. Srinivasan, P. Baaske, A. Simeonov, I. Katritch, F. A. Melo, J. E. Ladbury, G. Schreiber, A. Watts, D. Braun and S. Duhr, *Methods*, 2013, **59**, 301.
77. K. Retra, M. Geitmann, J. Kool, G. Smit, I. J. P. de Esch, U. H. Danielson and H. Irth, *Anal. Biochem.*, 2010, **407**, 58.
78. A. J. Thomson, M. H. P. Verheij, R. Leurs, I. J. P. de Esch and S. C. R. Lummis, *Biotechniques*, 2010, **49**, 822.
79. G. L. Card, L. Blasdell, B. P. England, C. Zhang, Y. Suzuki, S. Gillette, D. Fong, P. N. Ibrahim, D. R. Artis, G. Bollag, M. V. Milburn, S. H. Kim, J. Schlessinger and K. Y. Zhang, *Nat. Biotechnol.*, 2005, **23**, 201.
80. V. L. Nienaber, P. L. Richardson, V. Klighofer, J. J. Bouska, V. L. Giranda and J. Greer, *Nat. Biotechnol.*, 2000, **18**, 1105.
81. S. B. Shuker, P. J. Hajduk, R. P. Meadows and S. W. Fesik, *Science*, 1996, **274**, 1531.
82. H. J. Boehm, M. Boehringer, D. Bur, H. Gmuender, W. Huber, W. Klaus, D. Kostrewa, H. Kuehne, T. Luebbers, N. Meunier-Keller and F. Mueller, *J. Med. Chem.*, 2000, **43**, 2664.

83. M. S. Congreve, D. J. Davis, L. Devine, C. Granata, M. O'Reilly, P. G. Wyatt and H. Jhoti, *Angew. Chem., Int. Ed.*, 2003, **42**, 4479.
84. J. A. Christopher, S. J. Aves, K. A. Bennett, M. Congreve, A. Jazayeri, S. A. Khan, J. C. Patel, B. Tehan, G. Wiggin and F. H. Marshall, *Poster presentation, 247th ACS National Meeting Dallas*, Texas, 16–20 March 2014.
85. H. Wu, C. Wang, K. J. Gregory, G. W. Han, H. P. Cho, Y. Xia, C. M. Niswender, V. Katritch, J. Meiler, V. Cherezov, P. J. Conn and R. C. Stevens, *Science*, 2014, **344**, 58.
86. A. S. Doré, K. Okrasa, J. C. Patel, M. Serrano-Vega, K. Bennett, R. M. Cooke, J. C. Errey, A. Jazayeri, S. Khan, B. Tehan, M. Weir, G. R. Wiggin and F. H. Marshall, *Nature*, 2014, **511**, 557.

CHAPTER 6

Fragment-Based Lead Discovery Applied to Protein–Protein Interactions

RODERICK E. HUBBARD*[a,b]

[a]Vernalis (R&D) Ltd, Granta Park, Abington, Cambridge, CB21 6GB, UK; [b]York Structural Biology Laboratory, University of York, Heslington, York, YO10 5DD, UK
*E-mail: r.hubbard@vernalis.com

6.1 Introduction

The continuing advances in molecular, cellular and systems biology have identified many potentially important mechanisms in disease that rely on the interaction between two (or more) proteins. The interaction surfaces between the protein partners tends to be relatively large and flat, with a number of often shallow binding pockets where side-chains from the partners achieve the recognition and affinity of interaction.[1] Although analysis of mutational and peptide scanning data suggest these pockets can in some cases provide potentially druggable hotspots,[2] the sites are often quite distant from each other, making it a real challenge to identify and optimise suitable small molecules as inhibitors which retain drug-like properties.

Fragments provide an exciting opportunity for identifying initial start-points for drug discovery against such protein–protein interfaces. The main advantage is that it is possible to find fragments that bind to most binding

RSC Drug Discovery Series No. 47
Fragment-Based Drug Discovery
Edited by Steven Howard and Chris Abell

Published by the Royal Society of Chemistry, www.rsc.org

sites in most proteins. However, there are substantial additional challenges in all aspects of the fragment-based lead discovery (FBLD) process compared to FBLD against conventional targets such as kinases – including producing suitable protein, configuring robust assays, characterising fragment binding and achieving a balance between affinity, cellular activity and compound properties to identify suitable clinical candidates. This chapter will review progress in FBLD against such targets. At Vernalis, we have conducted fragment- and structure-based discovery against various protein–protein interaction (PPI) targets. To date, none of this work is published; disclosed is the discovery of pre-clinical candidates for inhibition of the protein Bcl-2, and our experiences with initial experiments on MDM2 and β-catenin. Our experience with these and other targets will be combined with this literature review to summarise the challenges and issues.

6.1.1 Definition of a Protein–Protein Interaction

An initial task is to define what a PPI is, and define the scope of the chapter. Many of the conventional enzyme targets (such as kinases, proteases, phosphatases, *etc.*) rely on recognition of one protein by another as an essential part of their mechanism. However, in those cases, it is either the catalytic machinery or a co-factor binding site that is usually targeted by inhibitors. The definition I will use to limit the discussion is that I will consider drug-discovery projects which identify small molecules that bind to a non-catalytic binding site on one protein; the binding of such a small molecule prevents the binding of another protein. This excludes some fascinating recent stories such as allosteric effectors of the activity of proteases[3] and kinases.[4,5]

6.1.2 Why is Discovery of Inhibitors of PPIs a Challenge?

Disruption of PPIs has been considered to be difficult, if not impossible, by many medicinal chemists for many years. There are three main factors that have contributed to that experience.

First, there is the extent and nature of the interacting surfaces between two proteins. As mentioned above, the interacting surface can have distinctive features, with recognisable hotspots and appropriate cavities and interaction points from which ligands could be developed. Examples are shown in Figure 6.1(a) and include the p53–MDM2 interaction,[6,7] the Bcl-2–peptide surface[8] and the Xiap system,[9] and for these, crystal structures of complexes (as well as peptide mapping studies) demonstrate they could be tractable. However, for some surfaces as shown in Figure 6.1(b), such as that of β-catenin interacting with Tcf-3 (ref. 10) or for hormone receptors such as growth hormone receptor interacting with growth hormone,[11] the surfaces are fairly flat and featureless, with any distinctive interactions too far apart for 'normal' sized small molecules to span.

Second, it has been difficult to generate hits from conventional high-throughput screening (HTS) for these targets. The chemical matter (even in large multimillion compound collections maintained by large companies) is not suitable for binding to such targets. The compounds are usually limited to experience against previous target classes, or the design of compounds for the collection are influenced by the established medicinal chemistry experience, which is predominantly against conventional enzyme targets.

Finally, there is the challenge of the nature of the interaction sites in many PPIs. They tend to be quite hydrophobic and extensive, which means that compounds that bind will themselves be quite extended and lipophilic – causing problems with compound properties. The embedding of rules about compound properties (such as the Lipinski rules[12]) into medicinal chemistry practices would not allow the large compounds required, such as ABT-737 for Bcl-2, to be taken forward (compound **6**) as they break all the Lipinski rules. In addition, there is emerging evidence that these binding interfaces can be quite plastic,

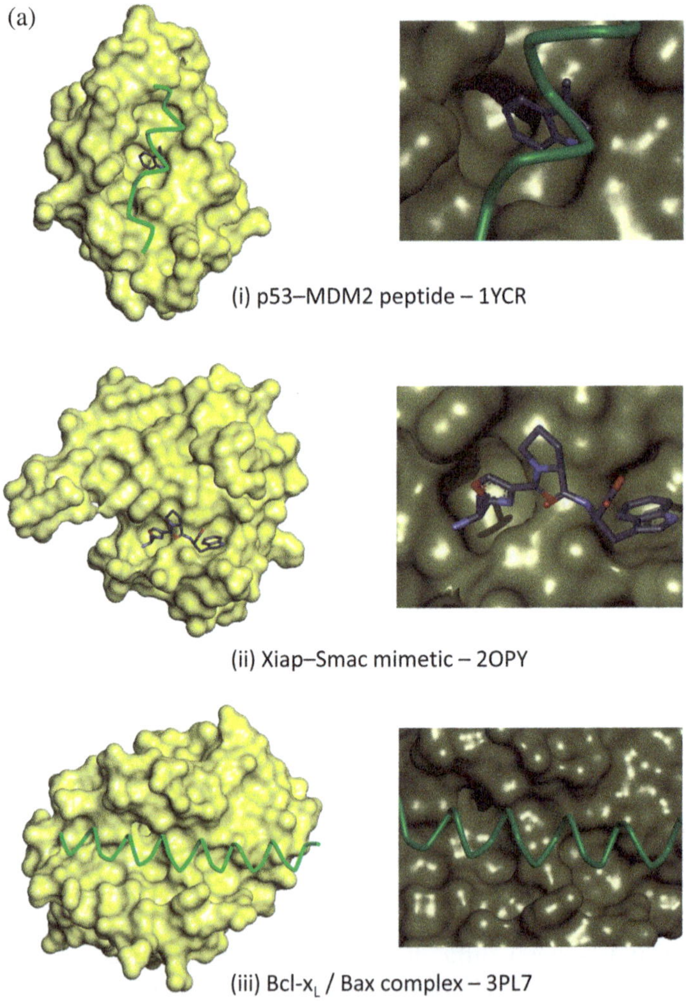

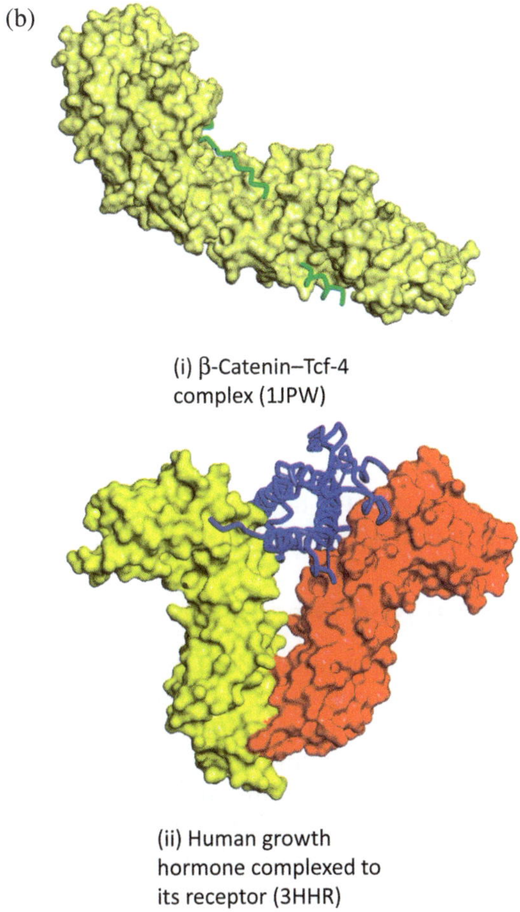

Figure 6.1 (a) Overview and detail of some PPI surfaces for which high-affinity small molecules have been identified. (i) The first structure (PDB:1YCR[6]) of MDM2 (surface) with a p53-derived peptide (green ribbon) bound, confirmed a hotspot for binding around a key tryptophan residue (stick representation). (ii) The first structure (2OPY[9]) of Bir2-Xiap protein (surface) complexed with a Smac-derived peptide mimic (atom-coloured stick representation). This demonstrates that affinity can be gained from ligands that mimic key side-chain interactions. (iii) The first structure of Bcl-xL (surface) with a pro-apoptotic peptide (Bax – green ribbon) bound (3PL7 (ref. 32)); although the recognition surface is quite extended, mutation studies demonstrate there are some key hotspot residues in the centre of the cleft that binds the helix. (b) Structure of more challenging PPI surfaces (i). β-Catenin–Tcf-4 complex (1JPW[10]); only portions of Tcf-4 (green ribbon) are visible in the electron density, but this is an extended and shallow surface with no distinctive hotspots. (ii) Human growth hormone (blue ribbon) in complex with two molecules (red and yellow surfaces) of the extracellular domain of its receptor (3HHR[11]); this is an extended surface across two domains.

that is they can undergo conformational change on challenge with ligands[13] as we will see later in this chapter. Although this has led to the development of attractive 'tethering' strategies,[14,15] it does make conventional design difficult based on structure–activity relationships (SARs) or structure-guided design. In addition, many of the attractive PPI targets (such as Bcl-2 binding to pro-apoptotic peptides, β-catenin binding to Tcf-3) exist as very high-affinity complexes – often with dissociation constants (K_D) less than 1 nM, again posing a formidable challenge and perceived hurdle to successful drug design.[16]

There have, however, been some successful projects to develop ligands against such types of interactions for some targets. Many of the more recent cases have been inspired by or derived from fragment-based discovery methods. As well as showing how to be successful, these examples illustrate the approaches and issues commonly associated with PPI targets.

The next section is a summary of the key elements of fragment-based discovery (perhaps reprising some of the material elsewhere in this book). This is followed by a review of some recent discovery projects that have generated lead compounds against such PPI targets. Finally, there is a discussion of the general lessons learnt from the published work and personal experiences.

6.2 Fragment-Based Discovery

The central feature of fragment methods is that the drug-discovery process begins with identification (usually by biophysical methods) of small (<250 Da), weakly binding (affinity of 100s of μM) compounds which are then optimised to drug candidates by structure-guided design. This section will summarise the essential steps in the overall process (summarised in Figure 6.2), with comments on how the methods have varied (in work published so far) for tackling PPIs rather than conventional targets.

6.2.1 Fragment Library Design

There is an extensive literature on fragment library design. Most libraries begin as selections of commercially available compounds that meet a set of criteria for molecular weight and physical properties such as solubility, absence of reactive or toxic groups, presence of suitable 'handles' for medicinal chemistry and chemical diversity. One of the earliest descriptions was the library at Vernalis for ligand-observed NMR screening;[17] this has subsequently been revised through medicinal chemistry experience, synthesis of fragments which present novel scaffolds and the long-term stability and solubility of the compounds.[18] Many of the larger pharmaceutical companies have published descriptions of their fragment libraries, such as Novartis,[19] AstraZeneca[20] and Pfizer.[21] A recent, useful review of library design was provided by the group at Monash[22] – this summarises the characteristics of most of the published libraries to date.

Many libraries (particularly those from commercial suppliers), rely on available compounds. The analysis by Mok *et al.*[23] found that biologically active medicinal chemistry substructures are not well represented in commercially available fragment sized compounds. However, most pharmaceutical companies rely on the medicinal chemists to select the compounds for library construction. This can bring their subjective experience of both chemical classes and synthetic tractability, which has usually evolved from working with conventional targets. This increases the coverage of bioactive space as seen by the early analysis by Vernalis of its library compared to the World Drug Index. However, this does mean that the libraries are somewhat biased towards previously seen or known substructures. Overall, this increases the challenge for PPIs as the fragment library is constrained to some extent by such past experience of the medicinal chemists.

The proposal that non-planar molecules are more drug-like[24] has recently generated a lot of interest in the idea of 3D fragment libraries[25] and the notion that fragments with more 3D shape may be more attractive for extended binding sites, such as those found in PPIs. As yet, the hypothesis has not been tested. There are no analyses published on shape bias in fragment hits against different classes of targets. However, even if there were, the results could be misleading as the constraint of synthetic tractability has limited the number of fragments of sufficient 3D character actually included in current libraries.

6.2.2 Finding Fragments that Bind

Finding fragments that bind to most targets is relatively straightforward. There are extensive reviews of the different techniques that have been developed and refined for screening fragment libraries (see for example ref. 26). Interestingly, most of the successful fragment screening campaigns described later in this chapter used NMR-based methods for hit identification. There are probably two reasons for this. The first is that this class of protein does not usually form crystals suitable for soaking of fragments. Secondly, NMR is the most robust technique for measuring fragments binding in the mM range, which can often be the case for PPI targets with such extended and feature-light binding surfaces.

There is wide variation in the hit rates reported for fragment screening by different groups ranging from 0.2% for Abbott[27] (the average for 23 screens) to over 20% for focussed screens (GSK kinase) and the Monash library (summarised in ref. 22). These hit rates reflect the sensitivity of the screening techniques and, certainly for ligand-observed NMR screening, differences in the rather subjective analysis of the NMR spectra, where small changes in peak intensity need to be interpreted by hand.

Figure 6.3(a) shows the validated hit rate for NMR screening against various targets at Vernalis over the past 10 years; the x-axis is the characterisation of the target binding site (we will call it *ligandability*) using the Sitemap method.[28] There has been variation in the composition and size of the fragment library, the sensitivity of the NMR instrumentation (with and without cryo-probe and improvements in the electronics of the console) and in the operator (as picking hits from changes in peak height is to some extent

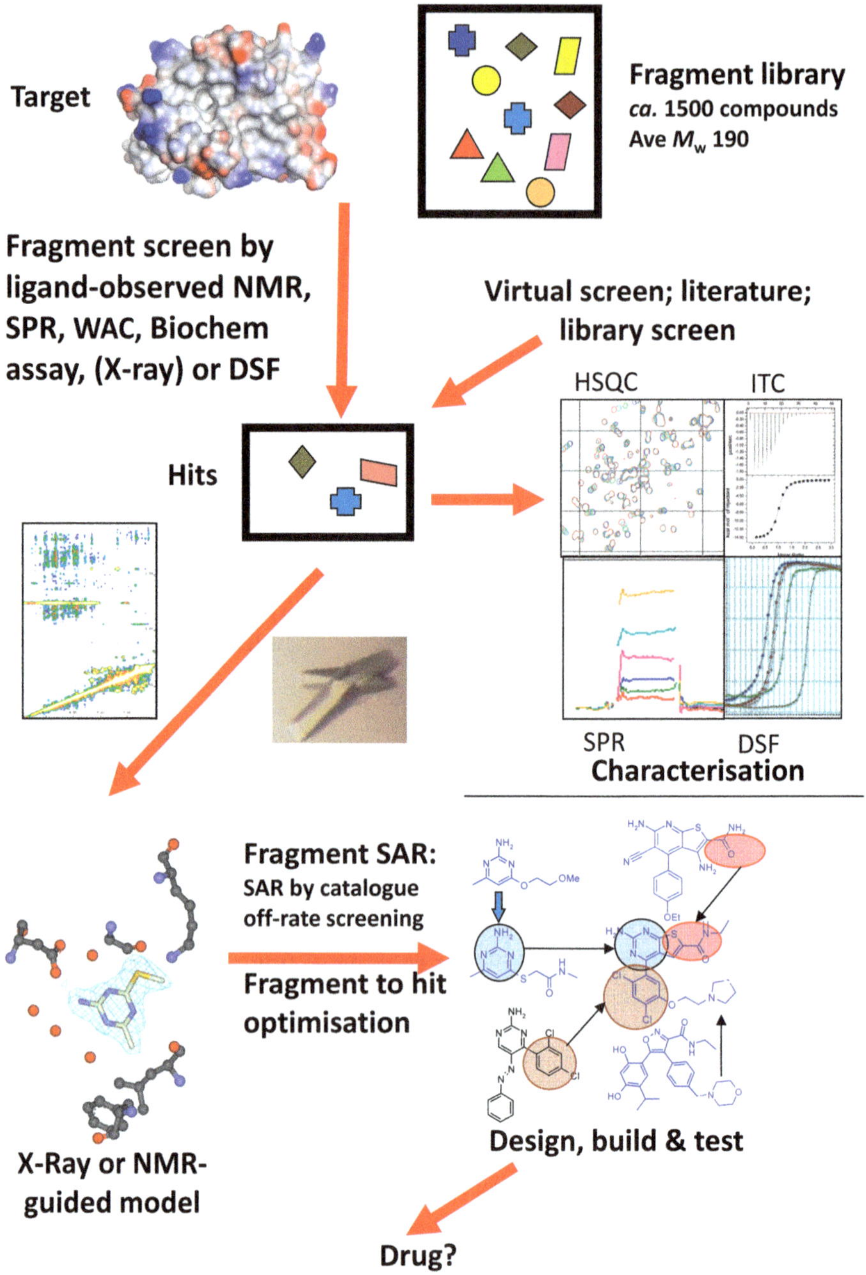

Figure 6.2 Overview of the FBLD process. More details can be found in ref. 26 and 57. A number of different biophysical or biochemical assays are used to identify fragments that bind to the target; this information is combined with hit compounds from HTS, virtual screening or the literature. A range of biophysical techniques are used to characterise the binding of compounds in more detail while attempts are made to generate a structure, preferably by X-ray crystallography but if needed, by NMR methods. Selected fragments are optimised to explore SARs either by purchase of near-neighbours (SAR-by-catalogue) or limited library synthesis. The fragments are then optimised by conventional

structure-guided medicinal chemistry. The abbreviations and brief summaries of the different experimental techniques are: NMR – nuclear magnetic resonance; ligand-observed NMR signals can be used to detect when a ligand has interacted with a protein. Experiments typically require 500 μM ligand and 10 μM protein and can detect binding affinities over a large dynamic range (K_D of 100 nM to 5 mM). Screening by NMR requires relatively large quantities of protein (10s of mg) and typically takes 2–3 days of machine time and 1–2 weeks of operator time to screen a library of 1500 fragments against a target. More details can be found in Chapter 3. SPR – surface plasmon resonance is a technique which monitors change in molecular weight. One component, either the protein or ligand, is immobilised to a surface, whose optical properties are sensitive to the weight of what is attached, and then the other component is flowed over the surface. Binding affinities, with a limit of 500 μM or better, can be measured. SPR requires relatively small amounts of protein and can be configured for relatively high throughput (100s of compounds per day). Further details on SPR can be found in Chapter 2. ITC – isothermal titration calorimetry is a technique which directly measures the change in enthalpy (ΔH) on binding of one component to another. Usually, the ligand is titrated onto the protein and, if solubility allows, the resulting saturation enables stoichiometry to be determined. The mid-point of the binding curve is taken to define the equilibrium constant from which ΔS, and thus ΔG, can be determined. The main limitations are that the interaction should have a sufficient heat change to be measured and that quite large amounts of protein are used up in each titration so it cannot be used for high throughput or screening. It is, however, one of the more reliable techniques for driving the early stages of a drug-discovery project, being less prone to artefacts from misbehaviour of protein and ligand. X-Ray – X-ray crystallography is discussed elsewhere in this chapter. HSQC – heteronuclear single quantum coherence is an NMR experiment typically using ^{15}N- (or sometimes ^{13}C) labelled protein to provide a 2D experiment showing the individual nitrogen or carbon peaks. Chemical shift perturbations caused by ligand binding can be used to identify binding sites and determine binding constants. Further details are described in Chapter 3. DSF – differential scanning fluorimetry is a technique which measures the change in stability of a protein to heat denaturation on binding of a ligand. The experiment can be run in 96-well plates on a PCR-type instrument, where the temperature is increased and the unfolding of the protein monitored as the change in fluorescence as an added dye (usually Sypro orange) binds to the unfolded protein. It is important to first optimise the buffer conditions. The method is attractive as it requires relatively small amounts of protein and ligand, compared with NMR and X-ray crystallography. It has been quite successful for screening for the binding of larger ligands. There are more challenges for measuring the small effects of fragment binding and false positives and negatives are common. However, it is a relatively fast and cheap method and can provide a route to identify fragments when other screening methods are not available, bearing in mind that there may be false positives. WAC – weak affinity chromatography; this is a relatively newly developed technique where the protein target is immobilised on a Sephadex bead in a high-performance liquid chromatography (HPLC) instrument. The retention time of ligands (often applied in a mixture) is related to the binding affinity. There can be issues with the stability of the protein on the beads, and buffer conditions need to be optimised to avoid false results (there can be a correlation between log*P* of the compounds and retention time for example). However, the technique has shown some recent promise.[58]

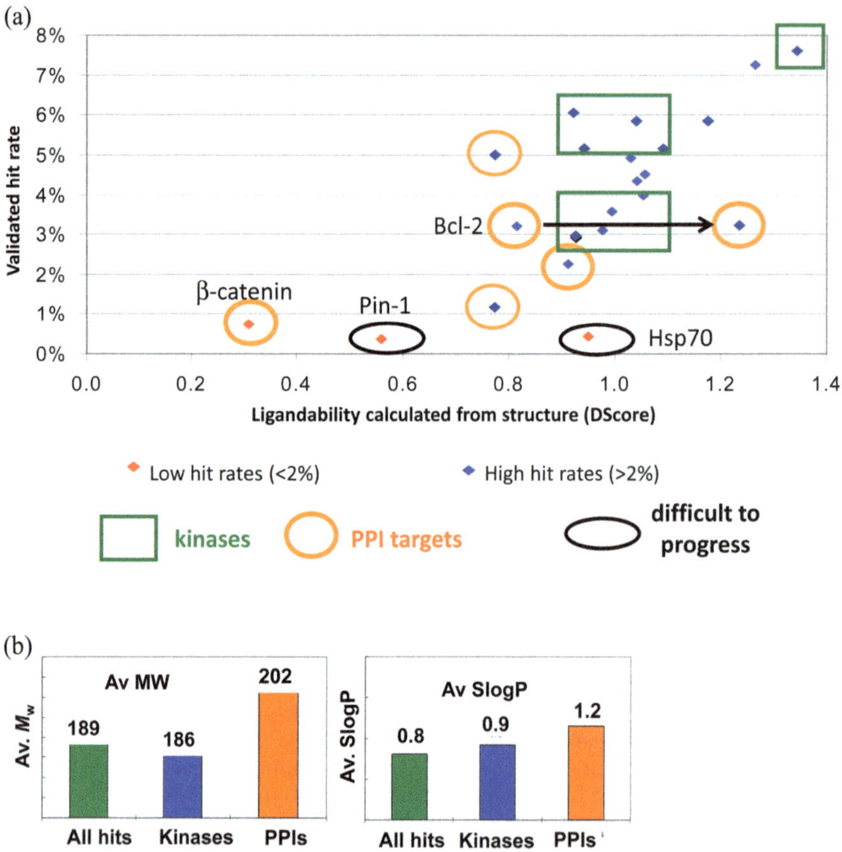

Figure 6.3 (a) Hit rates reported by Vernalis for screening of fragments by ligand-observed NMR methods against different targets. The *x*-axis is a measure of the ligandability of the active site of the protein calculated with the program Sitemap.[28] Ligandability is from a representative crystal structure (apo when available) – a value above 0.6 is usually taken as ligandable. The *y*-axis shows the validated hit rate from a number of ligand-observed NMR screens. These were conducted over the last 10 years with different libraries, instrumentation and operators, so the relative hit rates are approximate only. Red dots are for low hit rates (<2%); blue for higher hit rates. The targets ringed in green are kinases; targets in orange are PPI targets and those ringed in black are targets where hits were found to be difficult to progress/optimise. For Bcl-2, the arrow shows how the ligandability of the surface changes in response to the binding of a high-affinity ligand. This is an updated version of the analysis originally presented in 2009.[18] (b) Average M_w and SLogP of the hits found for different classes of target (data taken from the analysis presented by Vernalis in 2009).[18]

subjective). However, the analysis provides a useful set of data against which to discuss fragments for PPIs. In general, the fragment hit rate reflects the ligandability, as reported by others previously;[27] interestingly some of the PPI targets, such as Bcl-2, have a good hit rate with some 40 or so fragments found to bind. The low hit rate for β-catenin reflects the challenge of that particular interaction – long and fairly featureless. The two targets for which the hit rate is below that expected from analysis of the structure, are Pin-1 and Hsp70. In both cases, it took a lot of effort to progress hits to a reasonable affinity.

This notion of calculated ligandability can therefore be used for an initial characterisation of the potential binding sites on protein surfaces. The method (as implemented in DScore) uses analysis of the shape and chemical characteristics of known binding sites to assign a score to pockets on the surface of a protein. Another variant of this same idea is to use virtual fragment screening to assess the protein surface.[29] The challenge for all these computational methods is that the protein surface can change its character to accommodate a ligand through simple side-chain rotations and more so with adjustments in main chain or secondary structure elements. This idea of protein surface plasticity has been developed extensively by proponents of the tethering approach.[14] The protein surface can change substantially on challenge with a growing ligand. An example is some as yet unpublished results from a Servier/Vernalis collaboration that has generated pre-clinical candidate compounds for Bcl-2 (a target discussed later in this chapter). Here, the ligandability of the binding surface was seen to vary considerably in the crystal structures determined of ever more potent ligands, as illustrated on Figure 6.3. This phenomenon has also been commented on previously by Hajduk at Abbott.[30]

Many of the groups analyse the difference in the overall properties of the fragments that have been hits in screening campaigns compared to the overall properties of the fragment library. Abbott analysed the preference for different functional groups[31] and found a high hit rate for compounds containing a biphenyl. Novartis,[19] AZ,[20] Pfizer[21] and the Monash[22] group reported the difference in various physico-chemical properties between hits and non-hits. In general, the molecular weight and the hydrophobicity (either ALogP or SLogP) is slightly higher in the hits, as seen by Vernalis.[18] However, the Monash group see little difference in these particular properties, although, as this is the most recent analysis, they have profiled the 3D character (measured as the percentage of sp^3 hybridised atoms, fsp3). Interestingly this suggests that fsp3 is lower on the hits than the average in the library (see comments on 3D fragments below).

However, the only group to have further analysed the nature of the hits by protein class is one from Vernalis.[10] Figure 6.3(b) shows the average molecular weight and SLogP for the hits found by ligand-observed NMR screening at Vernalis.[18] The molecular weight of the hits against PPI targets is generally higher than for other targets, and the hits themselves are more hydrophobic. The data is not shown here, but the PPI hits also contain on average more rings than the hits for other targets and have a lower affinity. This is not unexpected – the PPI surfaces tend to be more extended and shallow and more hydrophobic.

As discussed above, although most of the fragment libraries have been constructed to be fairly generic, they are heavily biased by the previous experience of the medicinal chemist. This increases the percentage of flat, heterocyclic compounds. In addition, there are quite high solubility requirements for fragment screening, which limit the number of hydrophobic groups that are present. Given the size, shape and hydrophobicity of the sites being targeted for PPIs, it could be argued that it is appropriate to revisit the design of fragment libraries specifically to enrich for fragments for PPIs. However, introducing fragments with more 3D shape and increased hydrophobicity will be a challenge to meet the criteria for screening and synthetic tractability for evolution. However, the current interest in 3D fragments and PPIs will hopefully generate new libraries which will allow the hypothesis to be tested that PPIs favour more 3D compounds.

6.2.3 Characterising Fragment Binding

In a conventional medicinal chemistry project, it is usually possible to identify series of compounds where small changes in the molecules generate changes in the binding affinity and allow SARs to be established. From these SARs, it is then possible to build a virtual model of how the compounds are binding to the target and then design modifications that improve potency and/or drug-like properties.

It is very difficult to build such SARs for fragments. Usually, most changes to the fragment result in compounds with no detectable affinity, giving no guidance to the next steps in compound design. This is particularly true for PPI targets where the initial hits themselves are usually binding with an affinity that is at the limit of sensitivity of the assays available. In addition, the likelihood that modifications to the fragment will give different binding modes is increased with such weak affinity binding to what can be rather plastic binding sites. Therefore, in fragment-based design it is particularly important to have available robust models for how the fragments are binding to the target, to give that initial guidance to the design of compounds with higher affinity.

The most detailed models of compound binding are provided by X-ray crystallography, and can reveal the detail of binding mode, binding site and the involvement of solvent molecules. The first requirement is pure, folded protein which is soluble at the high concentrations usually required for crystallisation (10 mg ml^{-1}) and homogenous in fold and post-translational modification.

It is possible to produce protein for most target classes; the main rate-determining step for structure determination is obtaining crystals of the protein which diffract and for which the crystal packing allows for soaking of ligands into the crystal or where co-crystallisation conditions can be found for individual compounds in complex with the protein. This can be particularly challenging for PPI targets. For some targets such as the Bcl-2 family,[32] crystal structures can be obtained for the target in complex with a peptide representing the other partner. Although not yet published, it is also possible to get crystal structures of the target with no ligand bound. Here, the packing in the crystal between different parts of the protein mimics the PPI. However, it can be difficult to obtain a structure in complex with a fragment – the weak affinity of the fragment is insufficient to overcome the higher affinity interaction in the crystal packing. It takes a more evolved compound (in our experience better than 100 nM) to establish a tight enough complex that will find a different crystal packing, as in the case of ABT-737 binding to Bcl-2.[8]

There are many different strategies for trying to obtain a crystal structure of a target through exploration of many different protein constructs. Examples include generating protein with different domain boundaries, mutations to remove mobile surface amino acids or loops, use of homologues or orthologues (for example, much of the initial work on Bcl-2 compounds used crystal structures of Bcl-xL), the use of a protein chimera taking portions of sequence from different species, or mutating the residues in the active site of a closely homologous protein which does crystallise. Some of these have been used for PPI targets – for example, the K-Ras stories summarised below have used different constructs.

Most practitioners attempt to determine crystal structures to characterise fragment binding but in addition, a few have used crystallography as the primary screening method. Although rather technology-intensive, it is feasible to collect and process data on many hundreds of crystals with shape-controlled mixtures of fragments soaked in within a few weeks. This approach has worked for conventional targets but to date, an example of this approach to screening for PPIs has not been published. This is probably because of the challenge to obtain a crystal form for PPI targets with a suitable open active site.

The next level of quality of model is possible from NMR spectroscopy – either full protein structure determination or from identifying a sufficient number of defined ligand–protein interactions from which an NMR-guided model can be computed. Here, the main constraint is the molecular weight of the target. Although recent developments have raised this limit somewhat, it remains challenging if not impossible to generate the quality of NMR spectra required for proteins larger than 30 kDa, primarily due to spectral overlap and the line broadening as larger proteins tumble more slowly and thus give variation in signal during the timescale of the NMR experiment. The main requirements for determination of structure by NMR are protein solubility (typically 500 μM is required), stability (the experiments can take 7–14 days) and preparation of protein with isotopic labelling, usually with ^{15}N and ^{13}C for full structure determination. Once acquired, the NMR spectra can be assigned by associating each NMR peak with a particular nucleus in the protein. It should then be possible to obtain distances between bound ligands and the protein (through NOE measurements) from which models of ligand binding can be generated. Further details on NMR experiments that can be used to characterise fragment binding are discussed in Chapter 3 and ref. 33.

It sometimes can take some time to establish structure determination. For this reason it is usual to validate and characterise binding by orthogonal biophysical techniques. This is particularly important for fragments binding to

PPI targets because of the extensive nature of the binding site, the plasticity of the site and the possibility of cryptic sites well away from the functionally relevant interaction site. Figure 6.2 summarises the various biophysical methods that can be used to characterise the binding of fragments. For example, if HSQC measurements can be made, then it should be possible to confirm that fragments are binding to a consistent site. Competition experiments with SPR could also be used.

Finally, it is possible that a model of fragment binding could be generated by molecular docking calculations; however, such predictive modelling is usually only possible for binding sites which do not undergo conformational change and where the modeller has been able to build up a detailed understanding of the binding site from crystal structures of other compounds binding to the target – in which case, X-ray crystallography of the fragment should be possible.

6.2.4 Optimising Fragments

There have been a number of studies in which reasonably potent inhibitors of PPIs have been dissected to their constituent fragments. These inhibitors had been discovered previously by HTS, structure-based or peptidomimetic-inspired optimisation programs against targets such as the p53-MDM2 (ref. 7) and von Hippel–Landau protein and Hif-1α.[34] Not surprisingly, the fragments are not optimal for binding on their own,[7] and in most cases it was not possible to measure binding unless the screening conditions were significantly improved to detect very weak binding.[34] This is similar to what has been seen for more conventional targets such as kinases. For example, deconstruction of a virtual screening hit against CDK2 generated a fragment which was seen to bind in a totally different position and orientation than the original hit.[35]

These studies bring out three features of fragment-based discovery for PPIs. The first is that fragments may, in general, bind more weakly to PPI targets and require more sensitive screening conditions for detection.[34] Secondly, it emphasises that the affinity of an inhibitor can come from the collection of interaction points made across the protein surface by the complete inhibitor.[7] These features are not limited to PPIs; earlier work (*e.g.*, ref. 35 and 36) dissecting inhibitors for more conventional targets has previously made these points, but they are perhaps more relevant for PPIs. Finally, these studies emphasise that the initial stage of a fragment campaign should attempt optimisation of the affinity of the fragment starting point by careful exploration of fragment SARs, by near-neighbour acquisition or limited synthesis.

6.2.5 Fragment to Hit-to-Lead Strategies

The literature has emphasised three distinct strategies in how to evolve fragment hits. The original method, as used in SAR-by-NMR[37] (see examples below) is to identify and optimise the individual fragments that bind to distinct binding pockets and then link them together to make a more potent ligand. The challenge for this approach is to conceive a synthetic linker that can join the two fragments, without perturbing their initial, optimal binding positions and orientations. Although many practitioners have had difficulties in applying this approach to identifying inhibitors for conventional binding sites,[38] it has been successful for PPI targets as we will see in the later examples, perhaps because the binding surfaces can be distinct and separated on the surface. A more widely used method is fragment growth, where structural information is used to guide the growth and optimisation of an initial fragment. Lastly, a hybrid of these two methods is called *fragment merging*, where features from different overlapping fragments are combined in one compound. Both fragment growth and fragment merging rely on detailed structural guidance, usually provided by multiple crystal structures. As discussed above, it can be challenging to obtain the structures of fragments bound to PPI targets, so it is not surprising that there are relatively fewer examples of this approach compared to conventional targets.

6.3 Case Histories

6.3.1 FKBP

FK506 is a compound isolated from a soil bacterium that has been used as an immunosuppressant. When administered, it complexes a protein called FK506 binding protein (FKBP) and disrupts the PPI with calcineurin, blocking T cell activation.[39] Such immunosuppressant drugs were a major area of drug discovery in the early 1990s, in the search for drugs for post-transplant surgery. The structure of the FKBP–FK506 complex is shown in Figure 6.4(a) – this is a relatively well formed but large surface.[40] The discovery of potent small-molecule inhibitors for FKBP was the first published fragment-based drug-discovery project using the fragment-linking technique, SAR-by-NMR.[37] In this example, a library of fragments was screened against a ^{15}N-labelled sample of FKBP using ^{15}N-HSQC NMR, resulting in the identification of compound **1**. A subsequent screen by HSQC NMR of FKBP with a library of fragments in the presence of compound **1** identified compound **2** binding to an alternate site. Structures determined by NMR identified the appropriate vectors for linking the two fragments and, based on this, compound **3** was synthesised leading to a significant increase in potency. The activity numbers are the reported K_D values from the NMR HSQC titrations. This paper introduced one of the essential concepts of fragment-based discovery: that linking two weakly binding compounds can lead to a high-affinity compound, re-emphasising the observation of Jencks[41] that the relationship between free energy and equilibrium constant ($\Delta G = -RT \ln K$) means that addition of binding energy gives multiplication of the binding constants. In addition, the linking of fragments provides a variant of the ideas about coverage of chemical space[42] – linking two fragments identified from screening a library of 10^4 compounds binding in different sites is equivalent to screening a combinatorial library of 10^8 compounds.

6.3.2 Bcl-2 Family

The Bcl-2 family of proteins are arbiters of programmed cell death. The family consists of some 25 members with anti-apoptotic (such as Mcl-1, Bcl-2, Bcl-xL) members and two classes of pro-apoptotic members.[43] The first class is Bax and Bak which aggregate to permeabilise mitochondria, releasing caspases that trigger apoptosis. The action of Bax and Bak is inhibited through binding to the anti-apoptotic Bcl-2 members; however, the second class of pro-apoptotic proteins, such as Noxa, Bim and Bad can also bind to the anti-apoptotic members, which release Bax and Bak. All of these proteins contain one or more so-called BH (Bcl-2 homology) helical domains (BH1–4). The structure of Bcl-xL with Bax was shown above in Figure 6.1(a). It can be seen that the BH3 helix of Bax binds into the long, relatively hydrophobic groove on the surface of Bcl-xL. Similar structural arrangements are seen in the binding of proteins such as Bim, Bax and Bad to Bcl-2 and Bcl-xL and for Noxa to Mcl-1, discussed later in this section (see Figure 6.4(c)). These anti-apoptotic proteins are over-expressed in many cancers, therefore, inhibition of this interaction by BH3 mimetic molecules has emerged as a persuasive mechanism for cancer therapy.

Production of large quantities of these proteins for biophysical studies is relatively straightforward; in addition, the domains are of a size that has allowed NMR structures of family members such as Mcl-1, Bcl-2 and Bcl-xL to be determined. Finding fragments that bind to these proteins is also

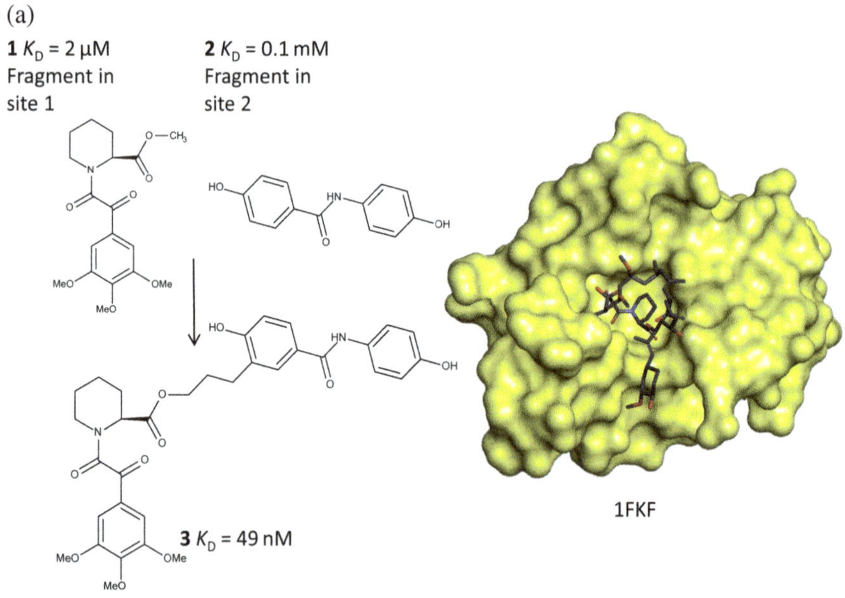

Figure 6.4a (a) The surface of FKBP is shown with the immunosuppressant FK506 in an atom-coloured stick representation (PDB:1FKF). This was the first example of SAR-by-NMR where fragments found in two sites (compounds **1** and **2**) were linked to give a high-affinity inhibitor, **3**.[18]

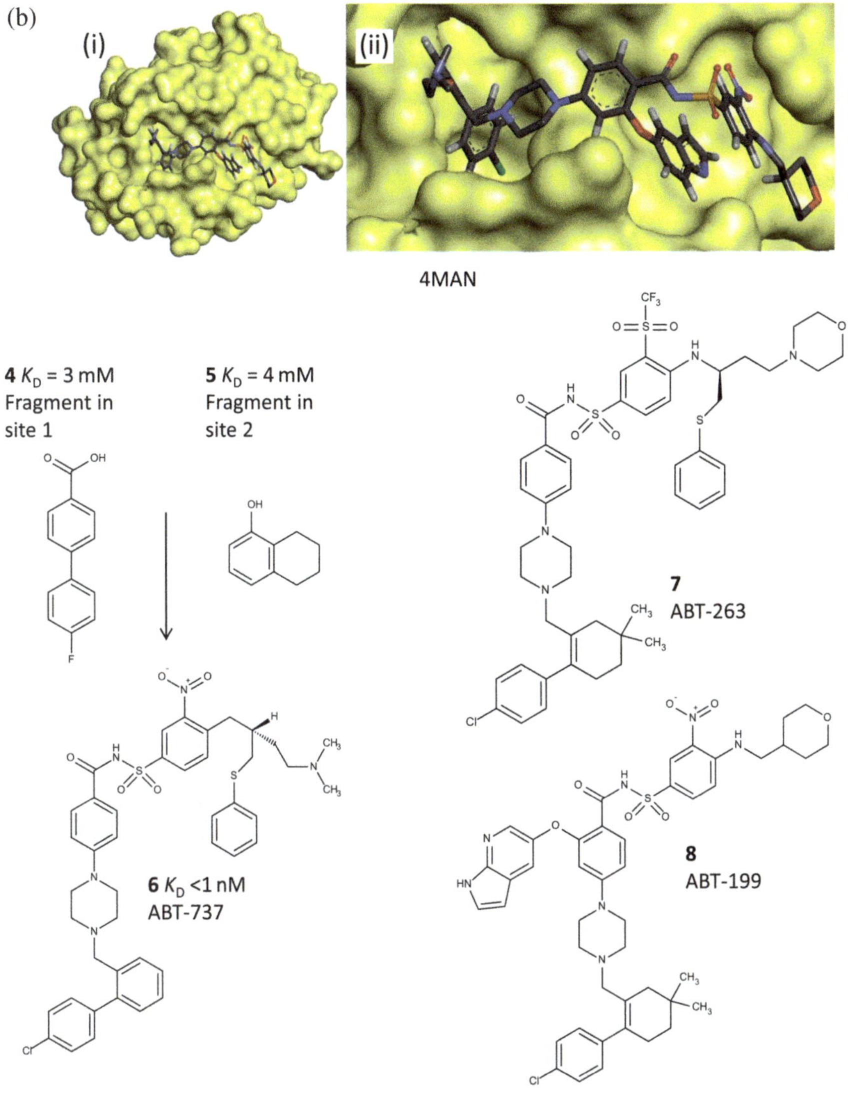

Figure 6.4b (b) Bcl-2–Bcl-xL. Overview (i) and detail (ii) of the surface of Bcl-xL is shown with the drug candidate, ABT-199 bound in an atom-coloured stick representation (PDB:4MAN). The initial drug candidate **6**, ABT-737,[8] was eventually identified by modification of initial hit compounds identified by linking fragments **4** and **5**. Subsequent optimisation identified **7**, ABT-263 with improved drug-like properties and then **8**, ABT-199, which is Bcl-2 selective.

relatively straightforward; where needed, peptides such as Bim for Mcl-1 and Bax for Bcl-2 can be used as competitors to confirm binding.

The first published drug-discovery campaign against these family members was the work of the Abbott group led by Fesik to develop ABT-737 (ref. 8) (**6**). This was an impressive demonstration of the power of fragments and

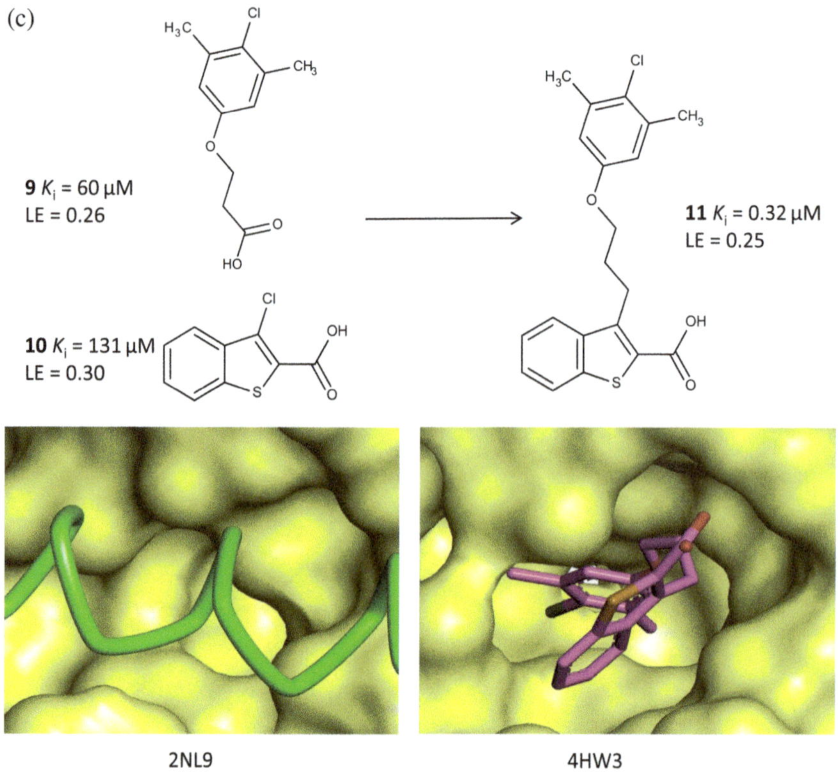

Figure 6.4c (c) Detail of the surface of Mcl-1 complexed (i) with the peptide Noxa (green ribbon, PDB:2NL9) and (ii) with the inhibitor **11** in an atom-coloured stick representation (purple carbon atoms) (PDB:4HW3) that was identified by linking fragments **9** and **10**. LE = ligand efficiency.

the SAR-by-NMR approach, requiring considerable commitment in moving from the initial fragment hits to the final compound. In brief, the initial hits were identified by monitoring shifts in the HSQC NMR spectrum of ^{15}N-labelled Bcl-xL The pattern of peak shifts allowed two hits to be identified that bound in different sites, the biphenyl acid **4** and a tetrahydronapthol **5**. The binding modes of these compounds were determined by NMR and a lengthy structure-guided design led eventually to ABT-737; this included the determination of how intermediate compounds bound to albumin in order to guide design of analogues with reduced plasma protein binding. The final ABT-737 compound, which is a potent inhibitor of both Bcl-2 and Bcl-xL, contains only echoes of the original fragments, and their relative position and orientation on the surface of the protein has changed during optimisation. Although the compound is no longer in clinical trials, it was important in establishing that fragment-based methods can generate potent compounds with effective biological efficacy. Subsequent work at Abbott (summarised in Figure 6.4(b)) identified the slightly smaller and more potent compound

(d)

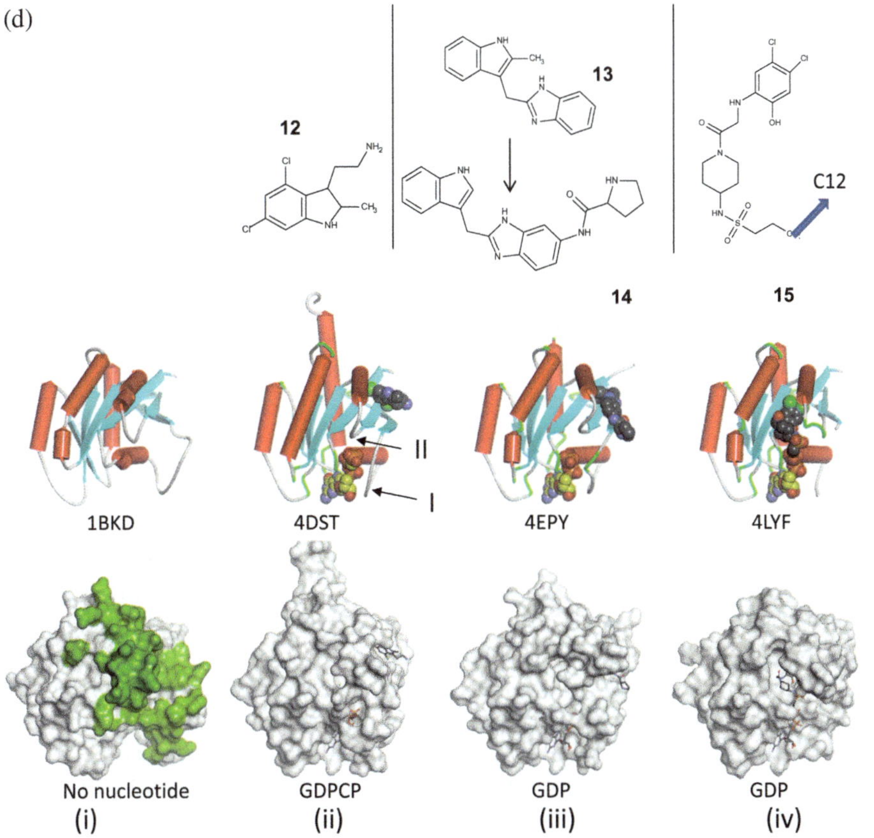

Figure 6.4d (d) Overview of compounds bound and the secondary structure and surface of (i) H-Ras with nothing bound showing, in green, the interacting surface with SOS, (ii) K-Ras with GDPCP (gold) and compound **12** (grey) bound with arrows pointing to the switch I and II regions, (iii) K-Ras with GDP (gold) and compound **14** (grey) bound and (iv) K-Ras with GDP (gold) and compound **15** (grey) bound. The four-letter codes are the PDB files from which the images were constructed. The blue arrow shows the point of covalent attachment of the ligand to cysteine residue 12 in K-Ras.

ABT-263 (ref. 44) (**7**); this progressed further in clinical trials before the Bcl-2 selective compound ABT-199 (ref. 45) (**8**) was produced which has shown impressive activity in clinical trials to date.

A comparison of the surface of Bcl-xL in Figures 6.1(a) and 6.4(a) is an example of the plasticity that can be seen in some PPIs.[13,16] This provides an opportunity, as the surface of the targets can become more druggable on challenge by the ligand, but it is also a challenge for structure-based design. The moving yellow circle on Figure 6.3(a) shows how the structure of Bcl-2 has changed in an as-yet-unpublished collaboration between Vernalis and Servier which has resulted in a compound now in clinical trials.

The development of Bcl-2 and Bcl-xL inhibitors demonstrated that the other pro-apoptotic family member, Mcl-1 could also be a very important target. Mcl-1 is responsible for conferring resistance to Bcl-2 inhibition and there are some cancers where the main driver is Mcl-1. This target has proven particularly challenging for drug discovery; the recent work by Fesik[46] (now at Vanderbilt University) has demonstrated that the methods can generate useful tool compounds for this target, although as yet, there are no reports of compounds entering clinical trials for Mcl-1. However, recent work by the Fesik group does provide an additional demonstration of the features of working with this class of protein.

The Vanderbilt Mcl-1 program is summarised in Figure 6.4(c). A relatively fast, new method of protein-observed ligand screening (SOFAST HMQC NMR) was used to screen 13 500 fragments against Mcl-1. This identified two series of hits (such as compounds **9** and **10**) which bind in adjacent pockets on the protein surface. NOE information was obtained to generate NMR-guided models of the fragment binding modes and a fragment-linking strategy was designed to generate higher affinity compounds such as **11**. This was a particularly well-behaved study as the SARs seen for the fragments binding to the different pockets were preserved in the merged compounds. In addition, the initial fragments and the final compounds showed good selectivity against Bcl-2 and Bcl-xL. However, the compounds are quite lipophilic and reflect the low ligand efficiency (0.25) often seen for compounds against such targets. This work emphasises again the plasticity of these protein surfaces – it can be seen in Figure 6.4(c) that the binding surface changes considerably between when the Bim peptide (PDB:2NL9 (ref. 47 and 48)) and compound **11** are bound. Conformation differences between the two structures were seen in both the protein backbone and amino acid side-chains.

6.3.3 K-Ras

The Ras family of proteins are small GTPases with a major role in cell signalling. With the assistance of proteins such as SOS1 (a guanine nucleotide exchange factor or GEF), Ras binds GTP, hydrolysing it to GDP and thus cycles between GTP-bound active and GDP-bound inactive states. In the active state, Ras interacts with high affinity with numerous effectors such as Raf and PI3K to stimulate growth. There are three Ras isoforms (K, N and H); K-Ras is the most frequently mutated and is present in more than 25% of human cancers.[49] Mutation leads to reduced GTPase activity, promoting K-Ras-dependent signalling and cancer cell survival or growth.

There have been extensive campaigns (mostly unpublished) within many organisations over many years using conventional HTS and medicinal chemistry approaches to identify small-molecule inhibitors of K-Ras. There has been little success reported. This is an extremely challenging target not least because the affinity of wild-type K-Ras for the native substrate (GTP/GDP) is in the pM range. Furthermore, the binding sites for effector proteins are relatively flat and featureless, plus the protein is quite dynamic in conformation.

It is the activated form (GTP-bound) of the mutant proteins that should be targeted, as most of the oncogenic mutations reduce GTPase activity, resulting in increased activation. However, there have recently been reports from three laboratories on the discovery of small molecules that inhibit mutant K-Ras with promising affinity – all were derived from fragment-based approaches.

Figure 6.4(d) introduces the protein and the various pockets. The structure shown in (i) is from an earlier study in which the structure of H-Ras (a close relative of K-Ras) complexed to the nucleotide exchange factor, SOS, was determined.[50] The structure was solved without a nucleotide bound and the green coloured surface shows the interaction point. The first report of fragment-based discovery work was from the group at Genentech.[51] All their structures were of the full-length K-Ras isoform 2b residues 1–188 with the G12D mutation and with R41S; the final C-terminal residues after the distinctive hexa-lys sequence were not defined. The structures were determined in complex with GDPCP – a non-hydrolysible GTP analogue which traps the protein in the active state with high affinity for the various effector proteins. A comparison of this structure with the others emphasises the importance of two loops (known as switch I and switch II) which change conformation depending on the GTP/GDP status of K-Ras. A fragment (DCAI, compound **12**) was seen to bind in a cleft sitting between the switch II loop and the central β-sheet region with a measured affinity in the 100s μM range. This compound was profiled in a range of functional assays and blocks nucleotide exchange, with some evidence of cellular activity. However, the study emphasises the challenge of drug discovery against such multifunctional targets – particularly that without potent tool compounds it is difficult to establish and validate even basic binding, as well as demonstrate that any cellular activity is associated with target engagement.

The second report of fragment-based discovery against K-Ras was from the Fesik group at Vanderbilt University.[52] The group used a shorter construct (1–169) of K-Ras, with various mutants (wild-type, G12D, G12V) and with C118S for stability in the NMR experiments. Protein-based NMR screening of 11 000 fragments against G12D, C118S K-Ras was performed with GDP bound, yielding 140 fragment hits that bound with between 1.3 and 2 mM affinity; a similar affinity seen for binding to all the mutant variants as well as to H-Ras. Subsequent X-ray crystal structure determination showed that the fragments (such as **13**) bind in the same region as the Genentech compound, DCAI (**12**). A first round of fragment evolution led to the highest affinity compounds (such as **14**), reported to have a K_D around 200 μM and showed inhibition of SOS-mediated exchanged at around 1 mM concentration.

The third report is from a group at UCSF[53] that exploits the disulfide tethering approach first developed by Sunesis. G12C is a common mutation in oncogenic Ras leading to decreased GTPase activity; the presence of the cysteine provides a reactive target for the ligand design. Four hundred and eighty disulphide-containing compounds were each incubated with (1–169) G12C K-Ras and the resulting mixtures screened using mass spectrometry. Hits from this initial screen were then progressed to X-ray crystallography which showed tethered compounds bound to a new cleft between switch I and switch II. Subsequent structure-guided elaboration identified more potent compounds (as measured by the time taken to link with the protein). Assessment of the compounds (such as **15**) in a range of binding and cellular assays demonstrated they had the potential to affect SOS-mediated nucleotide exchange, and also modified the relative affinity of GDP- *vs.* GTP-bound K-Ras for various effectors such as Raf. This translated into an effect on the viability of various cancer cell lines.

6.3.4 BRCA2/RAD51

The interaction of the tumour suppressor BRCA2 with the recombination enzyme RAD51 has two effects. The interaction stabilises the binding of RAD51 to single-stranded DNA, whilst also inhibiting the binding of RAD51 to double-stranded DNA. This has the effect of promoting RAD51-dependent strand exchange. It is proposed that small molecules that inhibit the BRCA2-RAD51 interaction should inhibit DNA repair. Studies led by Hyvonen[54] have identified fragments that are able to mimic the key binding motif (FXAA) in BRCA2 and the appropriate site in a model system of RAD51. Although no evidence has yet been presented of any functional effect of these weak inhibitors, or that inhibition of binding at such a site would indeed have a functional effect, the work does demonstrate that considerable care is needed in experimental design to identify such weakly binding compounds. First, a truncated RAD51 orthologue from a thermophilic organism was used, as monomeric human RAD51 is unstable. Second, a series of assays were used to identify fragment hits: the hits from an initial thermal shift assay were assessed in a competitive STD NMR experiment (the competitor was a peptide that covers the equivalent to the FXXA BRCA2 motif). These hits gave confidence to conduct a second NMR screen of a larger group of fragments. The resulting hits were confirmed by ITC and then finally a set of crystal structures were determined which showed the fragments binding to the expected site on the RAD51 orthologue.

6.3.5 Tethering Approaches

This approach has been pioneered by the scientists at Sunesis[14] and continues to be applied and developed by those associated with or influenced by the company.[13,53] The approach has two requirements. The first is a free cysteine close to the site on the protein where an inhibitor is to bind. This can either be naturally occurring, introduced by site-directed mutagenesis or a mutation that is present in a disease form of the protein (such as in K-Ras, see earlier). The second requirement is a library of compounds containing a disulphide bond. Under the screening conditions (usually pH 8), the non-covalent interaction of the compound with the binding site increases the local concentration of selected compounds, such that there is exchange between the disulphide bond of the compound and the free cysteine of the protein to 'tether' the ligand in the binding site. This can then be detected, usually by an electrospray mass spectroscopic approach.

The approach has been proposed as an effective way of evolving and identifying potent PPI inhibitors.[14,16] As discussed above, the interacting surfaces can sometimes undergo conformational change on binding of a ligand, so once tethered, the other end of the ligand can be modified to explore additional interactions. Once that other end has been optimised, then further optimisation of the end that was tethered is possible.

As well as the work on K-Ras summarised earlier,[53] another recent example of such a tethering approach applied to a PPI is the work of Arkin[13] which sought to identify inhibitors of the interaction between the cytokine, IL-2 and its receptor. The structure of an early, relatively weak inhibitor of this interaction binding to IL-2 (ref. 55) revealed some pockets in the surface that could be exploited. Introducing cysteine residues near these pockets allowed a tethering approach to identify various classes of compounds; although these have not (to date) been developed into potent inhibitors, and the story is rather involved, it does demonstrate the potential of the approach.

6.3.6 Computational Fragment Screening?

In this chapter, I have limited the detailed discussion to the use of experimental fragment screening methods. However, there have been some reports using computational methods to select fragments for subsequent experimental work. In one example,[56] virtual screening was used to identify binding hotspots on the Sec7 surface of the nucleotide exchange factor Arno – this has the potential for blocking interaction with Arf1, a guanine nucleotide binding switch protein in the Ras superfamily. This approach identified fragments that were then characterised experimentally. Although only weak inhibition of the interaction was found (2 mM), this case does highlight the potential use of computational methods for characterising potential functional binding sites in complex protein systems.

6.4 General Lessons

So – what are the general lessons about the challenges for fragment-based methods in drug discovery against PPI targets? The following includes a summary of key points described above.

Fragment Screening: in general, fragment hits bind more weakly to PPI targets than conventional targets, due to the shallower pockets. Also, modification to the screening conditions can be necessary in order to detect such weak binding and the screen tends to identify larger compounds from the library. The absence of a functional assay precludes a biochemical screen, and for most PPI targets, there are no readily available tool compounds from which a (usually fluorescent) suitably labelled analogue can be designed for use in a displacement assay. Most fragment-based approaches to PPIs cited in this chapter employed some form of NMR screen. For proteins less than about 30 kDa, HSQC with ^{15}N-labelled protein has proven very effective. This can be supplemented with titrations and/or structure determination. In other cases ligand-observed NMR (such as STD, WaterLOGSY) was used.

Characterisation of ligand binding: it is crucial to validate and characterise the binding of fragments before embarking on any optimisation, be that SAR-by-catalogue or chemical synthesis. This is particularly challenging for PPI targets as there can sometimes be a number of possible binding surfaces for such weakly interacting ligands (*e.g.*, K-Ras above). Crystal structures of fragments bound are often not possible to obtain (such as in the Bcl-2 family projects) as the main interaction site packs against itself in the crystal, requiring a more potent ligand (perhaps with a K_D of around 100 nM) before the crystal lattice can be disrupted and an alternate crystal form produced with the ligand bound. An alternate approach (for proteins below about 30 kDa) is to use a range of NMR techniques to either determine the full structure of the complex (as in the case of the Abbott work on Bcl-2 (ref. 8)) or to use techniques such as X-filtered NOESY to identify interactions between the fragment and the protein from which a model can be generated for ligand binding, such as the work on Mcl-1 by the Vanderbilt group.[46]

Fragment to hit-to-lead optimisation: the challenges here are two-fold. Firstly, the screening and characterising of compounds and their optimisation can be hampered by lack of a robust assay. Secondly, the proteins can adopt different conformations, as seen in the Bcl-2 family and, as seen with K-Ras, even a relatively small protein can exhibit many different faces. Without tool compounds, it can be a real challenge to validate the often weak binding obtained for initial fragments and establish a solid platform that gives the confidence needed for medicinal chemistry optimisation.

Compound properties: the size and character of most binding sites on PPIs are such that large and lipophilic compounds will be needed to achieve potency and affinity. Whereas the desired ligand efficiency (ΔG/number of heavy atoms) for a conventional target such as a kinase is above 0.35, a ligand efficiency of 0.25 at best is more typical for a PPI. Such large compounds often have difficulty achieving appropriate properties and it is essential to measure aggregation, cell penetration, solubility and plasma protein binding to understand the SARs when the cellular and primary assay data does not correlate.

Timescale: the Bcl-2 project from Abbott is the only PPI drug-discovery program from which a fragment-derived clinical candidate compound has been delivered. This project took a long time and a huge commitment and belief in the design hypotheses to realise the compounds. It is difficult in many organisations to justify spending so long on a single project. In our experience, achieving traction against a PPI target (perhaps equivalent to the exploratory work summarised above for Mcl-1 or K-Ras) takes a couple of years to build confidence in the project. Establishing that confidence can take many iterations of identifying and optimising compounds alongside developing and validating the assays until a firm foundation can be established for drug discovery. Once that is in place, then the initial series of compounds may, at best, provide tool compounds, and it can be necessary to start drug discovery again. What this requires is commitment to the project.

6.5 Conclusion

In summary, what is needed for fragment-based discovery against PPIs are three main qualities – patience, pragmatism and integration. It requires patience, as it can take a long time to make progress and to achieve a clinical candidate. For a conventional target (such as a kinase) one could hope to achieve this within 2–3 years; for a PPI, it can be double that time. It requires pragmatism as the nature of the targets will require the use of a whole range of assays and biophysical methods – it is not possible to rely on just one approach. Finally, it requires integration across the disciplines of medicinal chemistry, structural biology, assay technologies and modelling to recognise and solve the problems.

The first drug candidate for a PPI discovered by fragment-based methods was against Bcl-2, published in 2005, and there have been successive publications of compounds with differing profiles. It remains to be seen how long it will take for candidates to be identified for the other important targets reviewed in this chapter. However, the recently published progress on K-Ras and Mcl-1 suggests that such candidates will come from a fragment-based, or at least fragment-inspired, project.

References

1. D. C. Fry, *Curr. Protein Pept. Sci.*, 2008, **9**, 240–247.
2. D. Kozakov, D. R. Hall, G. Y. Chuang, R. Cencic, R. Brenke, L. E. Grove, D. Beglov, J. Pelletier, A. Whitty and S. Vajda, *Proc. Natl. Acad. Sci. U. S. A.*, 2011, **108**, 13528–13533.
3. J. Murray, A. M. Giannetti, M. Steffek, P. Gibbons, B. R. Hearn, F. Cohen, C. Tam, C. Pozniak, B. Bravo, J. Lewcock, P. Jaishankar, C. Q. Ly, X. Zhao, Y. Tang, P. Chugha, M. R. Arkin, J. Flygare and A. R. Renslo, *ChemMedChem*, 2014, **9**, 73–77.
4. D. A. Erlanson, J. W. Arndt, M. T. Cancilla, K. Cao, R. A. Elling, N. English, J. Friedman, S. K. Hansen, C. Hession, I. Joseph, G. Kumaravel, W. C. Lee, K. E. Lind, R. S. McDowell, K. Miatkowski, C. Nguyen, T. B. Nguyen, S. Park, N. Pathan, D. M. Penny, M. J. Romanowski, D. Scott, L. Silvian, R. L. Simmons, B. T. Tangonan, W. Yang and L. Sun, *Bioorg. Med. Chem. Lett.*, 2011, **21**, 3078–3083.
5. J. Zhang, F. J. Adrian, W. Jahnke, S. W. Cowan-Jacob, A. G. Li, R. E. Iacob, T. Sim, J. Powers, C. Dierks, F. Sun, G. R. Guo, Q. Ding, B. Okram, Y. Choi,

A. Wojciechowski, X. Deng, G. Liu, G. Fendrich, A. Strauss, N. Vajpai, S. Grzesiek, T. Tuntland, Y. Liu, B. Bursulaya, M. Azam, P. W. Manley, J. R. Engen, G. Q. Daley, M. Warmuth and N. S. Gray, *Nature*, 2010, **463**, 501–506.
6. P. H. Kussie, S. Gorina, V. Marechal, B. Elenbaas, J. Moreau, A. J. Levine and N. P. Pavletich, *Science*, 1996, **274**, 948–953.
7. D. C. Fry, C. Wartchow, B. Graves, C. Janson, C. Lukacs, U. Kammlott, C. Belunis, S. Palme, C. Klein and B. Vu, *ACS Med. Chem. Lett.*, 2013, **4**, 95–100.
8. T. Oltersdorf, S. W. Elmore, A. R. Shoemaker, R. C. Armstrong, D. J. Augeri, B. A. Belli, M. Bruncko, T. L. Deckwerth, J. Dinges, P. J. Hajduk, M. K. Joseph, S. Kitada, S. J. Korsmeyer, A. R. Kunzer, A. Letai, C. Li, M. J. Mitten, D. G. Nettesheim, S. Ng, P. M. Nimmer, J. M. O'Connor, A. Oleksijew, A. M. Petros, J. C. Reed, W. Shen, S. K. Tahir, C. B. Thompson, K. J. Tomaselli, B. L. Wang, M. D. Wendt, H. C. Zhang, S. W. Fesik and S. H. Rosenberg, *Nature*, 2005, **435**, 677–681.
9. A. D. Wist, L. Gu, S. J. Riedl, Y. Shi and G. L. McLendon, *Bioorg. Med. Chem.*, 2007, **15**, 2935–2943.
10. F. Poy, M. Lepourcelet, R. A. Shivdasani and M. J. Eck, *Nat. Struct. Biol.*, 2001, **8**, 1053–1057.
11. A. M. de Vos, M. Ultsch and A. A. Kossiakoff, *Science*, 1992, **255**, 306–312.
12. C. A. Lipinski, F. Lombardo, B. W. Dominy and P. J. Feeney, *Adv. Drug Delivery Rev.*, 1997, **23**, 3–25.
13. C. G. Wilson and M. R. Arkin, *Drug Discovery Today: Technol.*, 2013, **10**, e501–e508.
14. D. A. Erlanson, J. A. Wells and A. C. Braisted, *Annu. Rev. Biophys. Biomol. Struct.*, 2004, **33**, 199–223.
15. M. R. Arkin and J. A. Wells, *Nat. Rev. Drug Discovery*, 2004, **3**, 301–317.
16. J. A. Wells and C. L. McClendon, *Nature*, 2007, **450**, 1001–1009.
17. N. Baurin, F. Aboul-Ela, X. Barril, B. Davis, M. Drysdale, B. Dymock, H. Finch, C. Fromont, C. Richardson, H. Simmonite and R. E. Hubbard, *J. Chem. Inf. Comput. Sci.*, 2004, **44**, 2157–2166.
18. I. J. Chen and R. E. Hubbard, *J. Comput.-Aided Mol. Des.*, 2009, **23**, 603–620.
19. A. Schuffenhauer, S. Ruedisser, A. L. Marzinzik, W. Jahnke, M. Blommers, P. Selzer and E. Jacoby, *Curr. Top. Med. Chem.*, 2005, **5**, 751–762.
20. J. S. Albert, N. Blomberg, A. L. Breeze, A. J. Brown, J. N. Burrows, P. D. Edwards, R. H. Folmer, S. Geschwindner, E. J. Griffen, P. W. Kenny, T. Nowak, L. L. Olsson, H. Sanganee and A. B. Shapiro, *Curr. Top. Med. Chem.*, 2007, **7**, 1600–1629.
21. W. F. Lau, J. M. Withka, D. Hepworth, T. V. Magee, Y. J. Du, G. A. Bakken, M. D. Miller, Z. S. Hendsch, V. Thanabal, S. A. Kolodziej, L. Xing, Q. Hu, L. S. Narasimhan, R. Love, M. E. Charlton, S. Hughes, W. P. van Hoorn and J. E. Mills, *J. Comput.-Aided Mol. Des.*, 2011, **25**, 621–636.
22. B. C. Doak, C. J. Morton, J. S. Simpson and M. J. Scanlon, *Aust. J. Chem.*, 2013, **66**, 1465–1472.
23. N. Y. Mok, R. Brenk and N. Brown, *J. Chem. Inf. Model.*, 2014, **54**, 79–85.

24. F. Lovering, J. Bikker and C. Humblet, *J. Med. Chem.*, 2009, **52**, 6752–6756.
25. A. D. Morley, A. Pugliese, K. Birchall, J. Bower, P. Brennan, N. Brown, T. Chapman, M. Drysdale, I. H. Gilbert, S. Hoelder, A. Jordan, S. V. Ley, A. Merritt, D. Miller, M. E. Swarbrick and P. G. Wyatt, *Drug Discovery Today*, 2013, **18**, 1221–1227.
26. R. E. Hubbard and J. B. Murray, *Methods Enzymol.*, 2011, **493**, 509–531.
27. P. J. Hajduk, J. R. Huth and S. W. Fesik, *J. Med. Chem.*, 2005, **48**, 2518–2525.
28. T. A. Halgren, *J. Chem. Inf. Model.*, 2009, **49**, 377–389.
29. N. Huang and M. P. Jacobson, *PLoS One*, 2010, **5**, e10109.
30. S. P. Brown and P. J. Hajduk, *ChemMedChem*, 2006, **1**, 70–72.
31. P. J. Hajduk, M. Bures, J. Praestgaard and S. W. Fesik, *J. Med. Chem.*, 2000, **43**, 3443–3447.
32. P. E. Czabotar, E. F. Lee, G. V. Thompson, A. Z. Wardak, W. D. Fairlie and P. M. Colman, *J. Biol. Chem.*, 2011, **286**, 7123–7131.
33. B. Davis, *Methods Mol. Biol.*, 2013, **1008**, 389–413.
34. D. M. Dias, I. Van Molle, M. G. Baud, C. Galdeano, C. F. Geraldes and A. Ciulli, *ACS Med. Chem. Lett.*, 2014, **5**, 23–28.
35. C. M. Richardson, C. L. Nunns, D. S. Williamson, M. J. Parratt, P. Dokurno, R. Howes, J. Borgognoni, M. J. Drysdale, H. Finch, R. E. Hubbard, P. S. Jackson, P. Kierstan, G. Lentzen, J. D. Moore, J. B. Murray, H. Simmonite, A. E. Surgenor and C. J. Torrance, *Bioorg. Med. Chem. Lett.*, 2007, **17**, 3880–3885.
36. K. Babaoglu and B. K. Shoichet, *Nat. Chem. Biol.*, 2006, **2**, 720–723.
37. S. B. Shuker, P. J. Hajduk, R. P. Meadows and S. W. Fesik, *Science*, 1996, **274**, 1531–1534.
38. J. R. Huth, C. Park, A. M. Petros, A. R. Kunzer, M. D. Wendt, X. Wang, C. L. Lynch, J. C. Mack, K. M. Swift, R. A. Judge, J. Chen, P. L. Richardson, S. Jin, S. K. Tahir, E. D. Matayoshi, S. A. Dorwin, U. S. Ladror, J. M. Severin, K. A. Walter, D. M. Bartley, S. W. Fesik, S. W. Elmore and P. J. Hajduk, *Chem. Biol. Drug Des.*, 2007, **70**, 1–12.
39. J. Liu, J. D. Farmer Jr., W. S. Lane, J. Friedman, I. Weissman and S. L. Schreiber, *Cell*, 1991, **66**, 807–815.
40. G. D. Van Duyne, R. F. Standaert, P. A. Karplus, S. L. Schreiber and J. Clardy, *Science*, 1991, **252**, 839–842.
41. W. P. Jencks, *Proc. Natl. Acad. Sci. U. S. A.*, 1981, **78**, 4046–4050.
42. T. Fink and J. L. Reymond, *J. Chem. Inf. Model.*, 2007, **47**, 342–353.
43. A. Strasser, S. Cory and J. M. Adams, *EMBO J.*, 2011, **30**, 3667–3683.
44. C. Tse, A. R. Shoemaker, J. Adickes, M. G. Anderson, J. Chen, S. Jin, E. F. Johnson, K. C. Marsh, M. J. Mitten, P. Nimmer, L. Roberts, S. K. Tahir, Y. Xiao, X. Yang, H. Zhang, S. Fesik, S. H. Rosenberg and S. W. Elmore, *Cancer Res.*, 2008, **68**, 3421–3428.
45. A. J. Souers, J. D. Leverson, E. R. Boghaert, S. L. Ackler, N. D. Catron, J. Chen, B. D. Dayton, H. Ding, S. H. Enschede, W. J. Fairbrother, D. C. Huang, S. G. Hymowitz, S. Jin, S. L. Khaw, P. J. Kovar, L. T. Lam, J. Lee, H. L. Maecker, K. C. Marsh, K. D. Mason, M. J. Mitten, P. M. Nimmer, A. Oleksijew, C. H. Park, C. M. Park, D. C. Phillips, A. W. Roberts, D. Sampath,

J. F. Seymour, M. L. Smith, G. M. Sullivan, S. K. Tahir, C. Tse, M. D. Wendt, Y. Xiao, J. C. Xue, H. Zhang, R. A. Humerickhouse, S. H. Rosenberg and S. W. Elmore, *Nat. Med.*, 2013, **19**, 202–208.
46. A. Friberg, D. Vigil, B. Zhao, R. N. Daniels, J. P. Burke, P. M. Garcia-Barrantes, D. Camper, B. A. Chauder, T. Lee, E. T. Olejniczak and S. W. Fesik, *J. Med. Chem.*, 2013, **56**, 15–30.
47. E. F. Lee, P. E. Czabotar, M. F. van Delft, E. M. Michalak, M. J. Boyle, S. N. Willis, H. Puthalakath, P. Bouillet, P. M. Colman, D. C. Huang and W. D. Fairlie, *J. Cell Biol.*, 2008, **180**, 341–355.
48. P. E. Czabotar, E. F. Lee, M. F. van Delft, C. L. Day, B. J. Smith, D. C. Huang, W. D. Fairlie, M. G. Hinds and P. M. Colman, *Proc. Natl. Acad. Sci. U. S. A.*, 2007, **104**, 6217–6222.
49. S. Schubbert, K. Shannon and G. Bollag, *Nat. Rev. Cancer*, 2007, **7**, 295–308.
50. P. A. Boriack-Sjodin, S. M. Margarit, D. Bar-Sagi and J. Kuriyan, *Nature*, 1998, **394**, 337–343.
51. T. Maurer, L. S. Garrenton, A. Oh, K. Pitts, D. J. Anderson, N. J. Skelton, B. P. Fauber, B. Pan, S. Malek, D. Stokoe, M. J. Ludlam, K. K. Bowman, J. Wu, A. M. Giannetti, M. A. Starovasnik, I. Mellman, P. K. Jackson, J. Rudolph, W. Wang and G. Fang, *Proc. Natl. Acad. Sci. U. S. A.*, 2012, **109**, 5299–5304.
52. Q. Sun, J. P. Burke, J. Phan, M. C. Burns, E. T. Olejniczak, A. G. Waterson, T. Lee, O. W. Rossanese and S. W. Fesik, *Angew. Chem., Int. Ed.*, 2012, **51**, 6140–6143.
53. J. M. Ostrem, U. Peters, M. L. Sos, J. A. Wells and K. M. Shokat, *Nature*, 2013, **503**, 548–551.
54. D. E. Scott, M. T. Ehebauer, T. Pukala, M. Marsh, T. L. Blundell, A. R. Venkitaraman, C. Abell and M. Hyvonen, *ChemBioChem*, 2013, **14**, 332–342.
55. M. R. Arkin, M. Randal, W. L. DeLano, J. Hyde, T. N. Luong, J. D. Oslob, D. R. Raphael, L. Taylor, J. Wang, R. S. McDowell, J. A. Wells and A. C. Braisted, *Proc. Natl. Acad. Sci. U. S. A.*, 2003, **100**, 1603–1608.
56. J. Rouhana, F. Hoh, S. Estaran, C. Henriquet, Y. Boublik, A. Kerkour, R. Trouillard, J. Martinez, M. Pugniere, A. Padilla and A. Chavanieu, *J. Med. Chem.*, 2013, **56**, 8497–8511.
57. R. E. Hubbard, B. Davis, I. Chen and M. J. Drysdale, *Curr. Top. Med. Chem.*, 2007, **7**, 1568–1581.
58. E. Meiby, H. Simmonite, L. le Strat, B. Davis, N. Matassova, J. D. Moore, M. Mrosek, J. Murray, R. E. Hubbard and S. Ohlson, *Anal. Chem.*, 2013, **85**, 6756–6766.

CHAPTER 7

Fragment-Based Discovery of Allosteric Ligands

STEVEN HOWARD*[a]

[a]Department of Medicinal Chemistry, Astex Pharmaceuticals, Cambridge Science Park, Cambridge, CB4 0QA, UK
*E-mail: steven.howard@astx.com

7.1 Introduction

For many years it has been known that many enzymes contain secondary binding sites, beyond the main substrate (or orthosteric) site, where endogenous ligands, other proteins or drugs can bind and modulate enzyme function *via* an allosteric mechanism.[1–6] This chapter will explore how fragment-based approaches (FBAs) can be used to identify secondary binding sites and ligands that bind to them. In particular we will see how this contrasts with fragment-based approaches applied to orthosteric sites. Highlights from the recent literature will be used to illustrate screening techniques and strategies to tackle the particular challenges associated with identifying secondary binding sites. The chapter will show that fragment-based drug design (FBDD) is an excellent technique for identifying and targeting previously unknown allosteric pockets.

Historically, most drug discovery was based on exploiting the orthosteric sites of proteins, with rather limited effort consciously directed towards non-orthosteric (*e.g.*, allosteric) binding sites. However, some well-established drugs were found, retrospectively, to operate through an allosteric

RSC Drug Discovery Series No. 47
Fragment-Based Drug Discovery
Edited by Steven Howard and Chris Abell

Published by the Royal Society of Chemistry, www.rsc.org

mechanism. A good example is the sedative diazepam which acts *via* modulation of the $GABA_A$ (gamma-aminobutanoic acid) mediated ion channel. This drug was eventually discovered to bind not to the orthosteric GABA site, but to an alternative allosteric site.[7,8] Diazepam exerts its effect only in the presence of the endogenous orthosteric ligand (GABA) which leads to a different, more subtle, mode of action than could be achieved by simply blocking the endogenous agonist (GABA) site. Diazepam is a much safer medicine than many of the previously known sedatives.

These findings proved that identifying inhibitors that bind to allosteric pockets is of more than academic interest as it can lead to new medicines with unique modes of action which can have advantages over orthosteric drugs. Allosteric inhibitors can offer several therapeutic advantages over orthosteric compounds and these are discussed in the next section. In general, the term *secondary* will refer to any binding site known to be distinct from the orthosteric binding site, whether the site has any confirmed biological function or not. The term *allosteric* will refer only to those sites, or ligands, for which there is evidence of biological function.

7.2 Advantages of Allosteric Drugs

7.2.1 Target Selectivity

To date, 14 drugs targeting protein kinases have been approved for clinical use.[9] In the majority of cases, these compounds act by binding to the ATP pocket of the relevant kinase target.[10] However, as the ATP-binding site is largely conserved across the human kinome, it is unsurprising that many of these compounds, *e.g.*, sunitinib and sorafenib, also bind to additional *off-target* kinases.[11] This may be an undesirable feature as inhibition of certain off-target kinases has been suggested to be associated with cardiovascular, and other, side-effects.[12] While design of highly selective ATP-competitive compounds is possible, it is considered to be challenging. In contrast, there are a few examples of allosteric kinase inhibitors. Examples include the highly selective MEK (mitogen activate kinase) inhibitors, AZD-6244 and trametinib, which bind to an allosteric site of this enzyme.[13,14] The structural basis for selectivity is explained by the fact that this allosteric binding site is not observed in any other kinases. Interestingly, there is no known natural role for this particular allosteric pocket.

7.2.2 Targeting Proteins with Undruggable Orthosteric Sites

There are many therapeutic targets which, although well validated, are considered very challenging often due to the presence of poorly druggable orthosteric sites.[15] This problem can be circumvented if more druggable secondary (or allosteric) sites can be identified. A good example is provided by the CCR5 chemokine receptor; an important target for antiHIV therapy.[16] Viral particles gain entry to cells by first binding to the extracellular domain of the CCR5 receptor. However the recognition between viral peptides and the receptor is essentially a protein–protein interaction which is difficult to disrupt using small-molecule antagonists. Success against this target was ultimately achieved by the clinically approved drug maraviroc, which inhibits the CCR5 receptor *via* an allosteric mechanism.[17]

Another example is metabotropic glutamate receptor 5 (mGluR5) which is of interest as a target for treating various central nervous system (CNS) disorders. The challenge for this target has been identification of ligands for its polar orthosteric site (where the endogenous ligand is L-glutamate) which can also penetrate the blood–brain barrier. In the case of mGluR5, agonists and antagonists, with more attractive drug-like properties, have been identified which act *via* an allosteric mechanisms.[18]

7.2.3 Overcoming Drug Resistance

Resistance to therapy is a particular problem in oncology and anti-infectives. Mutation in the amino acid sequence of the drug-binding site is a common mechanism of resistance in both these therapeutic areas. In one example, leukaemia patients develop resistance to the kinase inhibitor, Imatinib, due to mutation (T315I) in the ATP-binding pocket of the target (bcr-Abl).[19] This mutation results in a larger, threonine, amino acid which disrupts the binding mode of the drug. In virology, patients with hepatitis C virus (HCV) typically develop resistance to HCV protease inhibitors *via* mutations in the active site of the viral HCV protease.[20] This is a frustrating situation because, in both cases, the enzyme remains a valid therapeutic target but the drugs cease to be effective. Encouragingly, scientists have recently discovered novel, previously unknown allosteric pockets for both these enzymes.[21–23] Work is at an early stage, but identification of inhibitors for these novel allosteric pockets offers a potential strategy to 'rescue' treatment for these patients. The case of HCV NS3 protein will be discussed in Section 7.6.

7.2.4 Target Activation

Activation of an enzyme target may be advantageous and, in some cases, this can be achieved with allosteric modulators. In an example described later, a fragment screen against pyruvate kinase M2 enzyme (PKM2), carried out at Astex, revealed L-serine bound to a previously uncharacterised pocket.[24] It was shown that L-serine binding to this allosteric site activates PKM2. Another good example, not discussed here, is the allosteric activators of glucokinase currently under clinical investigation.[25]

7.3 Key Concepts and Challenges for FBDD Applied to Allosteric Pockets

We have seen some examples which illustrate the potential benefits of identifying drugs acting *via* an allosteric mechanism. For the most part, this chapter will focus on the application of FBDD to identify previously unknown allosteric pockets in therapeutically relevant enzyme targets. The advantages offered by FBDD towards identifying allosteric sites and their ligands are, in principle, the same as for other classes of target.[26,27] However, the challenges associated with allosteric sites are different and this chapter will explore strategies to overcome these.

Fragment screening has proved itself to be a more efficient way of sampling chemical space than screening libraries of larger molecular weight compounds in a traditional high-throughput screening (HTS) format. Or to put it another way, screening fragment libraries typically provides higher hit rates than screening libraries of larger molecular weight compounds. This has been described extensively elsewhere.[28] In terms of fragment library design, there is no evidence to suggest that allosteric pockets, in general, demand any particular requirements. After all, allosteric pockets are not restricted to any particular family of proteins and there is nothing to suggest that allosteric pockets are fundamentally any different from any other protein pockets. The usual concepts of fragment library design, such as the *Rule of 3*, together with a suitably sensitive assay (*e.g.*, X-ray crystallography) should be fit for purpose.[29]

As we will find, FBDD is an excellent way to identify new, previously unknown, secondary pockets, as well as generating chemical hits for established allosteric binding sites. However, as mentioned at the beginning of this section, there are particular challenges to probing secondary pockets with FBDD. These include:

Challenge 1

Differentiating secondary site binders from orthosteric binders and/or false positives.

Challenge 2

Establishing the biological/functional relevance of secondary binding sites discovered as part of a fragment screen.

Challenge 3

Understanding of, and implications for, conservation of secondary sites within a protein class or across species. Section 7.6 on HCV NS3 inhibitors will illustrate this.[23]

Challenge 4

Allosteric pockets often lie outside the catalytic domain of enzymes. Therefore, screening using full-length protein constructs is likely to be necessary to maximise the chance of identifying novel, functionally relevant, allosteric pockets.[30]

The following sections will describe how different biophysical techniques, can address these challenges. Protein X-ray crystallography is a particularly powerful technique in addressing the first challenge as it can easily identify the binding mode of very low-affinity fragments. X-ray crystallographic screening can therefore very quickly differentiate fragments binding at secondary binding sites from those binding to orthosteric sites. The second challenge stems from the fact that fragment screening hits typically bind with low dissociation constants in the μM to mM range. Whilst detectable in more sensitive direct binding assays (*e.g.*, NMR, SPR, X-ray crystallography) they are typically too weak for detection by functional enzyme assays, let alone cellular assays. Fragment hits themselves are therefore inadequate as chemical tools for evaluating the functional relevance of the newly discovered secondary pockets. However, even at the early stages of a project, X-ray crystallography can provide evidence to support the functional relevance of a secondary binding site. For example, X-ray crystallographic data may show the fragment only binds to an inactive form of the protein. In other cases, identifying the binding mode of functional endogenous allosteric modulators (*e.g.*, amino acids) using X-ray crystallography can also support the biological relevance of pockets identified during fragment screening. The example of PKM2 is discussed in the next section. Ultimately, optimisation of fragments into more potent leads and subsequent biological profiling is necessary to fully characterise new binding sites and establish causality between structure and function.

7.4 Fragment Screening Techniques

7.4.1 Surface Plasmon Resonance

SPR is a well-established sensitive technique in fragment-based screening (FBS).[31] It is particularly well suited for high throughput and has relatively low requirements for amount of protein required. In recent years it has been applied even to G protein-coupled receptors (GPCRs), a most challenging class of membrane proteins. Chapter 2 discusses SPR in more detail: I will avoid re-iterating here and highlight only those features that are particularly relevant to probing secondary binding sites.

If known inhibitors (or endogenous ligands) are available for the orthosteric and/or allosteric site, SPR is well suited to probing for allosteric pockets. In a 2011 example, Danielson *et al.* used an SPR fragment-based screen to identify novel scaffolds for the allosteric inhibition of drug-resistant HIV-1 reverse transcriptase.[32] Fragment hits for this known allosteric pocket were validated by experiments demonstrating competition with the non-nucleoside RT inhibitor, nevirapine. The resulting hit (Figure 7.1, dissociation constant (K_D) = 270 μM, LE = 0.44) constituted a novel class of scaffold which was shown to be active against several nevirapine-resistant isozymes.

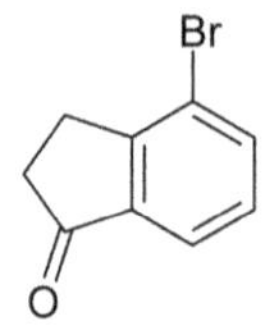

Figure 7.1 Fragment hit for HIV-1 reverse transcriptase identified *via* SPR-based fragment screening. LE = ligand efficiency, LLE = ligand lipophilicity efficiency.

7.4.2 Nuclear Magnetic Resonance

NMR is a very popular fragment screening technique and an excellent review can be found in Chapter 3. Like SPR, effectively screening for allosteric site binders generally requires the availability of a known orthosteric ligand which can be used in competition experiments to differentiate fragments binding at the orthosteric site from those binding at other secondary sites.

In the case of SPR and NMR, fragments binding at previously unknown allosteric sites can be identified. The challenge is then to optimise these fragments to a point at which they can provide chemical tools to validate the functional relevance of the secondary pocket. For this, X-ray crystallography is highly desirable.

7.4.3 X-ray Crystallography

We, and others, have shown that X-ray fragment screening is ideally suited to identify novel potential allosteric sites suitable for small-molecule drug development.[23,33] The technique is very sensitive (capable of detecting ligands with $K_D > 10$ mM, depending on the solubility of the compound) and can detect ligand-induced pockets, as well as pre-formed sites.[34] Information on binding mode is immediately available; allowing rapid differentiation of secondary site binders from those ligands bound to the orthosteric site. Structural information can then facilitate rapid structure-based optimisation of the fragment towards more potent leads and chemical tools in order to validate biological function. In some cases, there may be the caveat that secondary binding sites can be occluded due to the crystal packing arrangement. Conversely, the crystal packing could also result in artefactual secondary binding sites. However, close inspection of how symmetry related protein molecules interact within the crystal lattice can help mitigate these issues.

An excellent example of X-ray crystallographic fragment screening has been published by Arnold and colleagues. Their 2013 paper described X-ray crystallographic fragment screening against HIV-1 reverse transcriptase

(HIV-RT) (117 kDa) which identified fragments binding to, a remarkable, 16 new sites.[33]

Many inhibitors of HIV-RT are in clinical use. These fall into two categories; nucleoside HIV-RT inhibitors (NRTIs, *e.g.*, emtricitabine), which bind to the orthosteric site (oligonucleotide substrate site), and non-nucleotide HIV-RT inhibitors (NNRTIs, *e.g.*, rilpivarine) which bind to a secondary site and inhibit the enzyme *via* an allosteric mechanism. Rilpivarine binding is associated with a large protein movement leaving HIV-RT in an open, 'inactive', conformation. The Arnold group set out to identify fragments that could bind to this inactive conformation. Crystals of a HIV-RT variant (RT52A) in complex with rilvirapine, used to stabilise the open conformation, were screened against a library of ~800 fragments (in cocktails of 8). Fragments binding to seven new sites were identified (Figure 7.2).

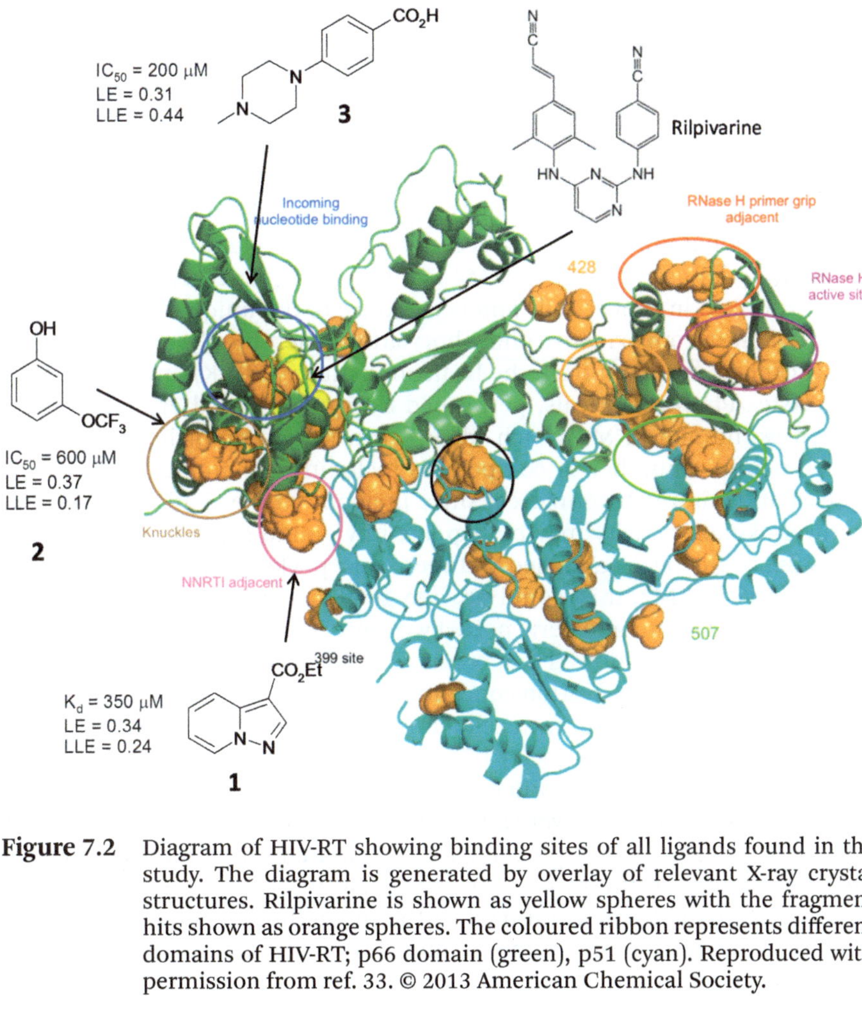

Figure 7.2 Diagram of HIV-RT showing binding sites of all ligands found in the study. The diagram is generated by overlay of relevant X-ray crystal structures. Rilpivarine is shown as yellow spheres with the fragment hits shown as orange spheres. The coloured ribbon represents different domains of HIV-RT; p66 domain (green), p51 (cyan). Reproduced with permission from ref. 33. © 2013 American Chemical Society.

The screen identified 34 fragments binding throughout HIV-RT at 16 sites, giving a hit rate of 4%. Interestingly, 4-bromopyrazole was observed at all 16 sites. Discounting pockets which bound only 4-bromopyrazole, this left seven sites. Of these, compounds binding at three sites were found to be inhibitory in an enzymatic assay. These included compounds **1** (K_D = 350 μM), **2** (K_D = 600 μM) and **3** (K_D = 200 μM) which showed weak activity in a functional enzyme assay.

X-ray crystallography easily deals with Challenge 1, *i.e.*, differentiating allosteric binders from orthosteric binders and/or false positives. The second challenge, demonstrating the biological relevance of these previously unknown pockets, is more difficult to answer without developing the hits into more potent compounds. The activity of compounds **1–3** in a functional enzyme assay is encouraging, but without structurally understood structure–activity relationships (SARs), it is difficult to discount alternative modes of action for these compounds. A more detailed biological validation of these pockets was outside the scope of this paper. However, this work does showcase the utility of X-ray crystallographic fragment screening to rapidly 'interrogate' a protein for secondary site binders. No other technique would have provided such detailed structural information on putative allosteric pockets and their ligands.

The next example shows how a combination of X-ray crystallography, together with identification of an endogenous ligand, was able to establish early functional relevance for a secondary binding site in the M2 isoform of pyruvate kinase (PKM2).[24] PKM2 is a tightly regulated enzyme which is over-expressed in cancer cells and is causative of the unique metabolic phenotype, a shift from oxidative phosphorylation to aerobic glycolysis, typical of many cancers.[35] A team from the Beatson Institute (CRUK), in collaboration with Astex Pharmaceuticals, performed a fragment-based crystallographic screen *vs.* human PKM2 which identified a novel non-orthosteric pocket which bound a number of compounds including the amino acid L-serine. This sparked immediate interest since previous studies had proposed that pyruvate kinase activity may be regulated by amino acids. If a causal relationship could be established between this pocket and serine-mediated regulation of PKM2, this would provide good validation of the pocket as a functionally relevant allosteric pocket worthy of further investigation. Initial *in vitro* experiments showed that serine activates recombinant PKM2 activity with a half-maximal activation concentration of 1.3 mM, a level that is within the physiological range of intracellular serine concentrations. Furthermore, isothermal titration calorimetry (ITC) was used to determine the K_D of serine as 0.2 mM and the titration curve was consistent with a 1 : 1 stoichiometry. No other amino acids were able to activate PKM2 in this way. Inspection of the PKM2–L-serine complex (Figure 7.3) showed serine forming multiple hydrogen bonds to the protein. In particular, interactions afforded by the hydroxymethyl side-chain rationalise the low affinity of glycine for PKM2 and may contribute to the unique ability of serine to activate PKM2. To complete this investigation, a recombinant PKM2 (H464A) mutant was produced

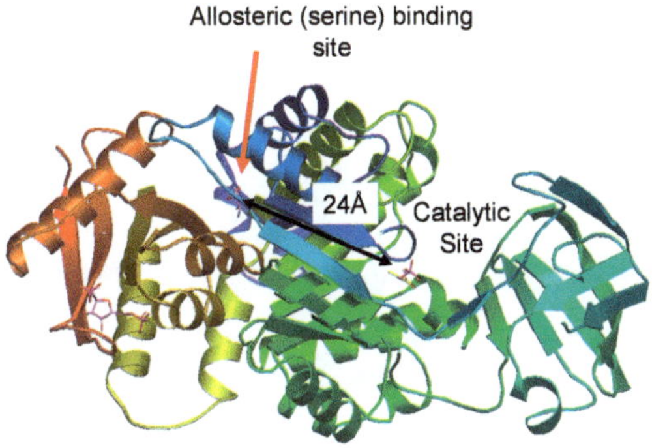

Figure 7.3 X-ray crystal structure of PKM2. The allosteric (serine) binding site (illustrated by the red arrow) is 24 Å from the catalytic site.

which should ablate serine binding by virtue of removing a key interaction within the binding site. PKM2 (H464A) retains functional activity but demonstrated neither direct binding to, nor activation by, serine. These data confirmed that the newly discovered pocket is a true allosteric binding site and provides encouragement that ligands for this site could be used to modulate PKM2 activity.

7.5 Lead Generation Case Studies

7.5.1 Soluble Adenylate Cyclase

In their 2014 paper on human soluble adenylate cyclase (hsolAC), Bethell *et al.* demonstrated the potential of FBS to both identify and exploit allosteric pockets for inhibitor design.[36] hsolAC catalyses the synthesis of the second messenger cAMP through cyclisation of ATP and is the only known mammalian enzyme to be directly regulated by bicarbonate. Bethell *et al.* described the first crystal structures of mammalian solAC, and the application of FBDD to identify small-molecule inhibitors of the human enzyme (Figure 7.4).[37] Bicarbonate binds to a putative allosteric pocket which is 12 Å from the substrate (ATP) pocket.

Before exploiting this putative allosteric pocket for inhibitor design, further reassurance was required that this was indeed the functionally relevant allosteric (bicarbonate) pocket. Closer inspection of the structure revealed that bicarbonate binding is associated with significant conformational rearrangement of the amino acid side-chains, resulting in changes to the ATP-binding pocket. Also, highly conserved residues around the bicarbonate-binding pocket are consistent with this being a functionally relevant allosteric site. hsolAC was screened against ~1600 fragments using a combination of X-ray crystallography and thermal shift assay. Interestingly, the majority of validated hits bound to the bicarbonate site rather than the adenine (substrate) site.

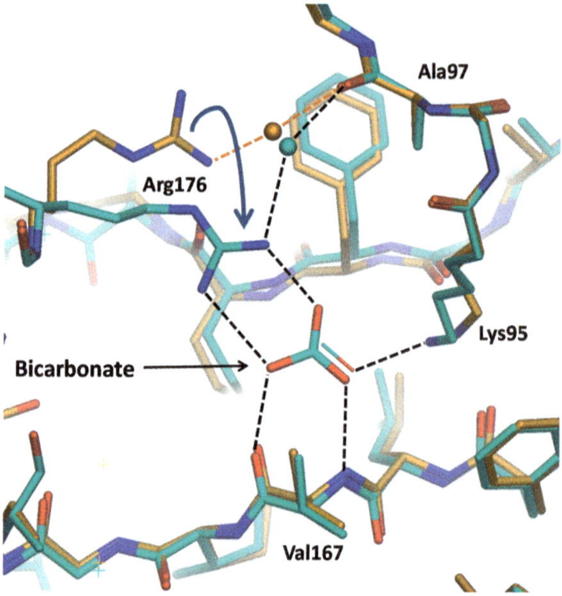

Figure 7.4 Comparison of the apo (orange) and bicarbonate-bound (cyan) hsolAC structures. The bicarbonate anion forms hydrogen bonds (black dashed lines) with the backbone carbonyl and NH groups of Val167, the guanidino group of Arg176 and amine side-chain of Lys95. The significant conformational change induced by bicarbonate is seen in the side-chain of Arg176 which undergoes an approximately 30° swing to hydrogen bond with bicarbonate.

The fragment hit **4** (IC_{50} > 1 mM) was identified as binding to the bicarbonate pocket albeit without meaningful affinity in a functional assay (Figure 7.5). A near-neighbour follow up screen of five-membered heterocyclic compounds identified the more potent analogue **5** (19 μM). The amino-oxadiazole mimics the hydrogen-bonding interactions made by bicarbonate. Structure-guided optimisation led to larger compounds which could span both the bicarbonate pocket and the proximal substrate (ATP pocket) resulting in sub μM inhibitor **6** (0.36 μM). Overall, this example demonstrates how X-ray crystallography can be used to identify hits for a highly flexible and previously uncharacterised allosteric site, and guide subsequent structure-based optimisation towards novel sub μM inhibitors.

7.5.2 Farnesyl Pyrophosphate Synthetase

The following example from Novartis illustrates how NMR screening, supported by X-ray crystallography, was used to find new chemical leads for an enzyme considered to have a poorly druggable orthosteric site.[38]

Jahnke and colleagues, reported the discovery of allosteric, non-bisphosphonate inhibitors of farnesyl pyrophosphate synthetase (FPPS).[38] Prior to

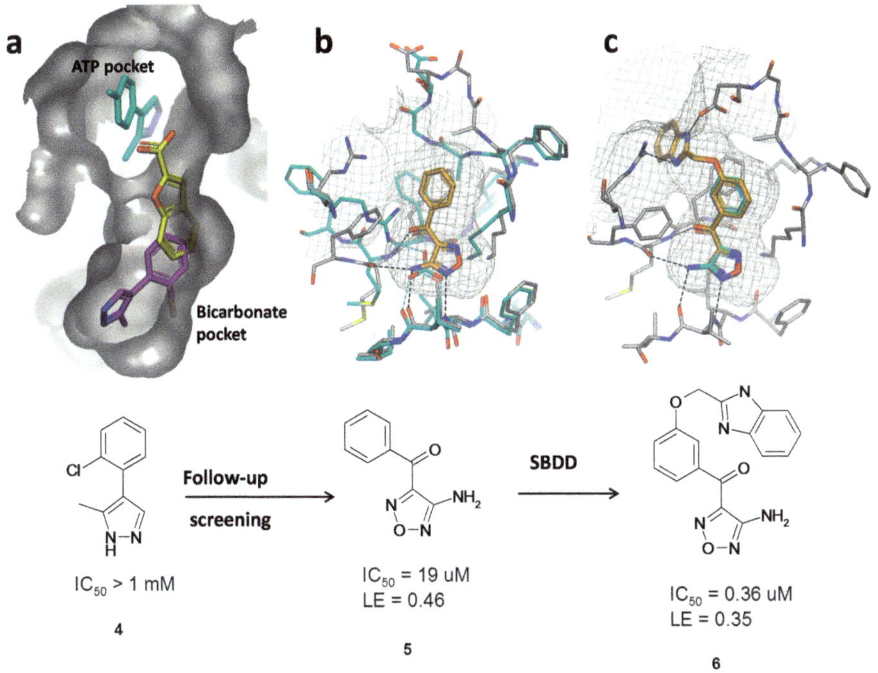

Figure 7.5 Fragment hit **4** and structure-based optimisation towards lead **6**. (a) Overlay of three different crystal structures showing fragments binding in multiple regions of the hsolAC catalytic and allosteric pockets. The phenyl pyrazole (cyan) binds in the ATP pocket, an aromatic carboxylic acid (yellow) induces a new pocket, and the pyrazole **4** (purple) binds in the allosteric bicarbonate pocket. (b) Overlay of the bicarbonate (cyan) and amino-furazane fragment **5** (orange) protein structures. Position 3 on the phenyl ring of compound **5** presents a good vector to access the ATP pocket. (c) Complex of hsolAC with compound **6** (orange) showing the benzimidazole motif extending into the ATP pocket. The structure of **5** is overlaid in cyan to illustrate the fragment-growing strategy.

this work, inhibitors of FPPS were dominated by the bisphosphonates such as zoledronate, which bind at the orthosteric (substrate) site (Figure 7.6). These compounds are highly efficacious in the treatment of bone diseases such as osteoporosis. In contrast to most drugs, bisphosphonates are highly charged, water-soluble compounds; properties that have a profound effect on their pharmacokinetics. On reaching the systemic circulation, approximately half the bisphosphonate dose is rapidly adsorbed to bone and the remainder is rapidly excreted unchanged *via* the kidney. Consequently, blood levels and soft tissue exposure are very low soon after administration.

In addition to bone disease, FPPS is emerging as a promising target for cancer therapy. However, the high affinity of bisphosphonates for bone mineral, and low blood levels, seem inappropriate for the direct treatment of soft-tissue tumours. Since effective orthosteric site binders seem to rely on

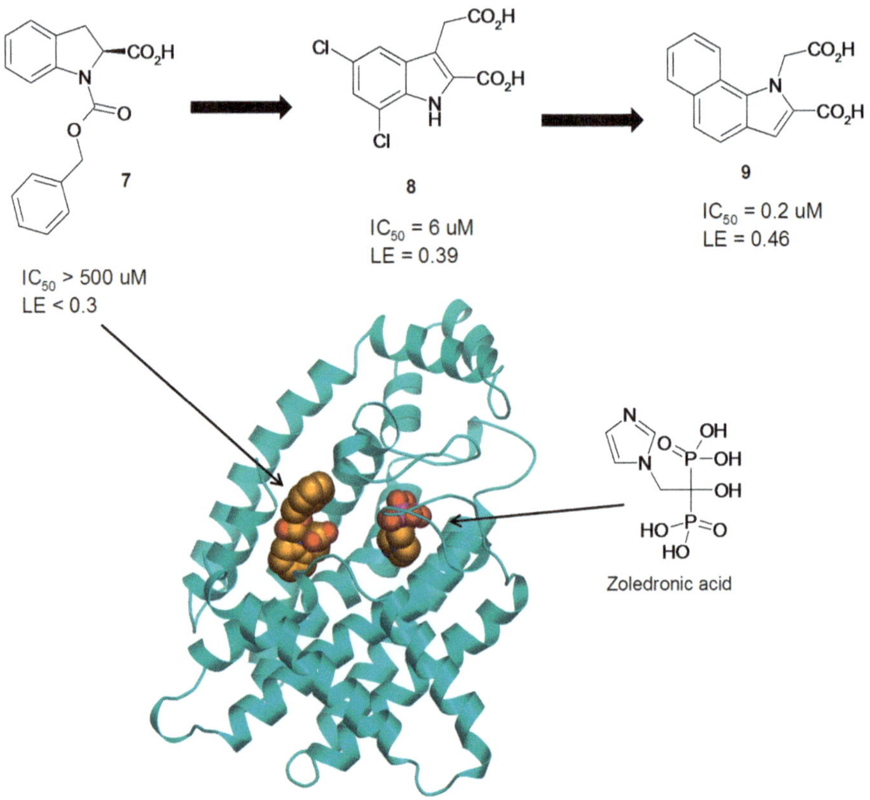

Figure 7.6 X-ray crystal structure of FPPS showing zoledronic acid bound to the substrate-binding pocket and compound **7** bound to the putative allosteric pocket. Optimisation of **7** led to compound **9**.

mimicking the highly polar substrate, dimethylallyl pyrophosphate (DMAPP), it is unsurprising that decades of research in this field have been unable to identify any alternative to bisphosphonate inhibitors.

In their paper, Jahnke *et al.* described their application of FBDD to this 'tough' target using NMR and X-ray crystallography. An NMR based-screen of 400 fragments identified the weakly binding indole 7 (IC_{50} > 500 μM). Competition binding experiments with zoledronic acid indicated an alternative binding mode which was confirmed by X-ray crystallography. The X-ray structure of 7 complexed with FPPS revealed the ligand bound in a previously unknown secondary pocket close to the protein C-terminus and remote from the substrate (DMAPP)/zoledronic acid binding site (Figure 7.6). Although binding with a dissociation constant in the μM range, the novel binding mode and the promise of a non-bisphosphonate series made 7 an interesting chemical starting point for medicinal chemistry optimisation.

Having differentiated the hit from orthosteric site binders, the Novartis team faced the challenge of establishing the biological and functional relevance of this uncharacterised allosteric pocket (Challenge 2). The low affinity of **7** meant any data from a functional bioassay would be unhelpful and so further compound optimisation was necessary. A follow up screen of close analogues from the compound collection quickly identified the more potent analogue **8** (IC_{50} = 6 μM, LE = 0.39) which exhibited K_D = 10 μM in a functional FPPS assay. Furthermore, **8** showed target-related activity in a cell-based assay. Based on this, structure-based design allowed the synthesis of new analogues, leading to the lead compound **9** which exhibited K_D = 0.2 μM in a functional biochemical assay. Importantly this was confirmed in a secondary ITC assay (K_D = 0.18 μM) and further validation was provided by X-ray crystallography which confirmed **9** binds to the same allosteric pocket.

In contrast to bisphosphonates, compound **9** showed low affinity for bone matrix, although its permeability was, unsurprisingly, low. Interestingly a subsequent HTS screen against FPPS, using the full-deck of company compounds, did not reveal any hits with affinities <5 μM. This illustrates the principal advantage of FBDD; as a more effective way of sampling chemical space.

The work also demonstrates the exquisite sensitivity of NMR, its ability to detect very low affinity (mM) fragments and also the utility of the technique in establishing binding mode through competition experiments. Competition experiments using NMR are discussed in more detail in Chapter 3. Last, but not least, this paper illustrates the importance of X-ray crystallography. Structural data was a salient feature of this work and was essential in driving the successful optimisation of **8** to **9**. This allowed the generation of structurally understood SARs which provided encouraging evidence of causality between the observed novel binding mode and biological action. The impact of the structural data on optimisation is evident from the increase in ligand efficiency of **9** over **8**.

7.6 Fragment to Allosteric Pre-clinical Candidate

In 2012, Astex described the use of X-ray crystallography to identify a novel allosteric pocket at the interface of the protease and helicase domain of hepatitis C virus (HCV) NS3 protein. NS3 is a bifunctional enzyme with a serine protease domain at the N-terminus and an ATP-dependent helicase domain at the C-terminus.[39] *In vivo*, the NS3 protein exists as a complex with NS4a, an essential co-factor required for full protease activity. The NS3 protein has been the focus of intense research as a drug-discovery target, leading to inhibitors of the protease activity being approved for the treatment of HCV infections.[40] However, as with all antiviral drugs, the emergence of resistance will accentuate the need for new drugs with different and complementary modes of action.

While it is possible to screen for inhibitors of protease activity using a protease-only domain, experience with other classes of protein suggests that the discovery of novel allosteric, or secondary, binding sites is more likely when screens are performed against full-length protein constructs, rather than the catalytic domain alone.[13,41] 176 Fragments were screened, in cocktails of 4, using crystals of the HCV NS3–NS4a full-length genotype

1b holoenzyme. Crystal structures of the protein in complex with a series of fragment hits revealed the existence of a new binding site at the end-face of the two domains (Figure 7.7(a) and (b)). This is a lipophilic site which is close to, but distinct from, the protease-binding site. Fragment hit **10** was chosen as the chemical starting point for hit optimisation (Figure 7.7(c) and Table 7.1).

Structure-guided optimisation was used to elaborate the low-affinity fragment into tight binding chemical leads suitable to investigate the functional relevance of this novel site. Two strategies to optimise the potency of the compound were immediately obvious from the structure (Figure 7.8). First two flanking *ortho*-fluorine substituents stabilised the biphenyl ether conformation. Second moving the aminomethyl group to the adjacent position on the benzene ring allowed creation of a salt–bridge interaction with Glu628. The resulting compound **11** showed good inhibition of protease activity in the full-length protein assay but was inactive against just the protease domain, consistent with the observed binding mode.

Further structure-based optimisation led to compounds **12–14** which enabled testing of this novel mode of action in the more realistic setting of a cell-based assay. The results showed a good correlation between the *in vitro* enzyme and cell-based assays. Compound **14** showed an EC_{50} of 83 nM in the cell-based antiviral assay (Replicon) and ~500-fold selectivity against the cellular cytotoxicity assay (Table 7.1). Further supporting evidence to the cellular mode of action was provided by enantiomer **15** which was inactive in the cell-based assay.

To establish an unambiguous causal link between antiviral activity and inhibition of the novel allosteric pocket, compound **14** was used to apply selective pressure on the Replicon until viral-resistance mutations emerged. Sequencing of these strains traced the cause of resistance to amino acid mutations (V630L) within the allosteric pocket. In particular, the lower affinity of **14** for the V630L mutant can be understood structurally; the leucine residue results in smaller pocket. The V630L mutant showed no resistance to the active-site protease inhibitor, telaprevir, further confirming the differential mode of action associated with **14**.

At this point it is worth re-iterating Challenge 3 regarding the question of whether an allosteric site is conserved across related proteins. When targeting a known allosteric site, *e.g.*, a kinase ATP site, one can be confident that this binding site will be, in a broad sense, conserved within the protein class and/or across species. This cannot be assumed for non-orthosteric sites. For example, the allosteric site in MEK is unknown in other kinases. Phylogenetic analysis of HCV NS3 protein showed the amino acids that line this pocket are highly conserved across different clinically relevant HCV genotypes. This suggests that the allosteric pocket is broadly conserved across viral genotypes and that a drug from this series would exhibit a broad spectrum of activity. Further optimisation of pharmaceutical properties led to compound **16** which was selected as a pre-clinical candidate for the treatment of HCV infection (Figure 7.9).[42]

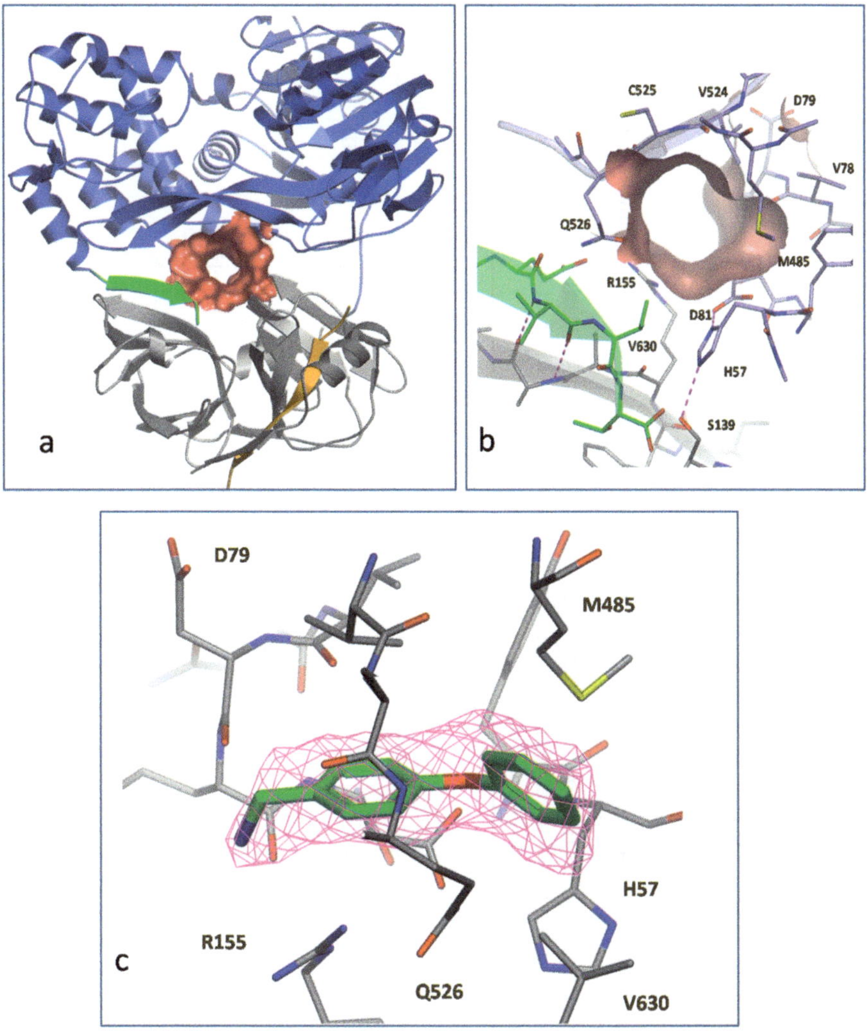

Figure 7.7 (a) X-ray crystal structure of full-length NS3–NS4a protein, illustrated as a ribbon diagram, with the novel allosteric site represented as a red surface. The protease domain is coloured in grey, the NS4a co-factor in orange, and the helicase domain in blue with the C-terminus of the helicase highlighted in green. (b) Magnified view of the protease–helicase interface and the newly discovered allosteric pocket. The six C-terminal residues of the helicase domain (D626–T631) bind in the active site of the protease domain and form backbone hydrogen-bonding interactions (dashed lines) which result in the formation of an antiparallel β-sheet. These interactions stabilise the protein in a compact conformation. The catalytic triad (S139, H57, D81) and other key residues are labelled. (c) Crystal structure of fragment hit **10** bound in the novel allosteric pocket of full-length NS3 protein. Binding is driven predominantly through van der Waals contacts. The pink mesh represents the electron density map (Fo–Fc omit map contoured at 3.9σ) for the ligands.

Table 7.1 *In vitro* activity profile of compounds binding at the allosteric binding site. K_D refers to the binding affinity of the ligand to the full-length protein measured by ITC; IC_{50} values reflect the inhibition of protease activity of the full-length and protease domain proteins respectively. LE is defined as the binding affinity in kcal per heavy atom, in this instance the IC_{50} values from the full-length protease assay have been used rather than the K_D values to allow comparison of all of the compounds. EC_{50} is the cell-based replicon activity of the compounds and CC_{50} reflects the cytotoxicity index.

Cmpd	Structure	K_D ITC/μM	Full-length protein IC_{50}/μM	LE	Protease domain IC_{50}/μM	EC_{50}/μM	CC_{50}/μM
10	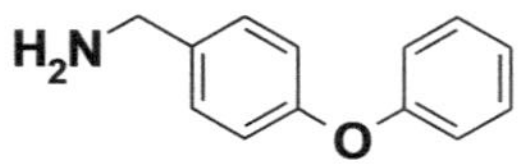	ND	~500	0.30	Inactive	ND	ND
11	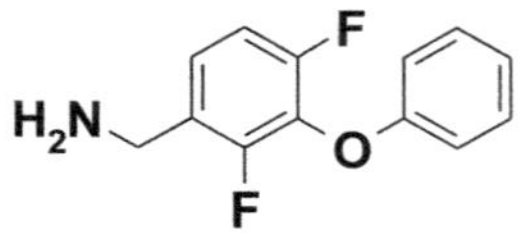	29	20	0.38	~1000	Inactive	40
12	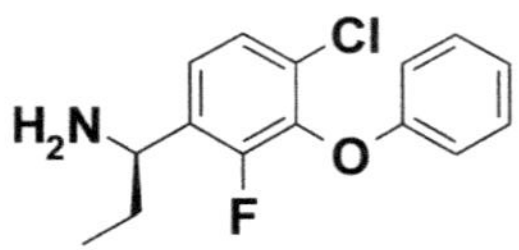	ND	1.3	0.42	Inactive	Inactive	11

13	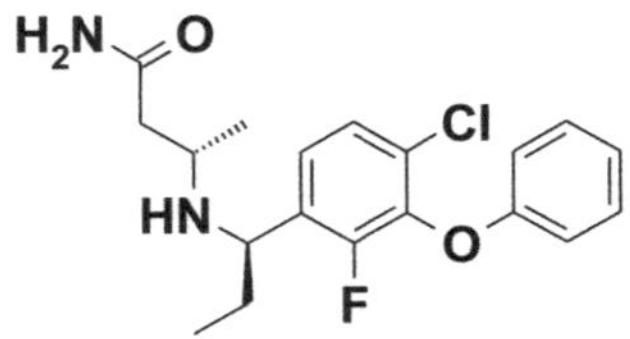	0.062	0.10	0.38	Inactive	0.4	>10
14	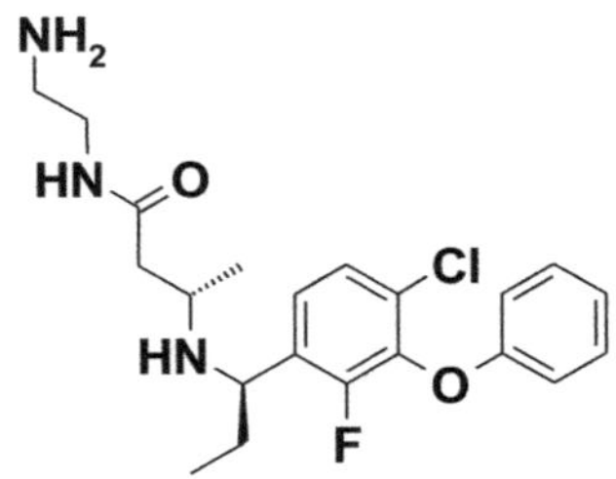	0.022	~0.01	~0.39	Inactive	0.0083	4.5
15	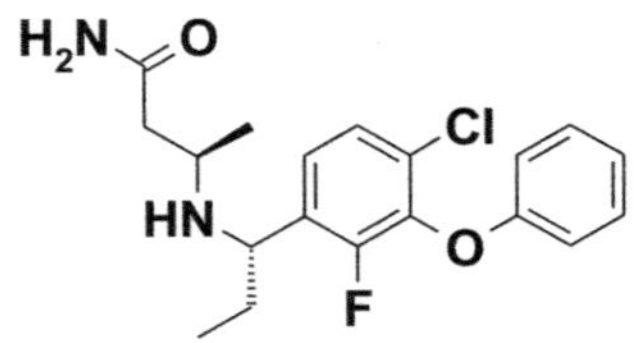	24	~300	<0.25	ND	Inactive	>30

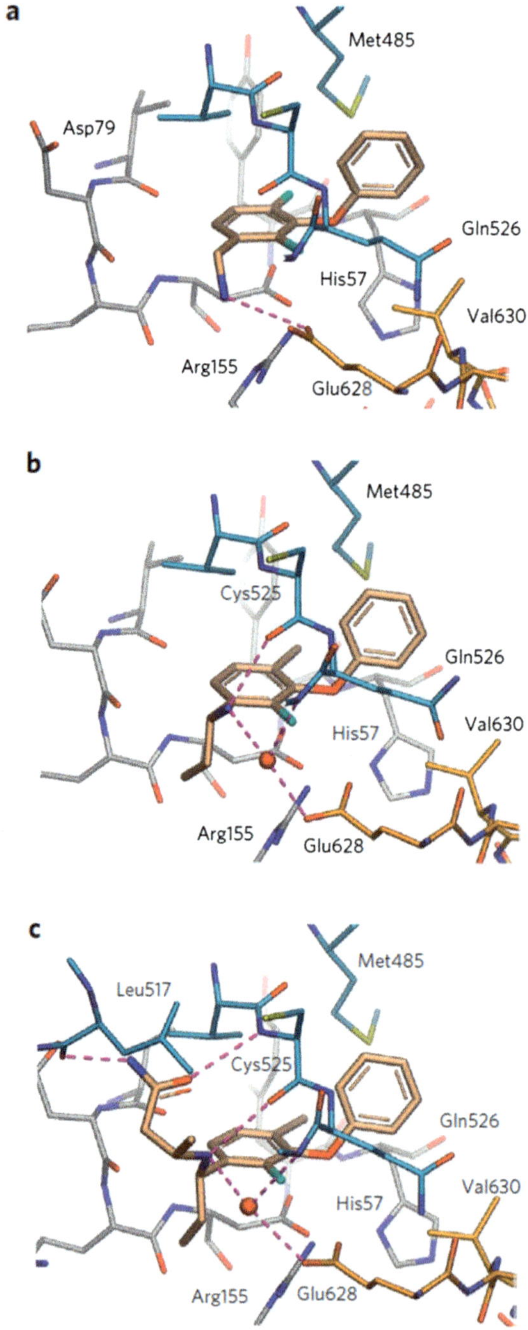
a
Met485
Asp79
Gln526
His57
Val630
Arg155
Glu628
b
Met485
Cys525
Gln526
Val630
His57
Arg155
Glu628
c
Met485
Leu517
Cys525
Gln526
Val630
His57
Arg155
Glu628

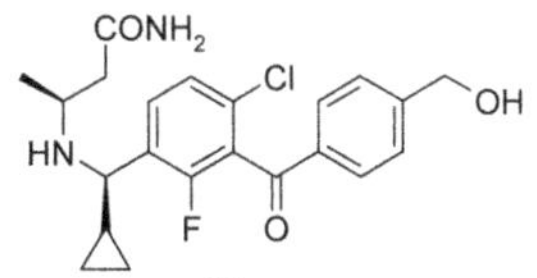

16

Activity		**Rat PK**	
IC_{50}	110 nM	Clb	7.4 mL/min/kg
EC_{50}	4 nM	$T_{1/2}$	1.2 h
clogP	2.0	Vss	0.5 L/kg
		$\%F_{PO}$	59%

Figure 7.9 Fragment-derived pre-clinical candidate **16**.

Overall, this work highlights the sensitivity of X-ray crystallography and its ability to detect low-affinity binding and immediately establish the binding mode. The work also illustrated the difficulties of working with novel pockets where no biological role or endogenous ligand is known. Fragment hits were inadequate to probe the biological relevance and significant effort was invested in optimising the series to a point where meaningful chemical tools were available. Using a chemical lead, identified by fragment screening and structure-based drug design (SBDD), we demonstrated that this site has a regulatory function on the protease activity *via* an allosteric mechanism (Figure 7.10).

7.7 Prevalence of Secondary Site Binders

In 2014, scientists at Astex set out to investigate the frequency of alternative binding sites both within their own database of X-ray crystal structures and within the PDB.[43] This analysis aimed to count the number of sites on a protein that had been shown to support the binding of a small molecule. Ligands

Figure 7.8 Protein–ligand co-crystal structures of compounds **11–13**, which were generated using structure-guided optimisation. (a) Compound **11**. Flanking fluorine substituents were introduced to stabilise the bound conformation of compound **10** and also improve hydrophobic interactions. The aminomethyl group was moved to the *meta*-position allowing the positively charged NH_3 group to interact with the acidic side-chain of Glu628. (b) Compound **12**. The 4-fluoro was replaced with a larger chlorine atom to maximise the van der Waals interactions. An ethyl group was introduced at the benzylic position to restrict the rotation of the aminomethyl side-chain, which now interacts with Glu628 *via* a water molecule and also forms a hydrogen bond to the backbone carbonyl of Cys525. (c) Compound **13**. The X-ray crystal structure of compound **12** was used to design a small set of compounds to probe the pocket formed by Tyr516, Leu517, Val524 and Cys525. This resulted in the identification of compound **13**, which makes key interactions with residues from the protease domain (Arg155 and His57) and the helicase domain (Leu517, Cys525, Gln526 and Met485) including the C-terminus (Glu628 and Val630). Hydrogen bonds are indicated by dashed lines, and conserved water molecules are indicated by red spheres.

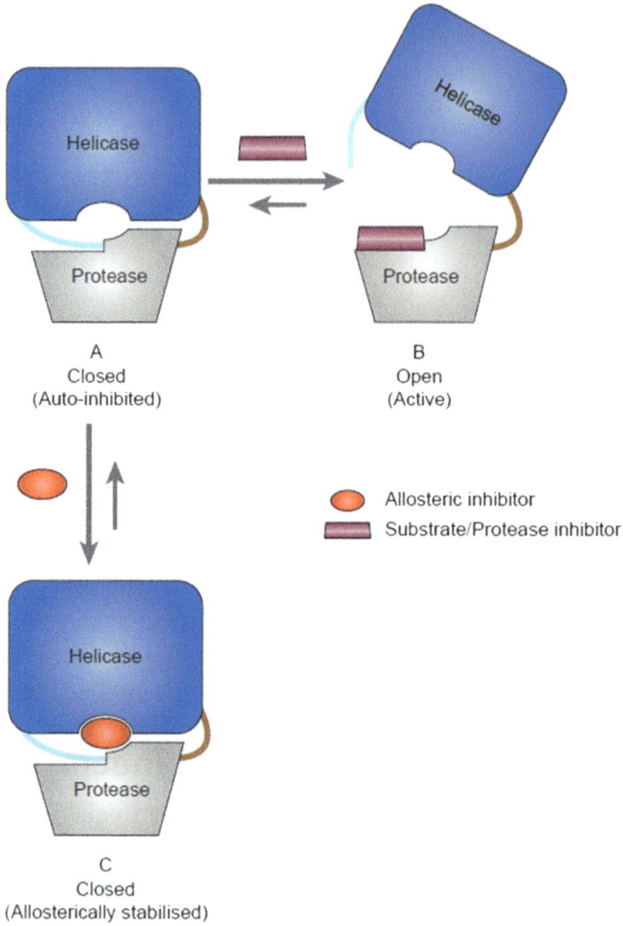

Figure 7.10 (A) Closed conformation: the C-terminus of the helicase domain occupies the protease active site and stabilises the protein in an auto-inhibited conformation. (B) Open conformation: required for proteolytic activity and also reportedly required for helicase activity. This conformation allows substrates to access the protease active site and is inhibited by peptidomimetic active site inhibitors such as Telaprevir and Boceprevir. (C) Closed conformation: compounds binding at the protease–helicase interface stabilise the auto-inhibited conformation of the protein and block catalytic function *via* an allosteric mechanism. Colour key: red oval – allosteric inhibitor; blue line – C-terminus of the helicase domain; brown line – flexible linker between the protease and helicase domains.

binding in interstitial sites, due to crystal packing, were first removed during this process, as were inorganic compounds, metal clusters, solvent molecules and covalently bound compounds.

The study included all X-ray structures of in-house protein targets for which at least 100 different fragments had been screened using X-ray crystallography. This provided 5590 X-ray structures representing 4950 distinct compounds

Table 7.2 Number of ligand-binding sites detected across 25 internal X-ray screening campaigns. In total, 54 sites were discovered, an average of 2.2 ± 1.3 sites per target.

Target	Number of sites observed	Target	Number of sites observed
Metap2	1	FGFR1	1
MELK	2	HspA2	4
bPKA-PKB	5	PKM2	5
PGDS (rat)	4	DNA ligase	3
Bace	4	Xiap	1
CDK2	2	Methyl transferase	2
Urokinase	1	Kinase	1
PARP1 (mouse)	1	PPI	1
HCV NS3	3	Target A	2
iNOSm	2	Target B	2
solAC[a]	1	Target C	2
JAK2	1	Target D	1
HSP90A	2		

[a]Although we did identify an allosteric bicarbonate site for SolAC, the methodology we used here for grouping binding sites only identifies a single site from our fragment screening campaign. This is because the fragments are observed to bind across the nucleotide and bicarbonate pockets. In ambiguous cases such as this we take a conservative approach and report a single site.

and 25 different protein targets. This work found that the majority of liganded protein structures have at least one additional ligand bound at a secondary site, with some having several additional binding sites (Table 7.2).

In one example, CDK2, two copies of the same ligand were seen; one binding to the ATP site and the other binding far away from the active site within the C-terminal lobe. In another case an indole fragment was observed bound to the DNA-binding groove of the helicase domain of HCV NS3 protein. This binding site was independently identified by a second group at Boehringer Ingelheim. Overall, 55 sites were observed across 25 targets, with an average of 2.2 ± 1.3 sites per target. It is possible that other sites exist, but were not observed due to occlusion in the particular crystal system being used. A similar analysis of a much wider range of protein classes within the PDB gave similar results.

7.8 Discussion

Fragment-based screening using a variety of biophysical techniques, such as SPR, NMR and X-ray crystallography, is a proven method for the identification of secondary binding sites on therapeutically important proteins. Of these biophysical techniques, X-ray crystallography is particularly well suited as it provides structural information on binding modes and facilitates structure-based optimisation of hits towards more potent inhibitors. In the case of the HVC NS3 protein, allosteric fragment hits were optimised towards a pre-clinical drug candidate for treatment of drug-resistant HCV infections.

Of course, identifying a secondary binding site is no guarantee of biological relevance. One could argue that there are many small hydrophobic niches within proteins that are capable of binding small molecules, but are of little interest for drug discovery. In the case of HIV-RT discussed in Section 7.4 (Figure 7.2), 3 alternative binding sites were shown to be associated with enzyme inhibitors but it would be very surprising if all the additional 9 pockets identified, which only bound 4-bromopyrazole, were biologically relevant.

7.9 Conclusion

Overall, a review of the literature, and our own experience, suggest that that secondary binding sites are present in more than half of the ligand–protein complexes for which X-ray crystallographic data is available. This brings us back to key Challenge 2; given a large number of putative allosteric sites with ligands bound, how does one prioritise which fragments are worthy of further optimisation. In this regard, X-ray crystallography is a useful technique which can help identify those binding sites which are more likely to be biologically relevant. For example, it can help identify pockets which are associated with an inactive form of the protein and/or which are conserved across a range of isozymes (*e.g.*, HCV NS3). Ultimately, establishing the true allosteric function of a secondary binding site requires more potent tool molecules which can be used in cell-based mode-of-action studies. Overall the X-ray crystallographic fragment-based approach has proven to be a powerful method to identify putative allosteric sites, molecular tools and lead molecules for drug discovery.

References

1. J. A. Hardy and J. A. Wells, *Curr. Opin. Struct. Biol.*, 2004, **14**, 706–715.
2. P. J. Conn, A. Christopoulos and C. W. Lindsley, *Nat. Rev. Drug Discovery*, 2009, **8**, 41–54.
3. R. Nussinov and C. J. Tsai, *Cell*, 2013, **153**, 293–305.
4. J. A. Lewis, E. P. Lebois and C. W. Lindsley, *Curr. Opin. Chem. Biol.*, 2008, **12**, 269–280.
5. D. Wootten, A. Christopoulos and P. M. Sexton, *Nat. Rev. Drug Discovery*, 2013, **12**, 630–644.
6. A. C. Kruse, B. K. Kobilka, D. Gautam, P. M. Sexton, A. Christopoulos and J. Wess, *Nat. Rev. Drug Discovery*, 2014, **13**, 549–560.
7. E. Sigel, *Curr. Top. Med. Chem.*, 2002, **2**, 833–839.
8. P. S. Miller and A. R. Aricescu, *Nature*, 2014, **512**, 270–275.
9. Z. O'Brien and M. Fallah Moghaddam, *Expert Opin. Drug Metab. Toxicol.*, 2013, **9**, 1597–1612.
10. J. J. Liao, *J. Med. Chem.*, 2007, **50**, 409–424.
11. M. A. Fabian, W. H. Biggs III, D. K. Treiber, C. E. Atteridge, M. D. Azimioara, M. G. Benedetti, T. A. Carter, P. Ciceri, P. T. Edeen, M. Floyd, J. M. Ford, M. Galvin, J. L. Gerlach, R. M. Grotzfeld, S. Herrgard, D. E. Insko, M. A. Insko, A. G. Lai, J. M. Lelias, S. A. Mehta, Z. V. Milanov, A. M. Velasco, L. M. Wodicka, H. K. Patel, P. P. Zarrinkar and D. J. Lockhart, *Nat. Biotechnol.*, 2005, **23**, 329–336.

12. H. R. Mellor, A. R. Bell, J. P. Valentin and R. R. Roberts, *Toxicol. Sci.*, 2011, **120**, 14–32.
13. T. C. Yeh, V. Marsh, B. A. Bernat, J. Ballard, H. Colwell, R. J. Evans, J. Parry, D. Smith, B. J. Brandhuber, S. Gross, A. Marlow, B. Hurley, J. Lyssikatos, P. A. Lee, J. D. Winkler, K. Koch and E. Wallace, *Clin. Cancer Res.*, 2007, **13**, 1576–1583.
14. A. K. Salama and K. B. Kim, *Expert Opin. Pharmacother.*, 2013, **14**, 619–627.
15. H. Zhao, *Drug Discovery Today*, 2011, **16**, 158–163.
16. W. T. Choi and J. An, *Exp. Biol. Med.*, 2011, **236**, 637–647.
17. J. Garcia-Perez, P. Rueda, J. Alcami, D. Rognan, F. Arenzana-Seisdedos, B. Lagane and E. Kellenberger, *J. Biol. Chem.*, 2011, **286**, 33409–33421.
18. J. P. Rocher, B. Bonnet, C. Bolea, R. Lutjens, E. Le Poul, S. Poli, M. Epping-Jordan, A. S. Bessis, B. Ludwig and V. Mutel, *Curr. Top. Med. Chem.*, 2011, **11**, 680–695.
19. D. Bixby and M. Talpaz, *Leukemia*, 2011, **25**, 7–22.
20. L. Shang, K. Lin and Z. Yin, *Curr. Pharm. Des.*, 2014, **20**, 694–703.
21. J. Zhang, F. J. Adrian, W. Jahnke, S. W. Cowan-Jacob, A. G. Li, R. E. Iacob, T. Sim, J. Powers, C. Dierks, F. Sun, G. R. Guo, Q. Ding, B. Okram, Y. Choi, A. Wojciechowski, X. Deng, G. Liu, G. Fendrich, A. Strauss, N. Vajpai, S. Grzesiek, T. Tuntland, Y. Liu, B. Bursulaya, M. Azam, P. W. Manley, J. R. Engen, G. Q. Daley, M. Warmuth and N. S. Gray, *Nature*, 2010, **463**, 501–506.
22. W. Jahnke, R. M. Grotzfeld, X. Pelle, A. Strauss, G. Fendrich, S. W. Cowan-Jacob, S. Cotesta, D. Fabbro, P. Furet, J. Mestan and A. L. Marzinzik, *J. Am. Chem. Soc.*, 2010, **132**, 7043–7048.
23. S. M. Saalau-Bethell, A. J. Woodhead, G. Chessari, M. G. Carr, J. Coyle, B. Graham, S. D. Hiscock, C. W. Murray, P. Pathuri, S. J. Rich, C. J. Richardson, P. A. Williams and H. Jhoti, *Nat. Chem. Biol.*, 2012, **8**, 920–925.
24. B. Chaneton, P. Hillmann, L. Zheng, A. C. Martin, O. D. Maddocks, A. Chokkathukalam, J. E. Coyle, A. Jankevics, F. P. Holding, K. H. Vousden, C. Frezza, M. O'Reilly and E. Gottlieb, *Nature*, 2012, **491**, 458–462.
25. J. M. Bowler, K. L. Hervert, M. L. Kearley and B. G. Miller, *ACS Med. Chem. Lett.*, 2013, **4**, 580–584.
26. C. W. Murray and D. C. Rees, *Nat. Chem.*, 2009, **1**, 187–192.
27. C. W. Murray, M. L. Verdonk and D. C. Rees, *Trends Pharmacol. Sci.*, 2012, **33**, 224–232.
28. M. M. Hann, A. R. Leach and G. Harper, *J. Chem. Inf. Model.*, 2001, **41**, 856–864.
29. H. Jhoti, G. Williams, D. C. Rees and C. W. Murray, *Nat. Rev. Drug Discovery*, 2013, **12**, 644–645.
30. W. I. Wu, W. C. Voegtli, H. L. Sturgis, F. P. Dizon, G. P. Vigers and B. J. Brandhuber, *PLoS One*, 2010, **5**, e12913.
31. A. M. Giannetti, *Methods Enzymol.*, 2011, **493**, 169–218.
32. M. Geitmann, M. Elinder, C. Seeger, P. Brandt, I. J. de Esch and U. H. Danielson, *J. Med. Chem.*, 2011, **54**, 699–708.
33. J. D. Bauman, D. Patel, C. Dharia, M. W. Fromer, S. Ahmed, Y. Frenkel, R. S. Vijayan, J. T. Eck, W. C. Ho, K. Das, A. J. Shatkin and E. Arnold, *J. Med. Chem.*, 2013, **56**, 2738–2746.

34. C. W. Murray and T. L. Blundell, *Curr. Opin. Struct. Biol.*, 2010, **20**, 497–507.
35. W. Yang, Y. Zheng, Y. Xia, H. Ji, X. Chen, F. Guo, C. A. Lyssiotis, K. Aldape, L. C. Cantley and Z. Lu, *Nat. Cell Biol.*, 2012, **14**, 1295–1304.
36. S. M. Saalau-Bethell, V. Berdini, A. Cleasby, M. Congreve, J. E. Coyle, V. Lock, C. W. Murray, M. A. O'Brien, S. J. Rich, T. Sambrook, M. Vinkovic, J. R. Yon and H. Jhoti, *ChemMedChem*, 2014, **9**, 823–832.
37. S. M. Saalau-Bethell, A. Cleasby, T. A. Sambrook, J. Coyle and M. Vinkovic, WO 2007/010285, 2007.
38. W. Jahnke, J. M. Rondeau, S. Cotesta, A. Marzinzik, X. Pelle, M. Geiser, A. Strauss, M. Gotte, F. Bitsch, R. Hemmig, C. Henry, S. Lehmann, J. F. Glickman, T. P. Roddy, S. J. Stout and J. R. Green, *Nat. Chem. Biol.*, 2010, **6**, 660–666.
39. R. K. Beran and A. M. Pyle, *J. Biol. Chem.*, 2008, **283**, 29929–29937.
40. M. P. Manns and T. von Hahn, *Nat. Rev. Drug Discovery*, 2013, **12**, 595–610.
41. M. A. Ashwell, J. M. Lapierre, C. Brassard, K. Bresciano, C. Bull, S. Cornell-Kennon, S. Eathiraj, D. S. France, T. Hall, J. Hill, E. Kelleher, S. Khanapurkar, D. Kizer, S. Koerner, J. Link, Y. Liu, S. Makhija, M. Moussa, N. Namdev, K. Nguyen, R. Nicewonger, R. Palma, J. Szwaya, M. Tandon, U. Uppalapati, D. Vensel, L. P. Volak, E. Volckova, N. Westlund, H. Wu, R. Y. Yang and T. C. Chan, *J. Med. Chem.*, 2012, **55**, 5291–5310.
42. D. Norton, mansucript in preparation.
43. F. Ludlow, mansucript in preparation.

CHAPTER 8

Fragment-Based Discovery of Antibacterials

CHRISTINA SPRY*[a,b] AND ANTHONY G. COYNE[a]

[a]Department of Chemistry, University of Cambridge, Cambridge, CB2 1EW, UK; [b]Research School of Biology, The Australian National University, Canberra, ACT 2601, Australia
*E-mail: christina.spry@anu.edu.au

8.1 Introduction

The increasing prevalence of drug-resistant bacteria Worldwide is a serious threat to our capacity to treat and control infectious diseases.[1] There is presently a shortage of novel classes of antibiotics in the pipeline and many large pharmaceutical companies have disengaged from antibiotic research (in part a result of the lack of financial incentives and increasing regulatory obstacles[2]). As a consequence, real concerns that we could return to a situation analogous to the pre-antibiotics era have emerged.[3,4] The discovery and development of new antibacterial agents with novel modes of action (to avoid susceptibility to pre-existing resistance mechanisms) is crucial. However, the development of a truly novel antibacterial agent presents a number of formidable scientific and technical challenges and is associated with a high risk of failure.[5]

Historically, Nature has been a valuable source of antibacterial agents, and indeed natural products (predominantly microbial metabolites) and natural product derivatives remain the mainstay of antibacterial chemotherapy today. Following the golden years of antibiotic research (1940–1960s), in

RSC Drug Discovery Series No. 47
Fragment-Based Drug Discovery
Edited by Steven Howard and Chris Abell

Published by the Royal Society of Chemistry, www.rsc.org

which the majority of antibiotic scaffolds currently on the market were discovered, the rate of discovery of novel antibacterial agents slowed dramatically. Interest in antibiotic research plummeted and correspondingly three decades passed in which no new classes of broad-spectrum antibiotics (with novel modes of action) were introduced to the market.[6,7]

The sequencing of the first bacterial genomes in the mid-1990s renewed an interest in antibacterial drug discovery as it enabled target-based approaches. Genome comparisons facilitated informed selection of potential broad-spectrum antibacterial targets absent or poorly conserved in humans, and essentiality in a given bacterium could be investigated using genetic tools. Many target-based high-throughput screening (HTS) campaigns were carried out, however, unfortunately, these largely failed to deliver candidates for antibacterial development; unexpectedly low hit rates – lower than those observed against targets in other therapeutic areas – and difficulties with optimising hits into cell-active leads were encountered.[5,8] Efforts turned to the alternate strategy of whole-cell HTS, as this offered the advantage that: (i) compounds with antibacterial activity (and hence the capacity to permeate the bacterial cell and evade efflux mechanisms) were selected for from the onset; and (ii) targets were assayed in the context of the cell. However, whole-cell HTS, which is inherently less sensitive than target-based HTS, requires effective means by which to: (i) distinguish between true hits and non-specific inhibitors; and (ii) identify the molecular target – two non-trivial tasks, and, as for the target-based HTS campaigns, the output of lead compounds from whole-cell HTS has so far been lower than anticipated.[5,9] The limited success of HTS in antibacterial drug-discovery projects to date has, to a large extent, been attributed to HTS collections lacking chemical diversity and being deficient in *antibiotic-like* compounds (marketed antibiotics are typically more hydrophilic and somewhat larger than the rule of five drugs active in other disease areas[10]), and highlights the need for alternative lead discovery mechanisms.

Fragment-based drug discovery (FBDD), an approach that combines an element of random screening with rational structure-based design,[11] offers key advantages over traditional HTS. In FBDD, inhibitors are developed from fragments (molecules of low molecular weight and complexity) that weakly, but efficiently, bind to the target protein. Fragments that bind the target are identified through the screening of a fragment library, typically using a combination of biophysical techniques to detect the inherently weak binding. Thereafter fragment *hits* are elaborated into high-affinity inhibitors with drug-like properties by an iterative process involving structure-based design, synthesis and binding affinity measurements. A key advantage of the fragment-based approach is that chemical space can be sampled much more comprehensively with fragments than with molecules of the size typically found in HTS libraries.[12] Additionally, by comparison with HTS, hit rates are higher, and although hits bind with lower affinity, they form higher quality, *ligand efficient* interactions with the target (a greater proportion of the atoms are involved in binding and fewer superfluous moieties are present that prevent the optimal binding modes of functional groups being achieved).[12]

Moreover the path from hits to high-affinity binders with the required physicochemical properties is typically clearer.

Now established in industry and academia, FBDD has proven to be a powerful and versatile approach for generating leads, even against targets previously deemed chemically intractable. Evidence of the success is the many compounds originating from FBDD projects that have entered clinical trials, including one that has now reached the market.[13] Although as yet this list does not include any candidate antibacterial drugs, there is an increasing number of examples of fragment-based approaches being applied to bacterial targets to be found in the literature, and in this chapter we highlight some of these. Enzymes predominate the list of bacterial targets being pursued with fragment-based approaches. However, in line with the more recent extension of fragment-based approaches to targeting protein–protein interactions (PPIs) in other therapeutic areas,[14–17] fragment-based approaches are also being applied to targeting bacterial PPIs and we present one example. The enzyme targets include several nucleotide-binding enzymes (*e.g.*, biotin carboxylase, DNA gyrase, DNA ligase and pantothenate synthetase), likely a reflection of the success with which kinases and other enzymes with nucleotide-binding pockets have been exploited with fragment-derived anticancer agents.[13]

In the studies highlighted here, fragment-based approaches have been employed (alone or in combination with more traditional approaches) to:

(i) identify novel leads against tractable targets – in some cases to identify new series that overcome the deficits of existing series (such as sensitivity to resistance mechanisms) or simply to increase and diversify the chemical matter in the armamentarium available against a given target;
(ii) tackle targets belonging to target classes for which traditional lead-finding mechanisms have been ineffective at yielding leads; or
(iii) identify binding hotspots and guide the optimisation of leads identified by other means.

8.2 Application of Fragment-Based Approaches to the Discovery of Broad-Spectrum Antibacterial Agents

8.2.1 Targeting Biotin Carboxylase

Biotin carboxylase (BC) forms part of the multifunctional enzyme complex acetyl-CoA carboxylase (ACCase) that catalyses the first committed step in fatty acid biosynthesis. BC catalyses the ATP-dependent carboxylation of a biotin molecule covalently attached to a second enzyme in the ACCase complex (BC carrier protein), and a third enzyme in the complex (carboxyl transferase) thereafter catalyses the transfer of the carboxyl group to acetyl-CoA to yield malonyl-CoA.[18] ACCase is essential for the growth of a wide range of bacterial pathogens and, like the other essential enzymes in the bacterial type-II fatty acid biosynthesis pathway, is widely considered a good target for antibacterial drug discovery.[19,20]

In a whole-cell screen of ~1.6 million compounds, Miller and colleagues[21] at Pfizer identified a series of pyridopyrimidines with antibacterial activity against a membrane-compromised, efflux pump-deficient strain of *Escherichia coli*. Antibacterial activity was demonstrated also against other Gram-negative bacteria, including clinical isolates of *Haemophilus influenzae* and *Moraxella catarrhalis*. The target of the pyridopyrimidines was subsequently determined to be the ATP-binding site of BC, and hence the activity of the pyridopyrimidines chemically validates this target. The lead pyridopyrimidine (**1**; Figure 8.1A) was also shown to have antibacterial activity *in vivo* (in mouse models of tissue-localised and systemic *H. influenzae* infection).

In an effort to identify additional inhibitors of BC, Mochalkin *et al.*[22] at Pfizer also used a fragment-based approach. Two fragment screens were performed in parallel. In one screen, a library of 5200 fragments was tested in pools of 10 compounds using an enzyme-coupled ACCase assay. Inhibitory compound pools were subsequently tested using ligand-observed NMR (saturation transfer difference NMR, known as STD NMR) to identify individual fragments binding to *E. coli* BC (*Ec*BC). This yielded 142 fragment hits, six of which inhibited ACCase activity with IC_{50} values below 95 μM. The second screen began with a virtual screen (guided by 3D shape and electrostatic similarity) of a library of 2.2 million compounds against *Ec*BC. This screen made use of the X-ray crystal structure of *Ec*BC in complex with pyridopyrimidine **1**. A prioritised set of 525 compounds was thereafter tested in the enzyme-coupled ACCase assay; IC_{50} values less than 10 μM were measured for 48 (9%) of these. Selected hits from the two fragment screens are shown in Figure 8.1A (compounds **2–6**); those shown (or close analogues of those shown) were, like pyridopyrimidine **1**, found by X-ray crystallography to bind to the ATP-binding site of *Ec*BC (Figure 8.1B). Also like pyridopyrimidine **1**, most of these fragments form hydrogen bonds with four active-site residues (Glu201, Lys202, Leu204 and Lys159) that form hydrogen bonds with the natural substrate (ATP).

With a number of different fragment hits to serve as starting points for elaboration strategies, there was the possibility for the development of a number of different series. The two elaboration methods chosen were merging and growing; fragment linking was not appropriate as the fragment hits were all observed to bind to the adenine-binding region of the ATP-binding site (Figure 8.1B). Docking, inhibitory activity in the enzyme-coupled ACCase assay, and co-crystal structures were used to guide fragment elaboration. Upon binding of pyridopyrimidine **1**, the dibromobenzene motif is accommodated in a new, induced, hydrophobic pocket in the ribose-binding region (Figure 8.1B). This provided an opportunity to elaborate fragment hits with motifs that could access this hydrophobic pocket in a similar fashion.

In one elaboration strategy fragment **2** (IC_{50} = 59 μM, ligand efficiency (LE)[23] = 0.44) was merged with pyridopyrimidine **1** (Figure 8.2A). The thiazole ring was replaced with an imidazole; next a 2,6-dichlorobenzyl substituent was introduced to mimic the 2,6-dibromobenzene group present in

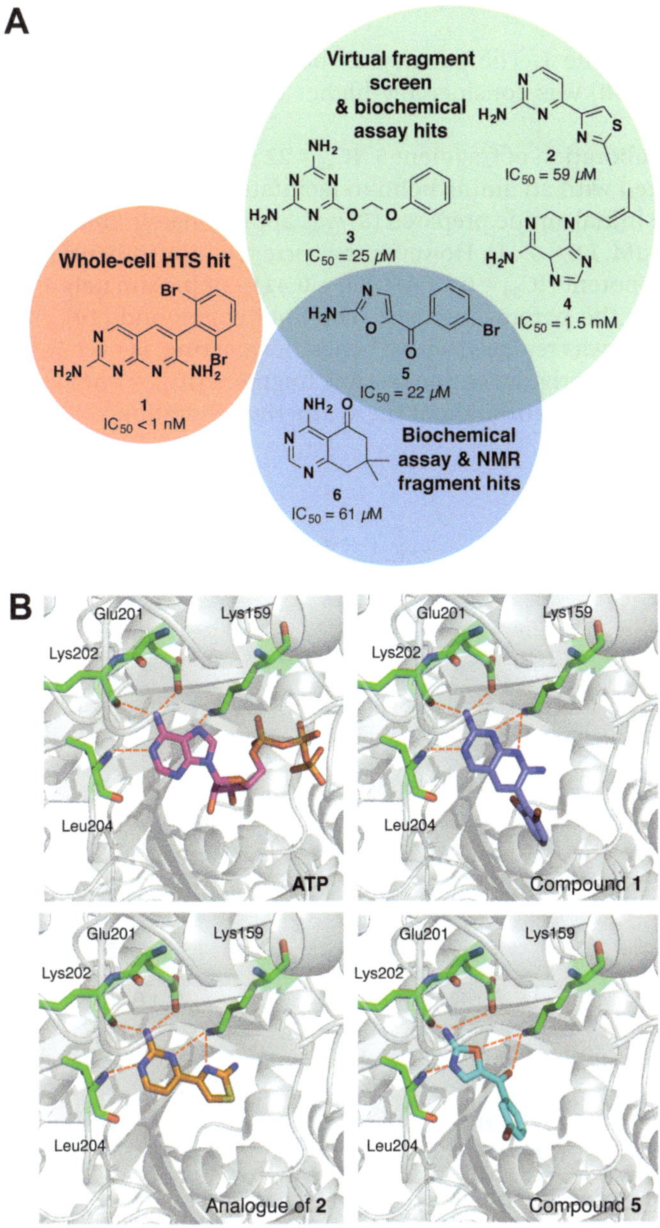

Figure 8.1 Biotin carboxylase (BC) fragment hits.[22] (A) Fragment hits identified in screens carried out against *Ec*BC. Hits were identified in a virtual fragment screen (and confirmed using a biochemical assay) and/or in a more conventional fragment screen involving both a biochemical assay and ligand-observed NMR. A bacterial BC inhibitor identified by whole-cell HTS[21] is shown for comparison (compound **1**). IC_{50} values measured in an *E. coli* ACCase enzyme-coupled assay are shown. (B) The binding poses of ATP, HTS hit **1** and two fragments, as determined from X-ray crystal structures [PDB:1DV2 (ATP), 2V58 (compound **1**), 2W70 (compound **2** analogue) and 2W6M (compound **5**)]. Key residues are shown in green, and dashed red lines represent hydrogen bonds.

pyridopyrimidine **1**. The potency of the merged compound (**7**; IC_{50} = 0.33 μM, LE = 0.40) was considerably higher than that of the original thiazole fragment.

In the elaboration of fragment **5** (IC_{50} = 22 μM, LE = 0.40) the ketone linker was replaced with an amide bond to facilitate subsequent synthesis (Figure 8.2B). The initial amide prepared (**8**) was approximately 20-fold more potent (IC_{50} = 1.3 μM, LE = 0.41). However, the corresponding dibenzylamide (**9**) was even more potent (IC_{50} = 0.13 μM, LE = 0.41) – approximately 200-fold more potent than the parent fragment. The final compound (**10**), in which one phenyl group was replaced with a benzodioxane ring, is over 3000-fold more potent than the original amino-oxazole fragment (IC_{50} = 7 nM *vs.* 22 μM). The LE was maintained throughout the synthetic elaboration.

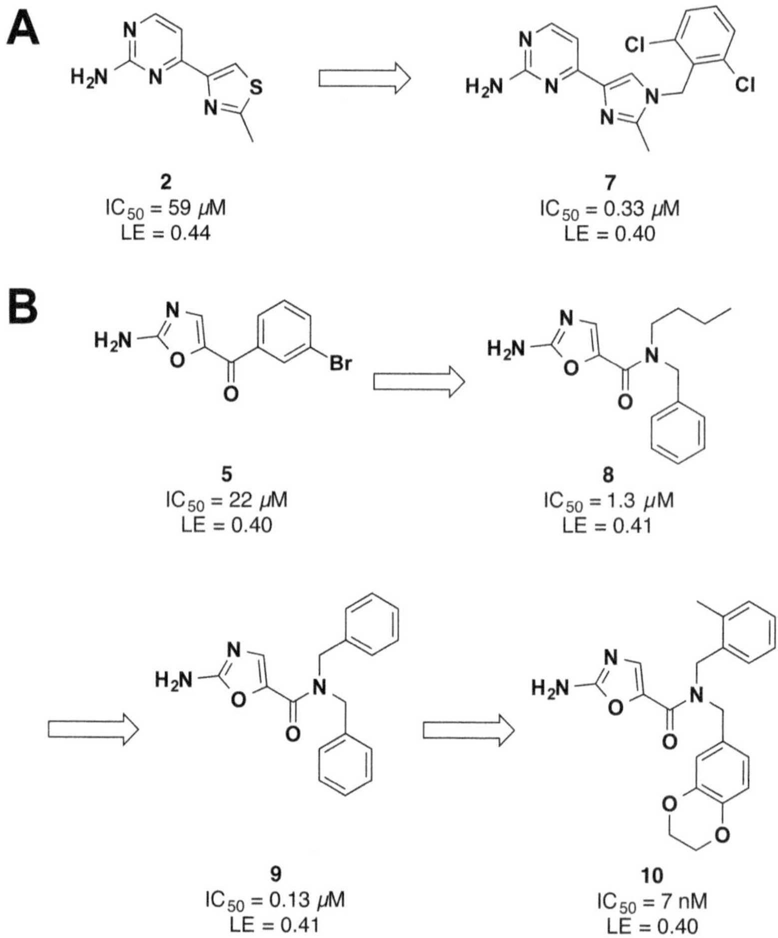

Figure 8.2 Elaboration of biotin carboxylase (BC) fragment hits.[22] Fragments **2** (A) and **5** (B) were elaborated using knowledge of the binding poses of the fragments and HTS hit **1**. IC_{50} values measured in an *E. coli* ACCase enzyme-coupled assay, and the corresponding LEs, are shown.

Both dibenzylamide **9** and benzodioxane **10** demonstrated activity against Gram-negative bacteria (a clinical strain of *M. catarrhalis* and efflux-pump-deficient strains of *E. coli* and *H. influenzae*), and the results of mode-of-action studies were consistent with the antibacterial activity resulting from inhibition of BC. Furthermore, despite structural similarity between bacterial BCs and human protein kinases, dibenzylamide **9** showed greater than 80-fold selectivity when tested against a panel of 40 different human protein kinases.

8.2.2 Targeting DNA Gyrase

DNA gyrase is a type-II topoisomerase that controls the topology of DNA. The enzyme introduces negative supercoils using energy from ATP hydrolysis. It is a tetrameric enzyme composed of two different subunits: DNA gyrase A (GyrA), which catalyses DNA cleavage and religation, and DNA gyrase B (GyrB), which catalyses ATP hydrolysis.[24] DNA gyrase is found only in prokaryotes, where its role in DNA replication is essential for viability. For these reasons it is widely regarded as an attractive antibacterial target.[25] Two classes of antibacterials that target DNA gyrase have been used clinically: aminocoumarins and quinolones. Classical aminocoumarins (*e.g.*, Novobiocin; Figure 8.3A) are *Streptomyces* metabolites that inhibit binding of ATP to GyrB. Although potent inhibitors of DNA gyrase, these compounds are relatively ineffective against Gram-negative bacteria, exhibit toxicity, and resistance towards them developed rapidly.[25,26] Quinolones (*e.g.*, Levofloxacin; Figure 8.3A), which inhibit DNA cleavage and religation, have enjoyed clinical success as broad-spectrum antibiotics.[25] However, the development of resistance is also compromising these DNA gyrase inhibitors[27] and novel inhibitors of this validated target, that are not sensitive to pre-existing resistance, are sought.

A number of independent studies describing the application of fragment-based approaches to the identification of DNA gyrase inhibitors can be found in the literature. In the three earliest studies, potential fragment hits were identified in *in silico* screens performed using the structure of an N-terminal fragment of *E. coli* GyrB (*Ec*GyrB) containing the ATP-binding domain.[26,28,29] The capacity to partake in key interactions with the ATP-binding site – *e.g.*, hydrogen bonds (with Asp73 and a conserved water molecule) that are known to be involved in the binding of the GyrB inhibitors Novobiocin and Cyclothialidine – was built into the screening constraints. A variety of techniques, including 2D ^{1}H–^{15}N heteronuclear single quantum coherence (HSQC) NMR and a DNA supercoiling assay, was used to verify experimentally which of the prioritised fragments bind GyrB.[26,29]

Following their screening efforts, Boehringer and colleagues[26] focussed on elaborating indazole (**11**; Figure 8.3B), a fragment observed to bind *Ec*GyrB with a dissociation constant (K_D value) of 10 mM. Guided by preliminary structure–activity relationships (SARs) gathered during screening, and knowledge of the active site structure and interactions important for binding of known inhibitors, the fragment was successfully grown at the C3 and C4 positions into a GyrB inhibitor (**12**) with >8000-fold improved potency against *E. coli* DNA gyrase (Figure 8.3B). In another study, Solmajer and colleagues[29] used a

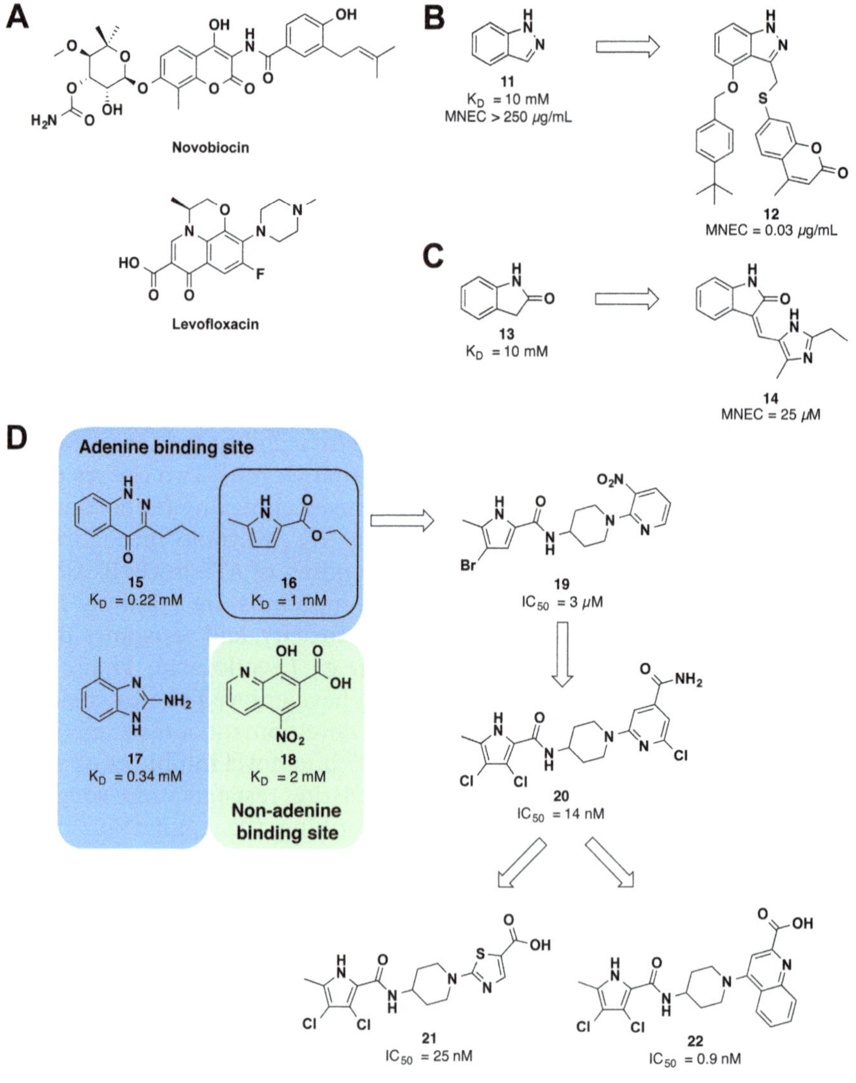

Figure 8.3 Fragment-based approaches to targeting DNA gyrase.[26,29,30] (A) The structures of two antibacterial DNA gyrase inhibitors used clinically: Novobiocin and Levofloxacin. (B)–(D) Fragment hits identified in independent fragment screens against DNA gyrase and the corresponding elaborated compounds. In (D), where multiple fragment hits are shown, an outline has been drawn around the fragment hit prioritised for elaboration. K_D values determined using NMR, maximal non-effective concentrations (MNECs) determined in an *E. coli* DNA gyrase supercoiling assay, and/or IC_{50} values determined in an *E. coli* DNA gyrase ATPase assay are shown.

linking strategy to elaborate one of the fragment hits identified in their *in silico* screen (indolin-2-one; **13**; K_D = 10 mM; Figure 8.3C). Additional *in silico* screens were performed to identify fragments predicted to bind in the ATP-binding site in close vicinity to the fragment, and the corresponding 'linked' indolin-2-one derivatives were tested. Subsequent SAR investigations yielded a derivative (**14**; in which the indolin-2-one moiety is linked *via* an ethylidene moiety to a substituted imidazole ring) that inhibits the DNA supercoiling activity of *E. coli* DNA gyrase at concentrations above 25 μM (Figure 8.3C).

In a subsequent study by Eakin and colleagues[30] at AstraZeneca, 1000 fragments, some of which are fragments of known GyrB inhibitors, were screened against the N-terminal ATP-binding domain of *Ec*GyrB using ^1H–^{15}N HSQC NMR. Fragments were initially screened in pools of ten and mixtures perturbing the chemical shifts of active-site residues were de-convoluted to identify individual hits. K_D values were determined from NMR titration experiments. The highest affinity hits caused chemical shift perturbations consistent with binding to the adenine-binding site of GyrB and displayed a common hydrogen-bond donor/acceptor motif also found in the adenine ring of ATP. Three of these fragment hits, with K_D values in the range of 0.2–1.0 mM, are shown in Figure 8.3D (compounds **15–17**). Interestingly, one of the most potent fragment hits (**15**) was a close analogue of a fragment hit identified in the *in silico*-based screen carried out by Solmajer and colleagues.[29]

Subsequently, a second NMR-based fragment screen carried out with the adenine pocket of GyrB blocked, yielded an additional fragment hit – a quinoline with a K_D value of 2 mM (**18**; Figure 8.3D) – that binds in a distal region of the ATP-binding pocket. This fragment hit highlighted the possibility of growing the higher affinity adenine-binding site fragments into a new region of the ATP-binding pocket.

Of the adenine-binding site fragment hits, compound **16**, a pyrrole ester (a group also present in the structure of the GyrB inhibitor Clorobiocin) was selected for elaboration. From docking experiments using the *Staphylococcus aureus* GyrB (*Sa*GyrB), this fragment was predicted to bind in a position partially overlapping with the adenine ring of ATP. A library of pyrrole-containing compounds was designed and prepared by coupling commercially available pyrrole carboxylic acids with a range of commercially available amines in a single synthetic step using standard peptide chemistry. Among those synthesised, the most potent pyrrolamide identified was compound **19**, a nitropyridine-bearing pyrrolamide that inhibits the ATPase activity of *E. coli* DNA gyrase with an IC_{50} value of 3 μM (as assessed using a malachite-green-based enzyme assay, Figure 8.3D). A crystal structure of this pyrrolamide in complex with the N-terminal ATP-binding domain of *Sa*GyrB revealed that the binding pose of the pyrrole was similar to that predicted in the docking experiments. The pyrrole group is positioned in the adenine-binding site and the pyrrole nitrogen and carbonyl substituent form hydrogen bonds with a conserved aspartic acid (Asp81) and water molecule, respectively. The pyridine moiety was observed to extend beyond the ATP-binding site; binding in a region similar to that occupied by the quinoline fragment hit **18**.

Although pyrrolamide **19** lacked antibacterial activity against the panel of bacteria tested, subsequent optimisation guided by X-ray crystal structures yielded compounds with improved affinity and antibacterial activity. Initially the nitro substituent on the pyridine ring, which was observed to orient towards solvent, was removed and amide and chloro substituents were introduced. At the same time chloro substituents were introduced to the pyrrole ring to pick up hydrophobic interactions in the adenine-binding pocket and reduce the pK_a of the pyrrole NH group, and thereby improve its hydrogen-bond-donating capacity. These modifications yielded a pyrrolamide with >200-fold improved potency (IC_{50} = 14 nM; compound **20**, Figure 8.3D) and antibacterial activity against Gram-positive bacteria (*S. aureus*, *Streptococcus pneumoniae* and *Enterococcus faecium*) and, to a lesser extent, against Gram-negative bacteria (*H. influenzae* and efflux-defective *E. coli*). Additional analogues of this pyrrolamide were synthesised; replacement of the pyridine moiety with a thiazole ring bearing a carboxylic acid substituent yielded a more soluble compound with a similar effect on enzyme activity (IC_{50} = 25 nM; compound **21**, Figure 8.3D) and bacterial growth, while replacement of the pyridine moiety with a quinoline bearing a carboxylic acid substituent yielded a compound that inhibits *E. coli* DNA gyrase with improved activity (IC_{50} = 0.9 nM; compound **22**, Figure 8.3D) and possesses enhanced antibacterial activity against Gram-positive bacteria.

The mode of action, selectivity, and *in vivo* efficacy of a representative of the pyrrolamide series (compound **21**) was also assessed.[30] Consistent with GyrB being the primary target of the pyrrolamides, laboratory generated *S. aureus* mutants showing resistance to the pyrrolamide contained single-point mutations in the gene encoding GyrB that would result in amino acid changes. The compound was found to be bactericidal, and to be 100-fold less active in inhibiting growth of mammalian and fungal cell lines than in inhibiting growth of *S. pneumoniae*. Importantly, this pyrrolamide also demonstrated efficacy in an immunocompetent-mouse pneumonia model with *S. pneumoniae*, albeit at relatively high oral doses.

8.2.3 Targeting DNA Ligase

DNA ligase (LigA) is a NAD^+-dependent enzyme that plays a fundamental role in DNA replication and repair.[31] The enzyme catalyses the formation of a phosphodiester bond between the 3′-hydroxyl of one nucleotide and the 5′-phosphate of another through a three-step mechanism that involves the formation of an adenylated intermediate. LigA is essential for bacterial growth, conserved among bacterial species, and is distinct from the ATP-dependent DNA ligases found in eukaryotes; for these reasons, compounds targeting DNA ligase hold promise as selective, broad-spectrum antibiotics.[31] In the last few years a number of small-molecule inhibitors of LigA have been identified, mostly from high-throughput screens.[31] As an alternative to HTS, Howard and colleagues[32] used a fragment-based approach to identify novel inhibitors of bacterial LigA.

S. aureus LigA (*Sa*LigA) was the bacterial LigA of choice in the fragment-based approach as it was possible to obtain a stable, de-adenylated, soakable form of this LigA that was suitable for screening by X-ray crystallography. Approximately 1500 fragments from an in-house library were screened using a combination of techniques: X-ray crystallography, ligand-observed NMR (water-ligand observed *via* gradient spectroscopy NMR, known as Water-LOGSY NMR), a fluorescence-based thermal shift (FTS) assay, and isothermal titration calorimetry (ITC) to measure the affinity of hits (Figure 8.4A). The result of the screen was 52 structurally validated hits. The K_D values of the hits ranged from 30 μM to 5.5 mM. One fragment hit, a chloro-substituted triazolopyrazine found to bind *Sa*LigA with a K_D value of 38 μM (**23**; Figure 8.4B), was selected for elaboration, in part due to its high LE (0.50). The fragment was observed by X-ray crystallography to bind in the AMP-binding site of the enzyme. Inhibitory activity in a time-resolved fluorescence resonance energy transfer (TR-FRET) DNA ligase assay was used to guide the elaboration process (summarised in Figure 8.4B).

Initially the pyrazine ring was replaced with a pyridine ring to facilitate subsequent growth at the pyridyl 4-position; the pyridine analogue was observed to bind with slightly improved affinity and LE (K_D = 10 μM, IC_{50} = 17 μM, LE = 0.54). Various replacements for the chloro substituent were explored due to the potential for reactivity; an analogue with a trifluoromethyl substituent (**24**) demonstrated the highest activity (IC_{50} = 16 μM, LE = 0.44). In an attempt to form a direct hydrogen bond between the ligand and the protein backbone, the 1,2,4-triazole moiety was replaced with a larger 6-azaindazole. This resulted in a 70-fold increase in potency; an IC_{50} value of 0.22 μM was measured for the elaborated compound (**25**; LE = 0.48). An ethanolamine substituent was introduced at the pyridyl 4-position in order to increase solubility and enable an X-ray crystal structure to be obtained. The structure of the resultant compound (**26**) bound to LigA confirmed that the nitrogen at the 6-position of the 6-azaindazole ring forms a hydrogen bond with the protein backbone as intended. It was hypothesised that the *Sa*LigA-bound conformation of the elaborated compound was not favourable because of steric and electrostatic clashes between the pyridine and azaindole rings, and in an attempt to stabilise the bound form, the pyridine ring was replaced with a pyrimidine. This gave rise to a compound with a K_D value of 25 nM and LE of 0.45 (**27**; Figure 8.4B).

This final elaborated compound showed antibacterial activity against a range of Gram-positive bacteria (*S. aureus*, *S. pneumoniae* and *Streptococcus pyogenes*; minimal inhibitory concentration (MIC) = 16–32 mg ml^{-1}) but typically lacked activity against wild-type Gram-negative pathogens. Antibacterial activity was observed against efflux mutant strains of Gram-negative pathogens such as *E. coli*, *Pseudomonas aeruginosa* and *Acinetobacter baumannii*, consistent with the lack of activity against wild-type Gram-negative pathogens resulting from efflux from the cell. A strain of *S. aureus* engineered to over-express LigA, and *E. coli* strains with mutations conferring resistance to a previously identified LigA inhibitor, were relatively less sensitive to the elaborated compound (as compared with the corresponding parent strains).

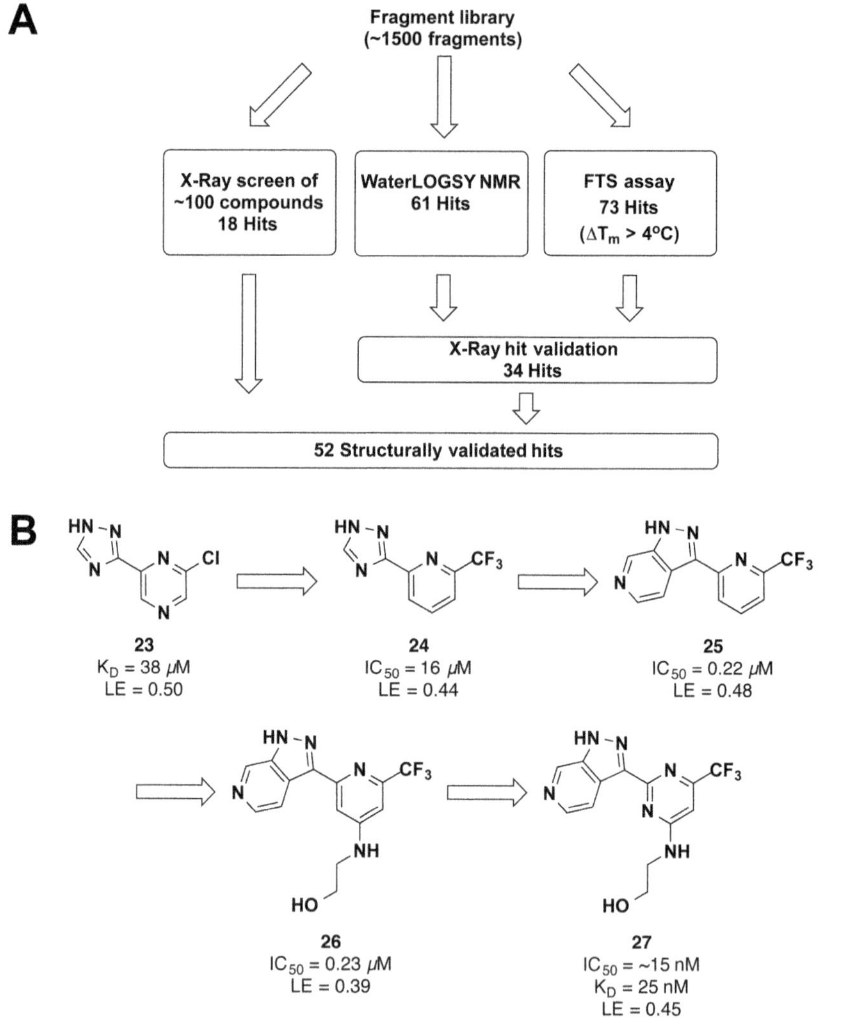

Figure 8.4 Fragment-based approach to targeting DNA ligase (LigA).[32] (A) Using X-ray crystallography, ligand-observed NMR and a FTS assay, a library of ~1500 fragments was screened against *Sa*LigA. The result of the screening was 52 structurally validated LigA hits. (B) Elaboration of a hit identified in the fragment screen against *Sa*LigA. K_D values measured by ITC and/or IC_{50} values measured in a TR-FRET assay, and the corresponding LEs, are shown. Where both K_D and IC_{50} values are given, the LE value was calculated using the K_D value.

This is consistent with the observed antibacterial effect against Gram-positive bacteria and efflux-deficient strains of Gram-negative bacteria resulting from LigA inhibition. Favourably, toxicity against a mouse lung lymphoma cell line was not observed at concentrations up to 100 μM (the highest concentration tested).

8.2.4 Targeting the ZipA–FtsZ Protein–Protein Interaction

PPIs have long been considered difficult targets for drug discovery.[33] This, in part, is due to the lack of well-defined pockets at protein–protein interfaces in which small-molecule inhibitors can bind. However, recently a number of PPIs have been successfully targeted using fragment-based approaches.[14–17] One PPI that has received interest as a potential antibacterial target is the ZipA–FtsZ interaction.[34] FtsZ is a conserved prokaryotic tubulin-like GTPase required for bacterial cell division. It polymerises to form a ring-like structure at the prospective site of cell division. In Gram-negative bacteria FtsZ recruits and binds the membrane-anchored protein ZipA, which in turn anchors FtsZ to the membrane.[35] FtsZ mutations that disrupt this interaction are lethal.[36]

The structure of a complex of the C-terminal interacting domains of FtsZ and ZipA has been solved and revealed that an α-helix of FtsZ binds in a broad, and predominantly hydrophobic, shallow cavity of ZipA.[37] The interaction is dominated by hydrophobic contacts.

With the aim of identifying small-molecule inhibitors of the ZipA–FtsZ interaction Tsao and colleagues[38] at Wyeth screened a library of 825 fragments against ZipA. To maximise diversity, no two fragments with the same ring scaffold were present in the library. The fragments were screened (initially in cocktails of up to six fragments) using $^{1}H–^{15}N$ HSQC NMR and ^{15}N-labelled ZipA. Of the 825 fragments screened, 16 were found to bind to ZipA (1.9% hit rate). Seven fragments perturbed the chemical shifts of peaks corresponding to residues involved in FtsZ binding, consistent with binding in the FtsZ-binding site (Figure 8.5A). Pyrimidines, indoles and aryl indazoles were among the scaffolds represented by these seven fragment hits (for examples, see compounds **28–31**, Figure 8.5B). All seven inhibited binding of a fluorescein-labelled FtsZ peptide to some extent when tested at a concentration of 50 μg ml^{-1} in a fluorescence-based polarisation (FP) assay, consistent with the NMR results. Fragment **30**, an indoloquinolizine that had also been identified in a previous FP screen for inhibitors of the FtsZ–ZipA interaction,[39] demonstrated the greatest inhibition in the FP assay and caused the greatest chemical shift perturbations in the HSQC experiments, and therefore was selected as the lead fragment for further investigations. An X-ray crystal structure of the indoloquinolizine bound to ZipA[40] showed that the fragment binds in the same shallow hydrophobic depression that the FtsZ peptide binds. The fragment is involved in a π-stacking interaction with Phe85 of ZipA, and the carbonyl group of the fragment hydrogen bonds with Asn63 *via* a water molecule.[38] The K_D value of this compound for ZipA was determined to be ~0.5 mM, as measured by $^{1}H–^{15}N$ HSQC NMR. By comparison, the FtsZ peptide binds with a K_D value of 7 μM.[39]

To explore SARs, 87 compounds similar to the indoloquinolizine fragment hit were selected and screened using $^{1}H–^{15}N$ HSQC NMR and an FP assay. However, none of the compounds – all of which contained a 6-6-5-6 ring system like the indoloquinolizine, or an analogous scaffold – showed a greater affinity for ZipA than the original fragment hit. X-Ray crystal structures

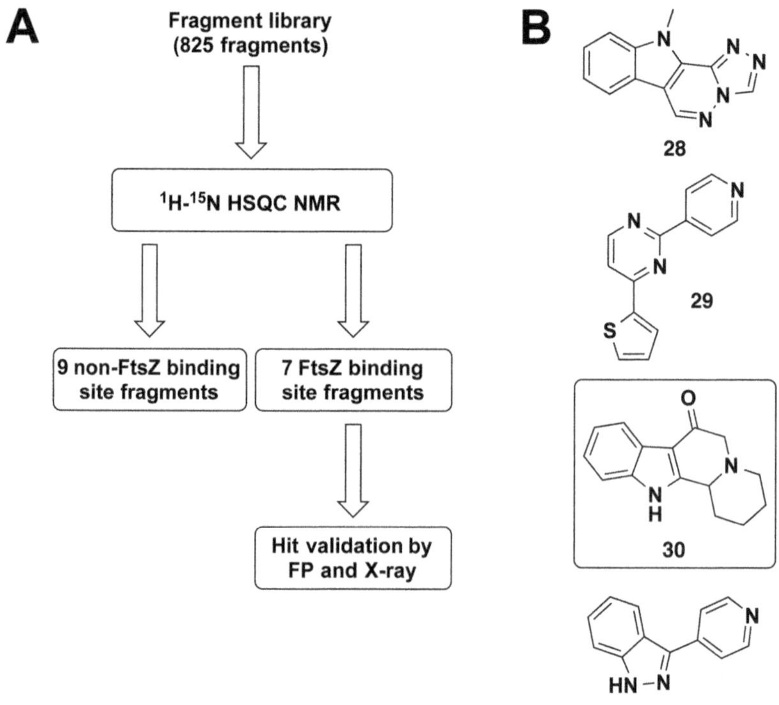

Figure 8.5 Fragment-based approach to targeting the ZipA–FtsZ protein–protein interaction.[38] (A) Using $^{1}H–^{15}N$ HSQC NMR a library of fragments was screened against ZipA. An FP assay and X-ray crystallography were used to confirm binding of fragments to the FtsZ-binding site. (B) Fragment hits found to bind in the FtsZ-binding site of ZipA. An outline has been drawn around the fragment found to cause the greatest chemical shift perturbations in the $^{1}H–^{15}N$ HSQC experiments and to have the greatest inhibitory effect in the FP assay.

revealed that, like the lead indoloquinolizine, the analogues were involved in a π-stacking interaction with ZipA, however, a variety of binding orientations were observed, which made structure-based design difficult.

In additional studies, the researchers at Wyeth directed efforts toward derivatising the indoloquinolizine fragment hit or an analogous scaffold.[40,41] The substituents were added to pick up new interactions and also to introduce groups that may improve solubility. A series of compounds based on merging of the indoloquinolizine fragment with an amino-oxazole-bearing HTS hit found to bind in an overlapping region of the FtsZ-binding site was also synthesised.[34] These efforts yielded additional analogues that bind to the FtsZ-binding site of ZipA, including a few that bind ZipA with slightly improved affinity relative to the lead fragment. A few compounds additionally demonstrated antibacterial activity against Gram-negative species of bacteria, however, whether the antibacterial effect was a result of inhibition of the FtsZ–ZipA interaction was not determined; antibacterial activity was also observed against Gram-positive bacteria, either pointing to the existence of a ZipA orthologue in these cells or off-target effects.[34] As the desired affinity for ZipA was not attained with analogues of the original indoloquinolizine fragment hit, researchers at Wyeth eventually pursued another hit identified in an HTS campaign.[41]

8.3 Application of Fragment-Based Approaches to Overcoming Pre-existing Resistance

8.3.1 Targeting β-Lactamases

The value of β-lactam antibiotics (penicillins, cephalosporins, carbapenems and monobactams) in the treatment of bacterial infections over the past 60 years cannot be overstated. However, the utility of these antibiotics is increasingly being compromised due to the development of bacterial resistance, primarily mediated by bacterial enzymes called β-lactamases that hydrolyse the β-lactam ring.[42] There exist hundreds of different types of β-lactamases; these can be divided into Classes A–D on the basis of amino acid sequence,[43] and further subdivided on the basis of substrate specificity profiles.[44] β-Lactam hydrolysis is catalysed either by an active-site serine (in the case of serine-β-lactamases, Classes A, C and D) or involves one or two active-site Zn^{2+} ions (in the case of metallo-β-lactamases, Class B).[42] One strategy to overcome β-lactamase-mediated resistance that has proved successful is the co-administration of β-lactam antibiotics with inhibitors of β-lactamases.[45] However, there is an increasing prevalence of β-lactamases that are not susceptible to inhibition by available β-lactamase inhibitors[42] and so if we are to restore the activity of β-lactams, novel β-lactamase inhibitors are needed.

HTS campaigns targeting β-lactamases have demonstrated mixed success (reviewed by Chen and colleagues[46]); while traditional HTS has delivered chemical leads for inhibiting metallo-β-lactamases, it has largely failed for serine β-lactamases. Fragment-based approaches are now being used to identify and advance new lead inhibitors of both serine and metallo-β-lactamases, and here we discuss some examples. For a more comprehensive review of the application of fragment-based approaches to the discovery of β-lactamase inhibitors, refer to the recent review by Chen and colleagues.[46]

8.3.1.1 Targeting Class A β-Lactamases

Chen and Shoichet[47] carried out a fragment screen against CTX-M β-lactamase, a Class A extended spectrum β-lactamase that confers resistance to third-generation cephalosporins. A set of 67 489 fragments was selected from the ZINC small molecule database and docked into the large open active site of a representative CTX-M β-lactamase (CTX-M-9). A set of 69 fragments was prioritised on the basis of docking scores and tested in a biochemical assay. Ten fragments with mM IC_{50} values were identified. Interestingly, by contrast, no compounds with inhibitory activity were identified from a prioritised set of 37 compounds selected in an *in silico* screen of a set of 1 147 326 lead-like molecules from the ZINC database carried out in parallel.[47]

X-Ray crystal structures were obtained for five of the fragment hits in complex with CTX-M-9 and the binding poses generally matched well with those predicted. Notably, most of the fragment hits identified contained either a carboxylate or a tetrazole (a common carboxylate isostere; *e.g.*, compounds **32–34**, Figure 8.6), which were found to bind to a similar region of the protein as the carboxylate of penicillins and cephalosporins. The high number of fragment hits with a tetrazole moiety (not a common moiety in known β-lactamase inhibitors) led Chen and Shoichet to search for elaborated analogues of the tetrazole fragment hits among the lower scoring lead-like docking hits not prioritised for testing. From four compounds tested, one tetrazole derivative with an inhibition constant (K_i value) of 21 μM was identified (**35**). Subsequent medicinal chemistry efforts directed toward introducing functional groups that could pick up interactions in two identified binding hotspots yielded a derivative with a K_i value of 89 nM (**36**).[48] The optimised inhibitor, which is the first non-covalent inhibitor of a Class A β-lactamase with nM affinity, was shown to increase the sensitivity of *E. coli* expressing CTX-M-9 to Cefotaxime 64-fold.

8.3.1.2 *Targeting Class C β-Lactamases*

Shoichet and colleagues[49] carried out a fragment screen against the β-lactamase AmpC, a Class C β-lactamase that hydrolyses a wide range of cephalosporins. As in the CTX-M fragment screen described above, molecular docking was used to prioritise fragments for testing; a library of 137 639 compounds (a mix of *Rule of 3* compliant fragments as well as some with more hydrogen-bond donors and/or acceptors) from the ZINC database was docked against the AmpC crystal structure. Forty eight of the top 500 scoring fragments were purchased and screened in a spectrophotometric assay, and some also in a surface plasmon resonance (SPR) assay. For 23 of these fragments inhibition constants below 10 mM were determined (K_i values between 0.7 and 9.2 mM, corresponding to LEs of 0.16–0.47). Interestingly, the fragment hits were structurally distinct from known AmpC inhibitors. X-Ray crystal structures were obtained for eight of the fragment hits in complex with AmpC. The predicted binding poses were mostly in agreement with those observed in the X-ray crystal structures, except for those fragments where fragment binding induced changes in the protein conformation. Interestingly, among the twenty three fragments found to bind AmpC, were two (including compound **32**, Figure 8.6) also among those found in the fragment screen carried out against the distinct CTX-M β-lactamase described above.

Shoichet and colleagues[50] used fragment-guided design to optimise a series of sulfonamide-bearing boronic acid β-lactamase inhibitors. Suitably substituted boronic acids have previously been shown to act as non-lactam,

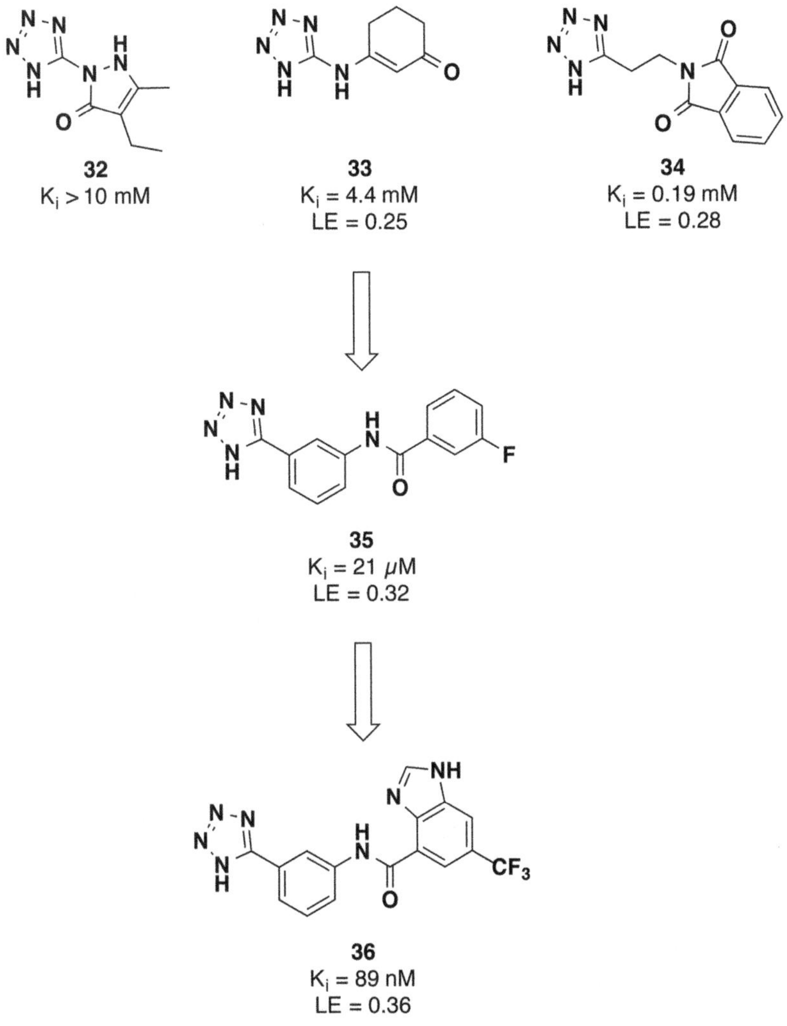

Figure 8.6 Fragment-based approach to targeting CTX-M β-lactamases.[47] A selection of fragment hits (compounds **32–34**) identified in a virtual screen and confirmed using a biochemical assay and X-ray crystallography are shown. Compound **35** was selected from docking hits on the basis of similarity to the fragment hits, and shown to inhibit CTX-M-9 β-lactamase. This compound was subsequently optimised into compound **36**, a high-affinity CTX-M-9 β-lactamase inhibitor.[48] K_i values measured in an absorbance-based assay (utilising the β-lactamase substrate nitrocefin), and the corresponding LEs, are shown.

transition state analogues that potently inhibit AmpC β-lactamase (K_i values in the nM range).[51] An example of one such boronic acid (compound **37**)[52] – a boronic acid bearing a sulfonamide moiety rather than the carboxamide group typically present in β-lactam mimics – is shown in Figure 8.7A. Although this and related sulfonamide boronic acids bind AmpC β-lactamase with high affinity, they display only modest activity in bacterial cell cultures, and efforts to increase β-lactamase affinity further had proved futile.[50,52] In an earlier study by this group,[53] two carboxylate-bearing fragments (fragments **38** and **39**) resulting from fragmentation of a known AmpC inhibitor (**40**) were observed to bind in a similar region of the AmpC crystal structure as the aryl carboxylate of boronic acid inhibitor **37** (Figure 8.7A and B). The same was true for fragment **32**, a tetrazole-containing fragment identified in the AmpC fragment screen described previously (Figure 8.7C). Interestingly, while the carboxylate of boronic acid inhibitor **37** forms a hydrogen bond only with one residue of AmpC (Ser212), the carboxylate/tetrazole moieties of the fragments hydrogen bond with two (Ser212 and Gly320). Therefore, with the aim of improving the affinity of the sulfonamide boronic acid inhibitors, attempts were made to form a hydrogen bond also with Gly320 by repositioning the aryl carboxylate or replacing it with a triazole. The strategy proved successful, yielding new compounds with 20–30-fold improved affinity (*e.g.*, compound **41**, Figure 8.7D). X-Ray crystal structures of the new compounds in complex with AmpC confirmed that the introduced carboxylate/tetrazole moieties of the new compounds closely recapitulate the binding poses of the corresponding moieties from the fragment hits and form the intended hydrogen-bonding interactions (Figure 8.7D). Further derivatisation of compound **41**, namely introduction of a trifluoromethyl substituent *meta* to the tetrazole, yielded an inhibitor with a K_i value of 0.05 nM (**42**; Figure 8.7E), corresponding to a 500-fold improved affinity relative to the initial lead boronic acid inhibitor (**37**; K_i value of 25 nM). Notably, this is the highest affinity β-lactamase inhibitor reported to date. An X-ray crystal structure of compound **42** in complex with AmpC was determined, and the compound was observed to form the same interactions with AmpC as compound **41**, the precursor molecule without the trifluoromethyl substituent. The trifluoromethyl substituent additionally was observed to pack with the backbone of Thr319 and the fluorine atoms appear to interact with the sulfonamide amide hydrogen.

The antiresistance activity of the developed compounds was investigated by comparing the MIC values of two third-generation cephalosporins (Cefotaxime and Ceftazidime) against a range of bacterial strains with high levels of resistance (as a result of AmpC over-expression) in the presence and absence of the boronic acid inhibitors. A decrease in MIC (typically >64-fold) was observed when Cefotaxime and Ceftazidime were combined with the optimised inhibitors, as compared to when the drugs were used alone. Interestingly, the new boronic acid derivatives also reduced the MIC values of Cefotaxime, and in some cases Ceftazidime, against a strain of *E. coli* that over-expresses a CTX-M β-lactamase, consistent with the new boronic acids also inhibiting Class A β-lactamases.

One compound (an analogue of compound **41** in which the phenyl group is replaced with pyridine) was also tested in a mouse model of bacteraemia and sepsis. Co-administration of the AmpC inhibitor with Cefotaxime to mice infected with a hospital-derived strain of *E. coli* that over-expresses AmpC,

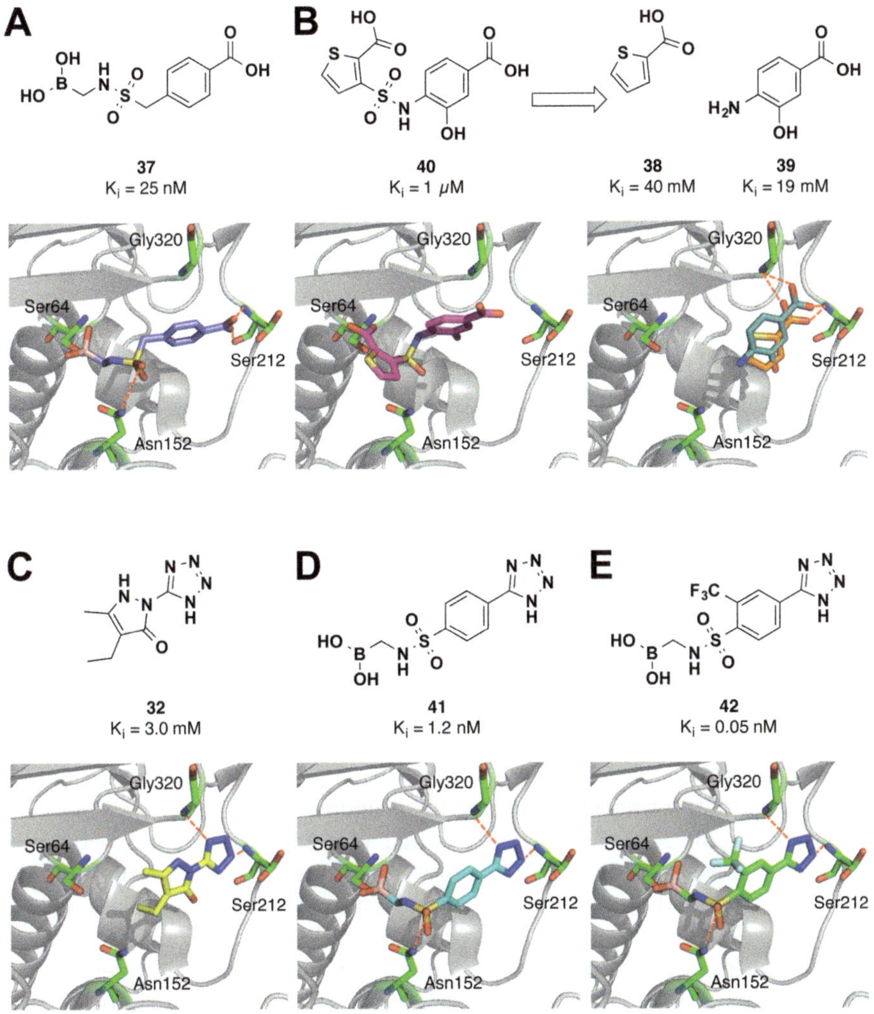

Figure 8.7 Fragment-guided optimisation of AmpC β-lactamase inhibitors.[50] (A) A lead sulfonamide boronic acid AmpC β-lactamase inhibitor.[52] (B) Carboxylate-bearing fragments resulting from de-fragmentation of a known AmpC inhibitor (compound **40**).[53] (C) A tetrazole-bearing fragment identified in a fragment screen against AmpC.[49] (D) and (E) Improved sulfonamide boronic acid AmpC inhibitors.[50] Optimisation of the boronic acid shown in (A) was guided by knowledge of the binding mode of the fragments shown in (B) and (C). The binding poses of the compounds, as determined from X-ray crystal structures, are shown [PDB:3O87 (A), 1XGJ, 2HDR, 2HDQ (B), 3GR2 (C), 4E3J (D) and 4E3N (E)]. Key residues are shown in green, and dashed red lines represent hydrogen bonds. K_i values measured in an absorbance-based assay (utilising the β-lactamase substrates nitrocefin or CENTA) are shown.

significantly prolonged survival of mice as compared with mice administered only with Cefotaxime.[50]

8.4 Application of Fragment-Based Approaches to Targeting *M. tuberculosis*

8.4.1 Targeting *M. tuberculosis* CYP121

M. tuberculosis cytochrome P450 enzymes (CYPs) are receiving increasing attention as potential targets for novel antituberculosis drugs.[54] CYPs are heme-containing enzymes that typically function to catalyse the monooxygenation of organic substrates, however a number of other functions have also been ascribed to these proteins.[55] The *M. tuberculosis* genome sequence[56] revealed there to be twenty putative CYP-encoding genes, an exceptionally high number considering the small size of the genome (4.4 Mb).

CYP121 is one of just a few *M. tuberculosis* CYPs to which a function has so far been ascribed. The enzyme has been shown to bind the cyclodipeptide cyclodityrosine (cYY) and catalyse the formation of an intramolecular C–C bond.[57] Although the physiological function of the reaction product is as yet unclear, Munro and colleagues[58] demonstrated disruption of the gene encoding CYP121 to be possible only in the presence of an additional copy, consistent with the protein being essential for viability of *M. tuberculosis in vitro*. In view of its essentiality and unique substrate specificity and catalytic mechanism, which are likely to facilitate selective targeting with small-molecule inhibitors, *M. tuberculosis* CYP121 (*Mt*CYP121) is considered a particularly attractive target for novel antituberculosis agents.

Abell and colleagues[59] recently described a fragment-based approach aimed at finding novel inhibitors of *Mt*CYP121. In this study, which represents the first example of a fragment-based approach aimed at this target class, an FTS assay was used to screen a commercially available library of 665 rule-of-three compliant fragments. Two complementary ligand-observed NMR techniques (STD and WaterLOGSY NMR) were subsequently used to confirm binding of hits identified in the primary screen, and also to identify those fragments that compete with the natural ligand (cYY) for binding and hence were likely to be active-site binders. Through this screening cascade (summarised in Figure 8.8A), 26 fragments that bind *Mt*CYP121 and compete with cYY for binding were identified. Eight of the 26 fragments (those most soluble under the crystallography conditions) were carried through to X-ray crystallography experiments, and four fragment-bound *Mt*CYP121 X-ray crystal structures were obtained. In each case, the fragments (with affinities between 0.4 and 3 mM, as measured by ITC, and LEs between 0.26 and 0.39) were observed in the *Mt*CYP121 active site, just as predicted from the NMR experiments.

Fragment hits **43** and **44** (Figure 8.8B) both possess an aryl amine moiety, and both demonstrate a type-II binding mode (they coordinate to the heme iron and in the process displace a coordinated water molecule); the heme iron coordination is through the aryl amine nitrogen (Figure 8.8B).

The binding poses of these two type-II binders were found to overlap, which presented an opportunity for fragment merging. The strategy to merge these fragments proved successful; a merged compound that retained the 1,2,4-triazole moiety of the first parent fragment and the 4-aminoquinoline scaffold of the second parent fragment (lacking only the methyl substituent to facilitate synthesis) was synthesised. This 1,2,4-triazoylquinoline (**45**, Figure 8.8B) was demonstrated to bind *Mt*CYP121 with a K_D value of 28 μM (as measured using ITC and a spectrophotometric heme-absorbance assay), corresponding to a LE matching that of the most ligand efficient parent fragment (0.39). The compound was also shown to inhibit the catalytic activity of *Mt*CYP121 with a K_i value of 50 μM in an absorbance-based activity assay. A number of structural analogues of compound **45** were also tested, but none demonstrated improved affinity. From analysis of the SAR it could be concluded that the heme-coordinating aryl amine was critical for binding. A crystal structure of compound **45** bound to *Mt*CYP121 revealed that the binding positions of the parent fragments were well preserved (Figure 8.8B).

The two additional fragment hits for which binding modes were determined by X-ray crystallography, lacked aryl amines, and were observed to bind away from the heme iron, towards the top of the active-site cavity. Two overlapping binding modes were observed for one of the two fragments (fragment **46**, Figure 8.8C). Attempts to develop higher affinity ligands based on merging two molecules of fragment **46** initially met with limited success; a biaryl-substituted triazole synthesised (**47**) was observed to bind *Mt*CYP121 (K_D = 2.8 mM, LE = 0.18; Figure 8.8C), but with lower affinity than the parent fragment. The latter was somewhat surprising as an X-ray crystal structure revealed that the two binding modes of fragment **46** were preserved in the merged compound. Quantum-mechanical-based conformational energy calculations, however, helped to shed light on why the affinity of this compound for *Mt*CYP121 was not higher; in the CYP121-bound conformation, the merged compound was conformationally strained due to a clash of the *ortho*-hydrogen atoms on the opposing phenol moieties. Consistent with this, the only structural analogue of compound **47** with improved affinity for *Mt*CYP121 (K_D = 500 μM) identified in this study contained an oxygen atom that increased the separation between the phenyl moieties.

In a follow up study, Abell and colleagues[60] identified the key interactions with *Mt*CYP121 that contributed to the strained *Mt*CYP121-bound conformation of compound **47**. They hypothesised that if they could increase the strength of interaction between compound **47** and a nearby Gln residue (Gln385; by forming a double hydrogen bond) it might be possible to shift the binding pose of the merged compound slightly toward this residue and slightly away from the residues thought to be restricting the rotational freedom of the molecule, and in this way sufficiently reduce the internal strain. An *in silico* docking screen was performed using a focussed library of biphenol analogues to identify new scaffolds that could facilitate the dual hydrogen bond with Gln385. Of the compounds that adopted the same general pose as biaryl-substituted triazole **47**, 3-aminopyrazole **48** was the compound predicted to interact most strongly with this residue

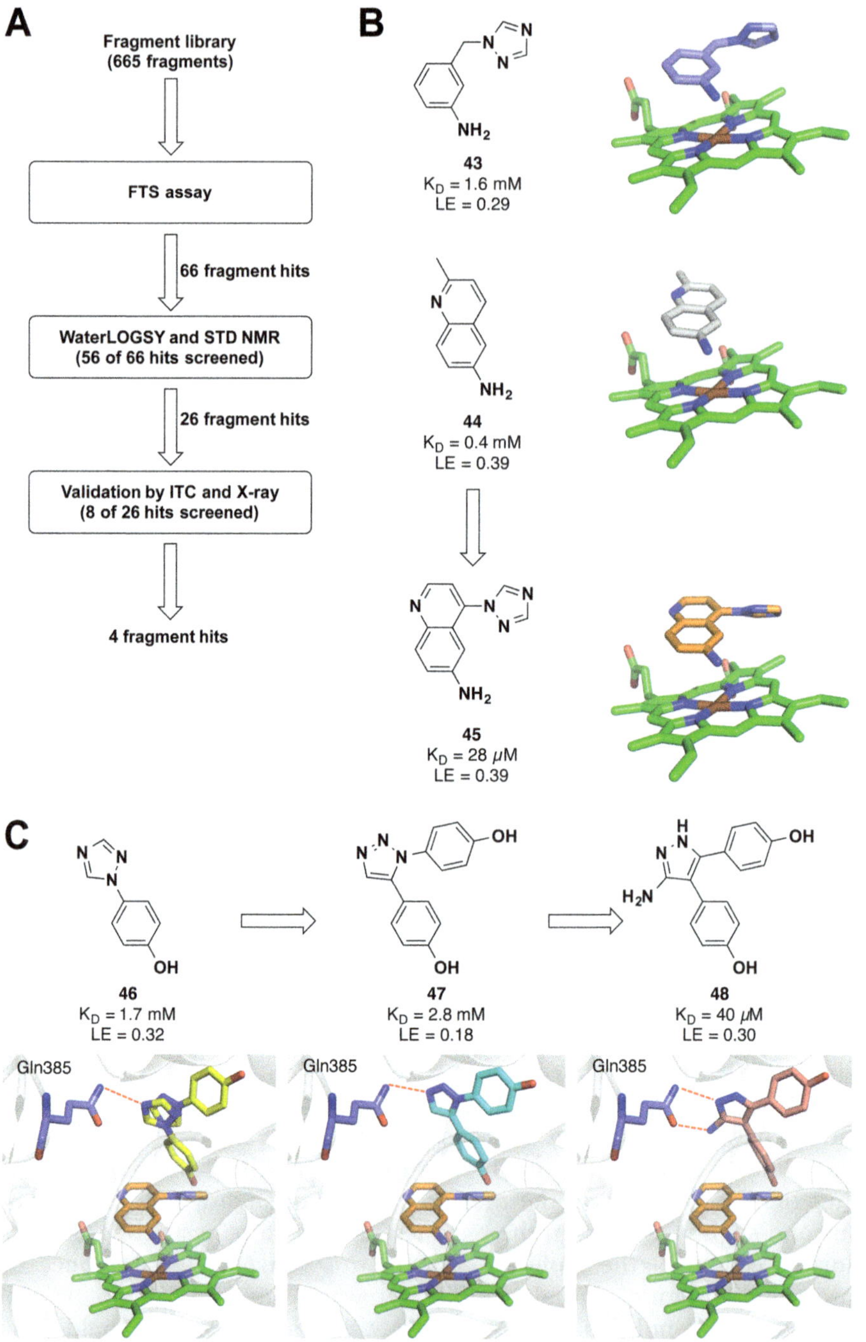
A
Fragment library
(665 fragments)
FTS assay
66 fragment hits
WaterLOGSY and STD NMR
(56 of 66 hits screened)
26 fragment hits
Validation by ITC and X-ray
(8 of 26 hits screened)
4 fragment hits
B
43
K_D = 1.6 mM
LE = 0.29
44
K_D = 0.4 mM
LE = 0.39
45
K_D = 28 μM
LE = 0.39
C
46
K_D = 1.7 mM
LE = 0.32
47
K_D = 2.8 mM
LE = 0.18
48
K_D = 40 μM
LE = 0.30
Gln385
Gln385
Gln385

(*via* a double hydrogen bond). Compound **48** was synthesised and showed a 70-fold higher affinity for *Mt*CYP121 (K_D = 40 µM, LE = 0.30), as compared to compound **47**. The X-ray crystal structure of aminopyrazole **48** bound to *Mt*CYP121 provided confirmation that the compound was positioned to form a double hydrogen bond with Gln385 and that the key interactions with compound **47** were retained (Figure 8.8C). Quantum mechanical calculations also revealed that the molecule was not strained like compound **47**.

To investigate selectivity, triazoylquinoline **45**, which was developed as part of the first merging strategy described above (K_D = 28 µM for *Mt*CYP121), was also tested against *M. tuberculosis* CYP125 (*Mt*CYP125). However, no heats of binding were observed by ITC.[59] Interestingly, a fragment screen against *Mt*CYP125 analogous to that performed against *Mt*CYP121, yielded just one common hit. These results are consistent with it being possible to identify isoform-selective ligands as early on as in the fragment screen. Triazoylquinoline **45** was also tested against a panel of three human hepatic CYPs, and with just one exception, less than 50% inhibition was observed at a triazoylquinoline concentration of 100 µM under the test conditions. Further evaluation of the *Mt*CYP121 ligands developed by Abell and colleagues and any higher affinity ligands to come to hand will be important for investigating the hypothesis that selective inhibition of *M. tuberculosis* can be achieved with *Mt*CYP121 inhibitors.

8.4.2 Targeting *M. tuberculosis* Pantothenate Synthetase

Pantothenate synthetase catalyses the ATP-mediated condensation of pantoate with β-alanine. The product of this reaction, pantothenate (vitamin B_5), is the universal precursor for the essential enzyme cofactor coenzyme A.[61]

Figure 8.8 Fragment-based approach to targeting *M. tuberculosis* cytochrome P450 121 (*Mt*CYP121).[59,60] (A) Through a fragment-screening cascade involving an FTS assay, ligand-observed NMR, ITC and X-ray crystallography, four fragment hits with K_D values between 0.4 and 3 mM were identified. (B) Two heme-coordinating aryl amine-containing fragment hits were merged to yield a *Mt*CYP121 inhibitor with a K_D value of 28 µM. (C) A merging strategy was also used to develop a high-affinity *Mt*CYP121 binder from a non-heme-coordinating fragment that was observed to bind in two different poses. The affinity of the initial merged compound was lower than that of the parent fragment. However, after modifying the scaffold to relieve conformational strain, the affinity for *Mt*CYP121 increased considerably. The binding poses, as determined from X-ray crystal structures, of the fragments and elaborated compounds are shown relative to the heme group (green) of *Mt*CYP121 [PDB:4G44, 4G45, 4G1X (B) and 4G47, 4G2G, 4KTF (C)]. The protein structure has been omitted from (B) for clarity. In (C), triazoylquinoline **45** (generated in the merging strategy) is superimposed for reference (orange) and Gln385 is shown in blue. Dashed red lines represent hydrogen bonds. K_D values measured by ITC and/or a heme-absorbance assay, and the corresponding LEs, are shown.

The importance of the gene encoding pantothenate synthetase (*panC*) for growth of *M. tuberculosis in vitro* has been demonstrated in independent studies.[62,63] Although access to exogenous pantothenate can render pantothenate biosynthesis in microorganisms non-essential,[64] pantothenate biosynthesis in *M. tuberculosis* has also been demonstrated to be important *in vivo*.[65] For these reasons, and because the pantothenate biosynthesis pathway is absent in humans,[64] *M. tuberculosis* pantothenate synthetase (*Mt*PS) has received much interest as a potential novel target for antituberculosis therapeutics.[61] Fragment-based approaches are among the strategies that have been used to identify inhibitors of *Mt*PS.

Abell and colleagues[66,67] screened a commercially available library of 1250 rule-of-three compliant fragments against *Mt*PS. The screen (summarised in Figure 8.9A), which involved a combination of FTS assays and ligand-observed NMR (STD and WaterLOGSY NMR), yielded 17 fragment hits. The binding of eight fragments (as measured by NMR) was reduced in the presence of saturating concentrations of ATP and binding of another eight was reduced in the presence of the pantoyl reaction intermediate (formed *in situ* from ATP and pantoate), consistent with binding in the ATP- and pantoate-binding pockets, respectively. The binding of 14 of the 17 fragments was detected by direct ITC, with K_D values ranging from 0.15–17 mM measured (corresponding to LEs of 0.18–0.37). Attempts were made to soak all 17 fragment hits into apo-*Mt*PS crystals that had been washed to remove a sulfate ion typically bound in the active site. A total of eight fragments were validated by X-ray crystallography, with four observed to bind in the ATP-binding site and four observed to bind in the pantoate-binding site.

Benzofuran-2-carboxylic acid (**49**), one of the fragment hits found to bind in the pantoate-binding site of *Mt*PS (K_D = 1 mM, LE = 0.34), and 5-methoxyindole (**50**), a fragment identified in an independent WaterLOGSY NMR screen of 52 randomly selected fragments,[66,67] became the basis of fragment elaboration efforts by Abell and colleagues (Figure 8.9B and C).[66] A methoxyindole **50**-bound *Mt*PS crystal structure showed that the fragment, with a K_D value of 1.1 mM (LE = 0.36), binds in the ATP-binding site of *Mt*PS, and, revealed possibilities for fragment growth.[66] An acyl sulfonamide was added at the C2 position of the fragment to serve as a handle for elaboration. This group had the required 'bend' to avoid steric clashes with the active-site wall and to fit into the binding pocket. An X-ray crystal structure of the elaborated fragment (**51**) bound to *Mt*PS showed that the sulfone oxygen was involved in a number of hydrogen-bonding interactions. Correspondingly the elaborated fragment showed an increase in affinity (K_D = 0.21 mM, LE = 0.28) relative to the parent fragment. The addition of a 4-methylpyridine ring on the sulfonamide (as in compound **52**) extended the fragment into the β-alanine-binding site and increased the binding affinity (K_D = 29 μM, LE = 0.26), as did the addition of an alkyl carboxylic acid chain at the N1 position of the indole (K_D = 1.5 μM, LE = 0.28). The final compound (**53**), which recapitulated the binding mode of the original methoxyindole fragment hit

(Figure 8.9B), was demonstrated to be an ATP-competitive inhibitor of *Mt*PS with a K_i value of 27 μM.

The binding sites of benzofuran **49** and methoxyindole **50** are separated by 3.1 Å and, as X-ray crystallography experiments revealed that the two fragments could bind simultaneously, a fragment elaboration strategy based on linking the two fragments appeared viable. Benzofuran **49** and methoxyindole **50** (substituted at the N1 position with an alkyl carboxylic acid moiety to pick up electrostatic interactions with the triphosphate-binding site, just like the 'grown' *Mt*PS inhibitor) were initially linked with a flexible alkyl ester. The flexible linker was chosen to allow the two fragments (which bind in different planes) the best opportunity to adopt their original binding poses. An increase in affinity for *Mt*PS was observed for the linked compound (K_D = 75 μM, LE = 0.20). A slight increase in affinity (K_D = 50 μM, LE = 0.20) was observed when an alkyl sulfonamide linker was used instead, but the greatest success was had when the two fragments were linked with a less flexible acyl sulfonamide linker to give compound **54** (K_D = 1.8 μM, LE = 0.26). Although the acyl sulfonamide introduced the bend in the molecule necessary to place the benzofuran fragment into the pantoate-binding pocket, and the moiety binds in a position to achieve the same hydrogen-bonding interactions as the parent fragment, the pose of the benzofuran moiety deviates slightly from that of the parent fragment, indicative of the optimum binding pose of this fragment not having been retained (Figure 8.9C). Like the elaborated compound of comparable affinity yielded from the fragment growing strategy, the linked compound was shown to inhibit the catalytic activity of *Mt*PS in a manner competitive with ATP (K_i = 9 μM).

Although X-ray crystallography was an integral part of the fragment-linking approach described above, Abell and colleagues[68] also showed that it was possible to drive a fragment-linking strategy based on methoxyindole **49** and benzofuran **50** primarily using ligand-observed NMR. Two-dimensional ^1H–^1H nuclear Overhauser effect spectroscopy (NOESY) experiments were performed to investigate the proximity and relative orientation of protein-bound fragments by observation of interligand Overhauser effects (ILOEs). Following introduction of both an alkyl carboxylic acid substituent at the N1 position of methoxyindole **50** to increase solubility, and a methyl group at the C2 position to increase sensitivity (as in compound **55**, Figure 8.10), a specific ILOE was observed between the methyl protons at the C2 position of **55** and the proton at the C3 position of compound **49**. This prompted the synthesis of amide-linked compound **56** (K_D = 0.86 μM, LE = 0.30; Figure 8.10), which binds *Mt*PS with an affinity approximately three orders of magnitude higher than the parent fragments and inhibits *Mt*PS activity in an ATP-competitive manner (K_i = 5.4 μM). A co-crystal structure confirmed that the binding modes of the parent fragments were conserved in the linked compound.

*Mt*PS inhibitors arising from the crystal structure-led fragment-elaboration efforts described above were subsequently tested for the ability to

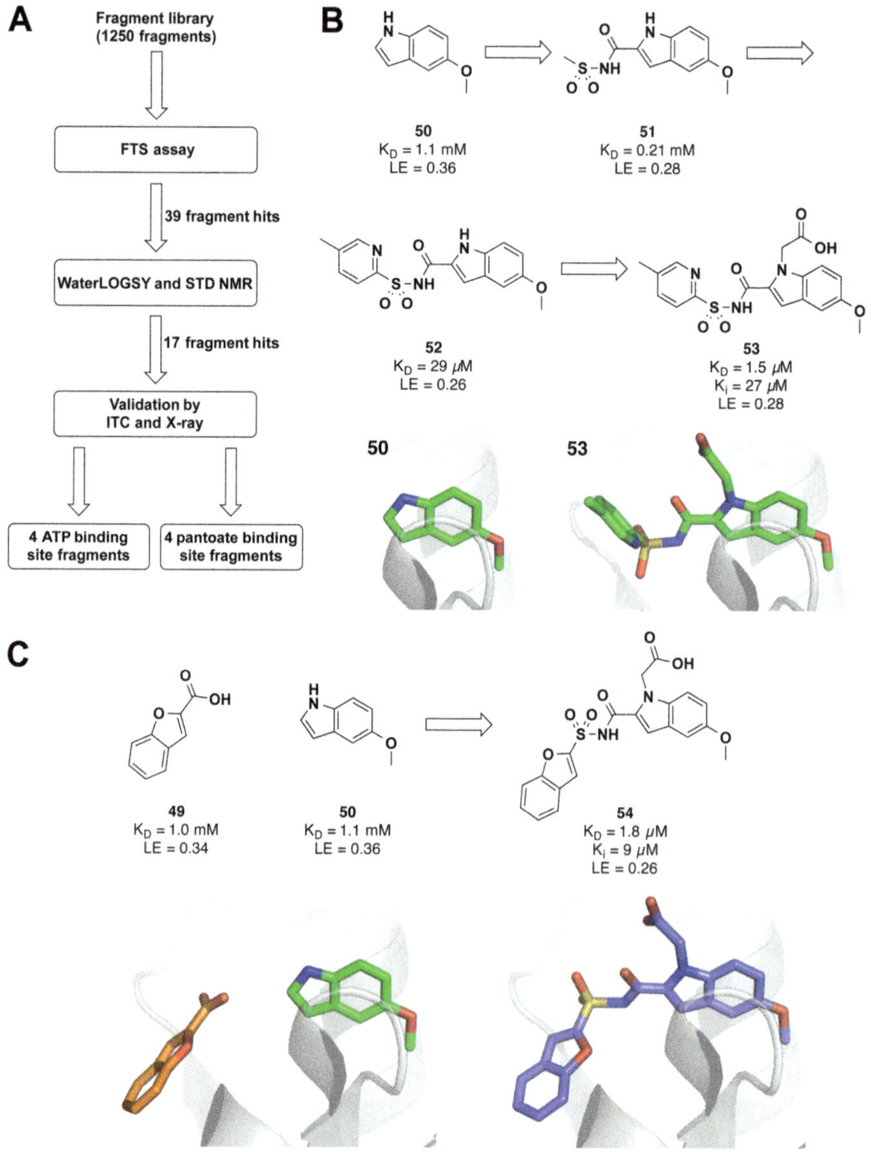

Figure 8.9 Fragment-based approach to targeting *M. tuberculosis* pantothenate synthetase (*Mt*PS).[66,67] (A) Through a fragment-screening cascade involving an FTS assay, ligand-observed NMR, ITC and X-ray crystallography, four fragments that bind to the ATP-binding site of *Mt*PS and four that bind to the pantoate-binding site of *Mt*PS were identified.[66,67] (B) Fragment **50** (a fragment identified in an independent WaterLOGSY-based screen that binds in the ATP-binding site of *Mt*PS) was elaborated stepwise into a higher affinity *Mt*PS binder using a fragment-growing strategy.[66] (C) Fragments **50** and **49** (a fragment identified in the screening cascade summarised in (A) that binds in the pantoate-binding site of *Mt*PS) were linked with an acylsulfonamide. The elaboration strategies were guided

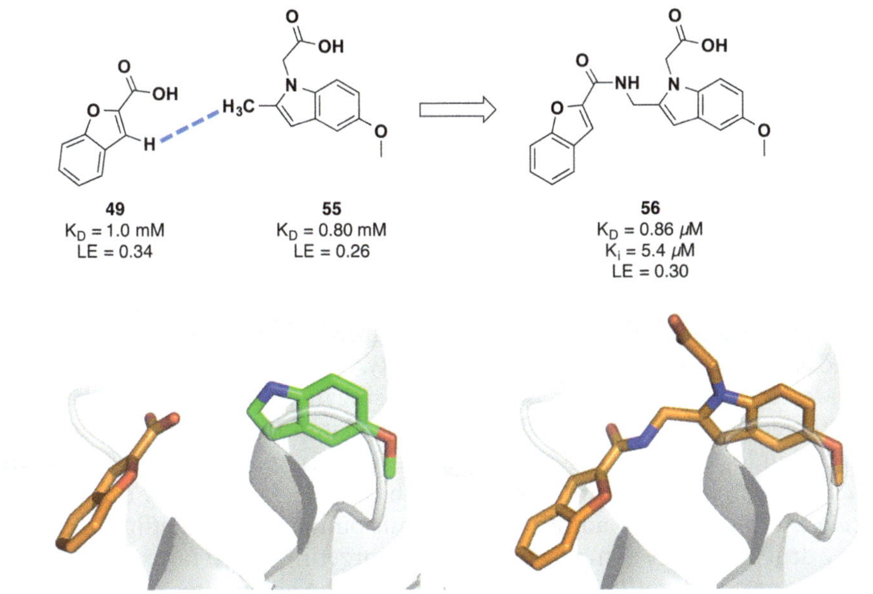

Figure 8.10 Elaboration of *M. tuberculosis* pantothenate synthetase (*Mt*PS) fragment hits guided by 2D ^{1}H–^{1}H NOESY.[68] An ILOE was observed between the two fragments shown (dashed blue line), which prompted synthesis of the amide-linked compound. The binding poses of the fragments and elaborated compound, as determined from X-ray crystal structures, are shown (PDB:3IMG, 3LE8). K_D values measured by ITC, and the corresponding LEs, are shown. A K_i value determined in an enzyme-coupled, absorbance-based assay is additionally shown for the elaborated compound.

inhibit growth of *M. tuberculosis in vitro*. Disappointingly, the compounds were without effect on growth of wild-type *M. tuberculosis* at concentrations up to 250 μM.[63] However, when tested against a mutant strain of *M. tuberculosis* under-expressing *panC*, sulfonamide-linked compound **54** (Figure 8.9C) demonstrated an inhibitory effect (MIC = 31–62 μM) that was alleviated upon supplementation with pantothenate.[63] The findings that the *Mt*PS-depleted cells were hypersensitive to the *Mt*PS inhibitor and that inhibition could be overcome with exogenous pantothenate is consistent with the compound acting through inhibition of pantothenate biosynthesis and demonstrates that failure of the compound to inhibit is not a result of the compound not being able to gain access to the target, but rather because greater inhibition of *Mt*PS is required.

---

by X-ray crystallography.[66] The binding poses of the fragments and elaborated compounds, as determined from X-ray crystal structures, are shown [PDB:3IMC, 3IUE (B) and 3IMG, 3IVX (C)]. K_D values measured by ITC, and the corresponding LEs, are shown. K_i values determined in an enzyme-coupled, absorbance-based assay are additionally shown for elaborated compounds.

8.4.3 Targeting *M. tuberculosis* Protein Tyrosine Phosphatases

Protein tyrosine phosphatases (PTPs) catalyse the de-phosphorylation of tyrosine residues in proteins and, in a coordinated manner with protein tyrosine kinases (the action of which they oppose), function in the transduction of signals that regulate fundamental cellular processes.[69] *M. tuberculosis* expresses two PTPs (MPtpA and MPtpB), both of which are secreted into the cytosol of the macrophage that plays host to the bacterium.[70–72] Although not essential for extracellular survival of *M. tuberculosis in vitro*, MPtpA and MPtpB are required for uncompromised growth in activated macrophages, and as such play a key role in pathogenicity.[70,72]

M. tuberculosis PTPs are receiving increasing interest as potential therapeutic targets.[73] This is due to their demonstrated importance for survival *in vivo*, and their localisation to the macrophage, which, circumvents the formidable task of designing inhibitors that are both capable of traversing the *M. tuberculosis* cell wall and evading *M. tuberculosis* pump-mediated efflux. However, any drug-discovery campaign chemically targeting a PTP faces a number of challenges due to: (i) the similarity between the active sites of PTPs, which means that selectivity can be hard to attain with small molecule inhibitors; (ii) the PTP active sites being highly positively charged, making it difficult to identify inhibitors with drug-like characteristics; and (iii) the PTPs being extremely susceptible to catalytic inactivation through oxidation, which can result in inhibitor screens yielding high numbers of false positives.[74]

A unique fragment-based technique known as substrate activity screening (SAS), first applied to the discovery of protease inhibitors,[75] is one approach that has proved effective for the discovery of selective *M. tuberculosis* PTP inhibitors. In SAS a customised fragment library is screened to identify fragment-sized enzyme substrates. The fragment substrates identified are thereafter optimised to improve affinity before being transformed into inhibitors by replacing the enzyme-reactive group with a non-metabolisable isostere. On the hunt for MPtpB inhibitors, Ellman and colleagues[76] prepared a library of 140 *O*-aryl phosphates from a diverse set of commercially available phenols less than 300 Da in size. To identify MPtpB substrates, the library was screened in an absorbance-based enzyme-coupled MPtpB assay in which the production of inorganic phosphate (P_i) that occurs upon MPtpB-catalysed phosphate hydrolysis is monitored. The practice of screening for substrates rather than inhibitors here cleverly avoids the false positives that can arise in inhibitor screens due to inactivation of the target enzyme through non-specific means. Several MPtpB fragment substrates were identified in the screen (*e.g.*, fragments **57–59**, Figure 8.11A), including some with Michaelis constants (K_M values) below 100 μM. Subsequent SAR explorations led to substrates with improved affinity (*e.g.*, compounds **60** and **61**). Replacement of the phosphate moiety with a non-hydrolysable mono-acidic phosphate isostere (isoxazole carboxylic acid) in turn yielded

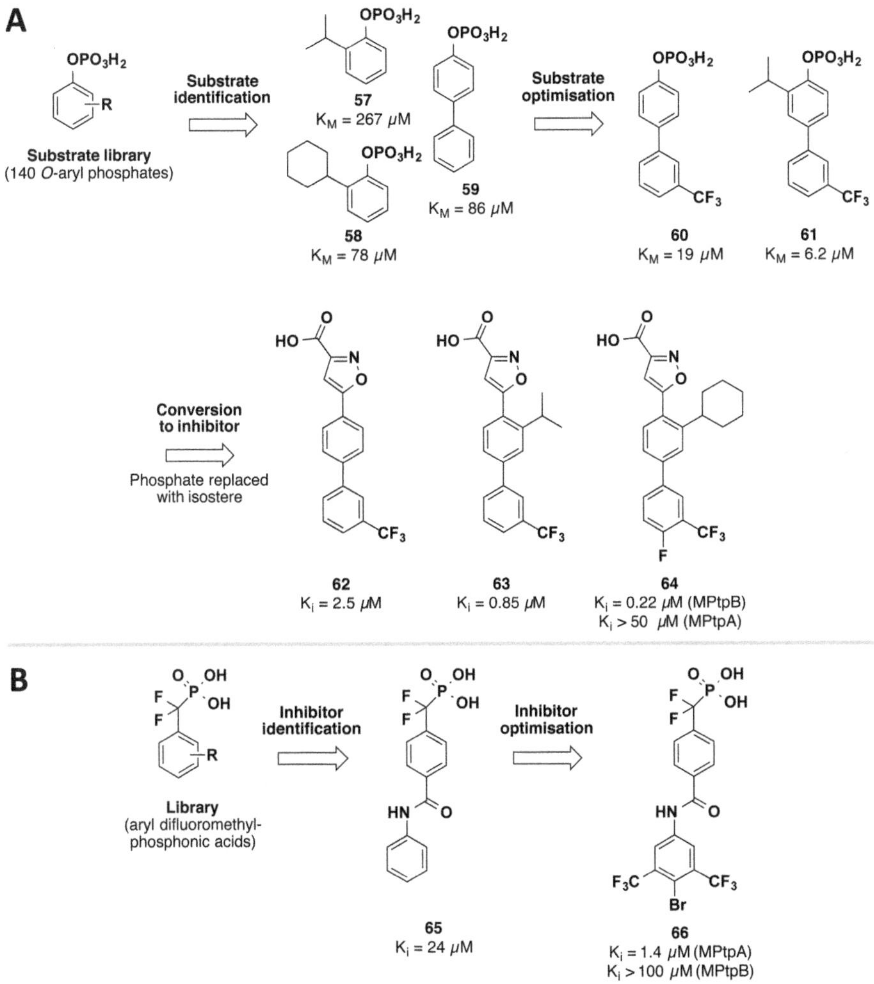

Figure 8.11 Fragment-based approaches to targeting *M. tuberculosis* protein tyrosine phosphatases (MPtps).[76,77] (A) A substrate-activity screening method was used to identify low-molecular-weight *O*-aryl phosphates that serve as substrates for MPtpB.[76] Following substrate optimisation, substrates were converted to inhibitors by the replacement of the phosphate moiety with a phosphate isostere. The approach yielded a selective inhibitor of MPtpB. (B) A library of low-molecular-weight aryl difluoromethylphosphonic acids was screened against MPtpA.[77] Optimisation of an initial hit yielded a selective inhibitor of MPtpA. K_M values determined using an enzyme-coupled, absorbance-based phosphate-release assay, and K_i values determined using the chromogenic substrate *para*-nitrophenyl phosphate, are shown.

inhibitors (*e.g.*, compounds **62–64**), the best compound (**64**) inhibiting with a K_i value of 0.22 μM (Figure 8.11A). Importantly this compound was shown to be selective for MPtpB, with K_i values >35-fold higher determined against MtPtpA and four human PTPs tested.

Ellman and colleagues[77] also used the library of 140 *O*-aryl phosphates as a starting point for MPtpA inhibitor identification. However, in this study a difluoromethylphosphonic acid was used as a phosphate mimic in place of the isoxazole carboxylic acid (Figure 8.11B). Informed by the results of a substrate screen against MPtpA, a small library of aryl difluoromethylphosphonic acids was prepared and tested. Among this library was a selective inhibitor of MPtpA with a K_i value of 24 μM (**65**; Figure 8.11B), which was selected for optimisation. The introduction of trifluoromethyl groups at the *meta*-positions and a bromo substituent at the *para*-position of the benzanilide ring subsequently yielded an MPtpA inhibitor with a K_i value of 1.4 μM (**66**). A K_i value >70-fold higher was measured against MPtpB, and K_i values at least 11-fold higher were measured against the additional six human tyrosine and dual-specificity phosphatases tested.

Rademann and colleagues[78] subsequently used a variation of fragment-based substrate activity screening to identify novel inhibitors of MPtpA. In this study, fragments binding to a secondary binding pocket next to the active site were sought. To identify such fragments, the fragment screen made use of a probe substrate (**67**; Figure 8.12) bearing an aldehyde functionality that could react with nucleophilic fragments binding in the secondary binding pocket. This in turn could increase the affinity of the substrate and thereby enhance substrate turnover. This strategy was termed *dynamic substrate enhancement*. The idea to target a pocket nearby to the active site stemmed from the finding that such pockets are less conserved among Ptps and therefore may provide the opportunity for selective targeting.[79] A library of 110 primary amine fragments was screened against MPtpA (as well as 3 human Ptps) in the presence of the aldehyde-bearing Ptp substrate, using the reagent malachite green to monitor P_i production and hence substrate turnover. In this screen, one fragment that enhanced the turnover of the substrate by MPtpA but not the other Ptps was identified (**68**; Figure 8.12). With the aim of converting this selective fragment hit into an inhibitor, analogues of the product predicted to result from the reaction of the aldehyde and the amine (**69**), in which the phosphotyrosine moiety was replaced with an isostere, were synthesised. This strategy yielded two compounds (**70** and **71**) that inhibit the activity of MPtpA and retain the selectively of the parent fragment; K_i values of 13 and 11 μM were measured for these inhibitors against MPtpA, while K_i values >300 μM were measured against the human Ptps tested (Figure 8.12).

Whether the MPtpA and MPtpB inhibitors identified in the studies reviewed here affect mycobacterial survival *ex vivo* in macrophages and indeed in animal models has not been reported. However, as the phosphate mimetics used in the studies performed by Ellman and colleagues were

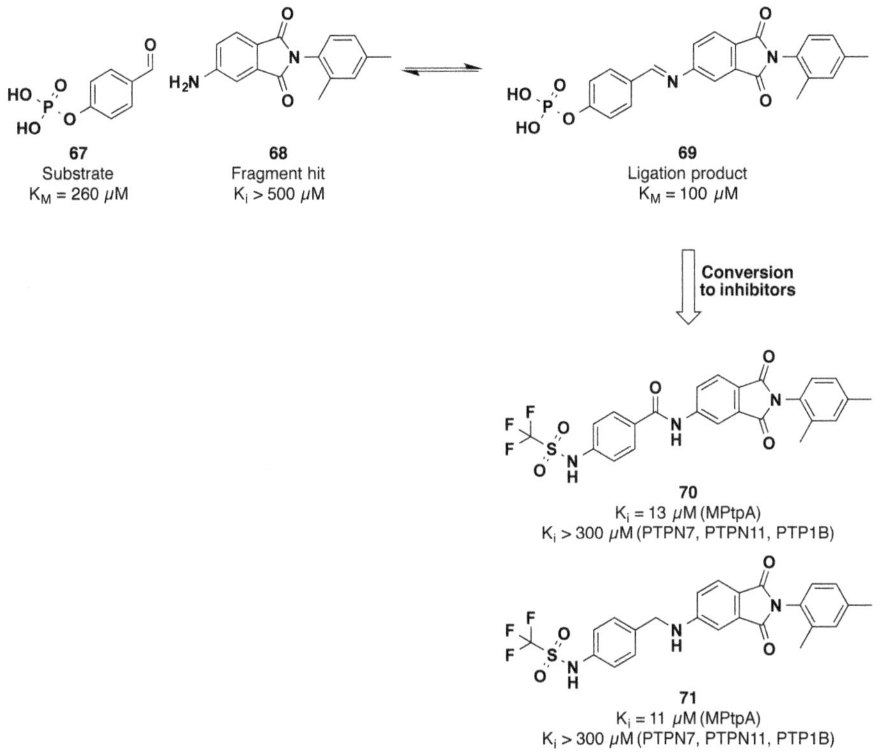

Figure 8.12 Targeting *M. tuberculosis* protein tyrosine phosphatase A (MPtpA) using a dynamic substrate enhancement approach.[78] Fragment **68**, which binds to a pocket adjacent to the active site of MPtpA, was identified by screening for fragments that increase MPtpA-catalysed turnover of an electrophilic *O*-aryl phosphate substrate (compound **67**) through template-assisted ligation.[78] Replacement of the phosphotyrosine moiety of the predicted ligation product with an isostere yielded selective MPtpA inhibitors. K_i values determined using the chromogenic substrate *para*-nitrophenyl phosphate, are shown.

selected on the basis that they form parts of molecules shown previously to exhibit good cell permeability,[80,81] the high-affinity Ptp inhibitors identified hold promise.

8.5 Conclusions

The examples presented herein demonstrate that fragment-based approaches are being utilised with success in antibacterial drug-discovery projects, both alone and in combination with other lead-finding mechanisms. Consistent with the previous success associated with targeting the ATP-binding sites of eukaryotic kinases using fragment-based approaches, fragment screens against bacterial BC, DNA gyrase, and LigA, and *Mt*PS (all enzymes with ATP-binding sites) yielded multiple ATP-binding site fragment hits that were successfully elaborated into high-affinity inhibitors (some with sub nM affinity). Importantly, the fragment screens generally provided novel scaffolds, offering the opportunity for (often multiple) novel series.

Fragment screens also yielded hits against the targets considered more challenging or that were previously unexploited. Difficulties were encountered in elaborating the ZipA fragment hits into higher affinity inhibitors, which is consistent with the general view that the optimisation of fragment hits against PPIs is more challenging than the optimisation of hits against more druggable target classes.[82] A more recent study by Oakley and colleagues[83] (published in the late stages of the preparation of this book chapter), did show more success in targeting a bacterial PPI. Fragment screening not only yielded fragments that bind to the protein-binding site of the bacterial sliding clamp (DNA polymerase III subunit β), but also led to the identification of higher affinity compounds that, as a consequence, inhibit DNA replication, and possess antibacterial activity. Hence, fragment-based approaches are also beginning to prove their worth in yielding inhibitors that can disrupt bacterial PPIs.

PTPs represent another challenging target class, and yet, as described, clever twists on a traditional fragment-based approach yielded new lead series of MPtp inhibitors with potent and selective activity. Fragment-based approaches also readily yielded two distinct novel inhibitors of *Mt*CYP121, a target for which selective inhibitors had not previously been identified. The new *Mt*CYP121 inhibitors will hopefully facilitate investigations into the druggability of this target.

Many of the fragment-based projects reviewed here not only yielded potent enzyme inhibitors, but also yielded compounds with whole-cell inhibitory activity against wild-type and/or membrane-compromised bacteria, that appear to be on-target. Furthermore, at least in the BC and DNA gyrase studies, some of the elaborated compounds also showed antibacterial activity *in vivo* in animal models of bacterial infection.

In the case of a lead *Mt*PS inhibitor, which lacked activity against wild-type *M. tuberculosis*, conditional *M. tuberculosis* mutants that under-express PS were used to show that the compound does indeed gain entry into the mycobacterium and inhibit the target. This result highlights the value of coupling fragment-based approaches with whole-cell screening against conditional mutants that under-express the intended molecular target; this can provide confirmation of on-target whole-cell inhibitory activity at an earlier stage of the development process.

The examples presented here not only highlight the value of fragment-based approaches in lead identification but also in late-stage lead optimisation; fragment-based approaches were shown to revitalise a lead series and deliver new β-lactamase inhibitors that could decrease the MIC values of third-generation cephalosporins to clinically relevant concentrations *in vitro*, with activity also *in vivo*.

Although many pharmaceutical companies have withdrawn from the area of antibacterial drug discovery, over half of the projects reviewed here were driven by, and/or involved contributions from pharmaceutical companies, presumably reflecting new hope that fragment-based approaches can also deliver in this area of drug discovery. With the growing awareness of the dire need for novel antibacterials and recognition of the advantages offered by fragment-based approaches, it seems likely that we will see an increase in the number of fragment-based antibacterial drug-discovery projects. As this transpires it will be interesting to consider the optimal composition of fragment libraries and the best suited fragment screening techniques, and whether these deviate from those that are yielding high-quality hits against eukaryotic targets. In view of the rich history of natural products in antibacterial drug discovery, fragments derived from the fragmentation of natural products (including 3D fragments) may also serve as valuable starting points for antibacterial lead discovery.

Acknowledgements

CS and AGC thank Professor Chris Abell and Dr Steven Howard for proofreading this chapter. AGC would like to thank the BBSRC for funding (BB/I019669/1). CS is funded in part by a NHMRC Overseas Biomedical Fellowship (1016357), and in part by a Grant from the Foundation for the National Institutes of Health with support from the Bill and Melinda Gates Foundation.

References

1. World Health Organization, *The Evolving Threat of Antibacterial Resistance*, 2012.
2. M. A. Cooper and D. Shlaes, *Nature*, 2011, **472**, 32.
3. O. Cars, A. Hedin and A. Heddini, *Drug Resist. Updates*, 2011, **14**, 68–69.
4. J. Carlet, P. Collignon, D. Goldmann, H. Goossens, I. C. Gyssens, S. Harbarth, V. Jarlier, S. B. Levy, B. N'Doye, D. Pittet, R. Richtmann, W. H. Seto, J. W. van der Meer and A. Voss, *Lancet*, 2011, **378**, 369–371.
5. D. J. Payne, M. N. Gwynn, D. J. Holmes and D. L. Pompliano, *Nat. Rev. Drug Discovery*, 2007, **6**, 29–40.
6. J. H. Powers, *Clin. Microbiol. Infect.*, 2004, **10**, 23–31.
7. K. Lewis, *Nat. Rev. Drug Discovery*, 2013, **12**, 371–387.
8. P. F. Chan, D. J. Holmes and D. J. Payne, *Drug Discovery Today: Ther. Strategies*, 2004, **1**, 519–527.
9. M. N. Gwynn, A. Portnoy, S. F. Rittenhouse and D. J. Payne, *Ann. N. Y. Acad. Sci.*, 2010, **1213**, 5–19.
10. M. Gualtieri, F. Baneres-Roquet, P. Villain-Guillot, M. Pugniere and J. P. Leonetti, *Curr. Med. Chem.*, 2009, **16**, 390–393.
11. D. E. Scott, A. G. Coyne, S. A. Hudson and C. Abell, *Biochemistry*, 2012, **51**, 4990–5003.
12. C. W. Murray and D. C. Rees, *Nat. Chem.*, 2009, **1**, 187–192.
13. M. Baker, *Nat. Rev. Drug Discovery*, 2013, **12**, 5–7.

14. N. Basse, J. L. Kaar, G. Settanni, A. C. Joerger, T. J. Rutherford and A. R. Fersht, *Chem. Biol.*, 2010, **17**, 46–56.
15. T. Oltersdorf, S. W. Elmore, A. R. Shoemaker, R. C. Armstrong, D. J. Augeri, B. A. Belli, M. Bruncko, T. L. Deckwerth, J. Dinges, P. J. Hajduk, M. K. Joseph, S. Kitada, S. J. Korsmeyer, A. R. Kunzer, A. Letai, C. Li, M. J. Mitten, D. G. Nettesheim, S. Ng, P. M. Nimmer, J. M. O'Connor, A. Oleksijew, A. M. Petros, J. C. Reed, W. Shen, S. K. Tahir, C. B. Thompson, K. J. Tomaselli, B. Wang, M. D. Wendt, H. Zhang, S. W. Fesik and S. H. Rosenberg, *Nature*, 2005, **435**, 677–681.
16. A. Potter, V. Oldfield, C. Nunns, C. Fromont, S. Ray, C. J. Northfield, C. J. Bryant, S. F. Scrace, D. Robinson, N. Matossova, L. Baker, P. Dokurno, A. E. Surgenor, B. Davis, C. M. Richardson, J. B. Murray and J. D. Moore, *Bioorg. Med. Chem. Lett.*, 2010, **20**, 6483–6488.
17. A. J. Potter, S. Ray, L. Gueritz, C. L. Nunns, C. J. Bryant, S. F. Scrace, N. Matassova, L. Baker, P. Dokurno, D. A. Robinson, A. E. Surgenor, B. Davis, J. B. Murray, C. M. Richardson and J. D. Moore, *Bioorg. Med. Chem. Lett.*, 2010, **20**, 586–590.
18. J. E. Cronan Jr and G. L. Waldrop, *Prog. Lipid Res.*, 2002, **41**, 407–435.
19. S. W. Polyak, A. D. Abell, M. C. Wilce, L. Zhang and G. W. Booker, *Appl. Microbiol. Biotechnol.*, 2012, **93**, 983–992.
20. Y. M. Zhang, S. W. White and C. O. Rock, *J. Biol. Chem.*, 2006, **281**, 17541–17544.
21. J. R. Miller, S. Dunham, I. Mochalkin, C. Banotai, M. Bowman, S. Buist, B. Dunkle, D. Hanna, H. J. Harwood, M. D. Huband, A. Karnovsky, M. Kuhn, C. Limberakis, J. Y. Liu, S. Mehrens, W. T. Mueller, L. Narasimhan, A. Ogden, J. Ohren, J. V. Prasad, J. A. Shelly, L. Skerlos, M. Sulavik, V. H. Thomas, S. VanderRoest, L. Wang, Z. Wang, A. Whitton, T. Zhu and C. K. Stover, *Proc. Natl. Acad. Sci. U. S. A.*, 2009, **106**, 1737–1742.
22. I. Mochalkin, J. R. Miller, L. Narasimhan, V. Thanabal, P. Erdman, P. B. Cox, J. V. Prasad, S. Lightle, M. D. Huband and C. K. Stover, *ACS Chem. Biol.*, 2009, **4**, 473–483.
23. LE = $-\Delta G$ (kcal mol^{-1}) per non-hydrogen atom = $-RT\ln(IC_{50}/K_i/K_D)$ (kcal mol^{-1}) per non-hydrogen atom; A. L. Hopkins, C. R. Groom and A. Alex, *Drug Discovery Today*, 2004, **9**, 430–431.
24. R. J. Reece and A. Maxwell, *Crit. Rev. Biochem. Mol. Biol.*, 1991, **26**, 335–375.
25. F. Collin, S. Karkare and A. Maxwell, *Appl. Microbiol. Biotechnol.*, 2011, **92**, 479–497.
26. H. J. Boehm, M. Boehringer, D. Bur, H. Gmuender, W. Huber, W. Klaus, D. Kostrewa, H. Kuehne, T. Luebbers, N. Meunier-Keller and F. Mueller, *J. Med. Chem.*, 2000, **43**, 2664–2674.
27. K. J. Aldred, R. J. Kerns and N. Osheroff, *Biochemistry*, 2014, **53**, 1565–1574.
28. S. Firth-Clark, N. P. Todorov, I. L. Alberts, A. Williams, T. James and P. M. Dean, *J. Chem. Inf. Model.*, 2006, **46**, 1168–1173.
29. M. Oblak, S. G. Grdadolnik, M. Kotnik, R. Jerala, M. Filipic and T. Solmajer, *Bioorg. Med. Chem. Lett.*, 2005, **15**, 5207–5210.

30. A. E. Eakin, O. Green, N. Hales, G. K. Walkup, S. Bist, A. Singh, G. Mullen, J. Bryant, K. Embrey, N. Gao, A. Breeze, D. Timms, B. Andrews, M. Uria-Nickelsen, J. Demeritt, J. T. Loch III, K. Hull, A. Blodgett, R. N. Illingworth, B. Prince, P. A. Boriack-Sjodin, S. Hauck, L. J. MacPherson, H. Ni and B. Sherer, *Antimicrob. Agents Chemother.*, 2012, **56**, 1240–1246.
31. S. Shuman, *J. Biol. Chem.*, 2009, **284**, 17365–17369.
32. S. Howard, N. Amin, A. B. Benowitz, E. Chiarparin, H. Cui, X. Deng, T. D. Heightman, D. J. Holmes, A. Hopkins, J. Huang, Q. Jin, C. Kreatsoulas, A. C. L. Martin, F. Massey, L. McCloskey, P. N. Mortenson, P. Pathuri, D. Tisi and P. A. Williams, *ACS Med. Chem. Lett.*, 2013, **4**, 1208–1212.
33. M. R. Arkin and J. A. Wells, *Nat. Rev. Drug Discovery*, 2004, **3**, 301–317.
34. A. G. Sutherland, J. Alvarez, W. Ding, K. W. Foreman, C. H. Kenny, P. Labthavikul, L. Mosyak, P. J. Petersen, T. S. Rush III, A. Ruzin, D. H. Tsao and K. L. Wheless, *Org. Biomol. Chem.*, 2003, **1**, 4138–4140.
35. A. I. Rico, M. Krupka and M. Vicente, *J. Biol. Chem.*, 2013, **288**, 20830–20836.
36. S. A. Haney, E. Glasfeld, C. Hale, D. Keeney, Z. He and P. de Boer, *J. Biol. Chem.*, 2001, **276**, 11980–11987.
37. L. Mosyak, Y. Zhang, E. Glasfeld, S. Haney, M. Stahl, J. Seehra and W. S. Somers, *EMBO J.*, 2000, **19**, 3179–3191.
38. D. H. Tsao, A. G. Sutherland, L. D. Jennings, Y. Li, T. S. Rush III, J. C. Alvarez, W. Ding, E. G. Dushin, R. G. Dushin, S. A. Haney, C. H. Kenny, A. K. Malakian, R. Nilakantan and L. Mosyak, *Bioorg. Med. Chem.*, 2006, **14**, 7953–7961.
39. C. H. Kenny, W. Ding, K. Kelleher, S. Benard, E. G. Dushin, A. G. Sutherland, L. Mosyak, R. Kriz and G. Ellestad, *Anal. Biochem.*, 2003, **323**, 224–233.
40. L. D. Jennings, K. W. Foreman, T. S. Rush III, D. H. Tsao, L. Mosyak, Y. Li, M. N. Sukhdeo, W. Ding, E. G. Dushin, C. H. Kenny, S. L. Moghazeh, P. J. Petersen, A. V. Ruzin, M. Tuckman and A. G. Sutherland, *Bioorg. Med. Chem. Lett.*, 2004, **14**, 1427–1431.
41. L. D. Jennings, K. W. Foreman, T. S. Rush III, D. H. Tsao, L. Mosyak, S. L. Kincaid, M. N. Sukhdeo, A. G. Sutherland, W. Ding, C. H. Kenny, C. L. Sabus, H. Liu, E. G. Dushin, S. L. Moghazeh, P. Labthavikul, P. J. Petersen, M. Tuckman, S. A. Haney and A. V. Ruzin, *Bioorg. Med. Chem.*, 2004, **12**, 5115–5131.
42. R. J. Worthington and C. Melander, *J. Org. Chem.*, 2013, **78**, 4207–4213.
43. K. Bush, G. A. Jacoby and A. A. Medeiros, *Antimicrob. Agents Chemother.*, 1995, **39**, 1211–1233.
44. D. M. Livermore, *Clin. Microbiol. Rev.*, 1995, **8**, 557–584.
45. S. M. Drawz and R. A. Bonomo, *Clin. Microbiol. Rev.*, 2010, **23**, 160–201.
46. D. A. Nichols, A. R. Renslo and Y. Chen, *Future Med. Chem.*, 2014, **6**, 413–427.
47. Y. Chen and B. K. Shoichet, *Nat. Chem. Biol.*, 2009, **5**, 358–364.
48. D. A. Nichols, P. Jaishankar, W. Larson, E. Smith, G. Liu, R. Beyrouthy, R. Bonnet, A. R. Renslo and Y. Chen, *J. Med. Chem.*, 2012, **55**, 2163–2172.

49. D. G. Teotico, K. Babaoglu, G. J. Rocklin, R. S. Ferreira, A. M. Giannetti and B. K. Shoichet, *Proc. Natl. Acad. Sci. U. S. A.*, 2009, **106**, 7455–7460.
50. O. Eidam, C. Romagnoli, G. Dalmasso, S. Barelier, E. Caselli, R. Bonnet, B. K. Shoichet and F. Prati, *Proc. Natl. Acad. Sci. U. S. A.*, 2012, **109**, 17448–17453.
51. E. Caselli, R. A. Powers, L. C. Blasczcak, C. Y. Wu, F. Prati and B. K. Shoichet, *Chem. Biol.*, 2001, **8**, 17–31.
52. O. Eidam, C. Romagnoli, E. Caselli, K. Babaoglu, D. T. Pohlhaus, J. Karpiak, R. Bonnet, B. K. Shoichet and F. Prati, *J. Med. Chem.*, 2010, **53**, 7852–7863.
53. K. Babaoglu and B. K. Shoichet, *Nat. Chem. Biol.*, 2006, **2**, 720–723.
54. S. A. Hudson, K. J. McLean, A. W. Munro and C. Abell, *Biochem. Soc. Trans.*, 2012, **40**, 573–579.
55. *Cytochrome P450: Structure, Mechanism, and Biochemistry*, ed. P. R. de Montellano, Kluwer Academic/Plenum Publishers, New York, 2005.
56. S. T. Cole, R. Brosch, J. Parkhill, T. Garnier, C. Churcher, D. Harris, S. V. Gordon, K. Eiglmeier, S. Gas, C. E. Barry III, F. Tekaia, K. Badcock, D. Basham, D. Brown, T. Chillingworth, R. Connor, R. Davies, K. Devlin, T. Feltwell, S. Gentles, N. Hamlin, S. Holroyd, T. Hornsby, K. Jagels, A. Krogh, J. McLean, S. Moule, L. Murphy, K. Oliver, J. Osborne, M. A. Quail, M. A. Rajandream, J. Rogers, S. Rutter, K. Seeger, J. Skelton, R. Squares, S. Squares, J. E. Sulston, K. Taylor, S. Whitehead and B. G. Barrell, *Nature*, 1998, **393**, 537–544.
57. P. Belin, M. H. Le Du, A. Fielding, O. Lequin, M. Jacquet, J. B. Charbonnier, A. Lecoq, R. Thai, M. Courcon, C. Masson, C. Dugave, R. Genet, J. L. Pernodet and M. Gondry, *Proc. Natl. Acad. Sci. U. S. A.*, 2009, **106**, 7426–7431.
58. K. J. McLean, P. Carroll, D. G. Lewis, A. J. Dunford, H. E. Seward, R. Neeli, M. R. Cheesman, L. Marsollier, P. Douglas, W. E. Smith, I. Rosenkrands, S. T. Cole, D. Leys, T. Parish and A. W. Munro, *J. Biol. Chem.*, 2008, **283**, 33406–33416.
59. S. A. Hudson, K. J. McLean, S. Surade, Y. Q. Yang, D. Leys, A. Ciulli, A. W. Munro and C. Abell, *Angew. Chem., Int. Ed.*, 2012, **51**, 9311–9316.
60. S. A. Hudson, S. Surade, A. G. Coyne, K. J. McLean, D. Leys, A. W. Munro and C. Abell, *ChemMedChem*, 2013, **8**, 1451–1456.
61. C. Spry, K. Kirk and K. J. Saliba, *FEMS Microbiol. Rev.*, 2008, **32**, 56–106.
62. C. M. Sassetti, D. H. Boyd and E. J. Rubin, *Mol. Microbiol.*, 2003, **48**, 77–84.
63. G. L. Abrahams, A. Kumar, S. Savvi, A. W. Hung, S. Wen, C. Abell, C. E. Barry III, D. R. Sherman, H. I. Boshoff and V. Mizrahi, *Chem. Biol.*, 2012, **19**, 844–854.
64. S. Y. Gerdes, M. D. Scholle, M. D'Souza, A. Bernal, M. V. Baev, M. Farrell, O. V. Kurnasov, M. D. Daugherty, F. Mseeh, B. M. Polanuyer, J. W. Campbell, S. Anantha, K. Y. Shatalin, S. A. Chowdhury, M. Y. Fonstein and A. L. Osterman, *J. Bacteriol.*, 2002, **184**, 4555–4572.
65. V. K. Sambandamurthy, X. Wang, B. Chen, R. G. Russell, S. Derrick, F. M. Collins, S. L. Morris and W. R. Jacobs Jr, *Nat. Med.*, 2002, **8**, 1171–1174.

66. A. W. Hung, H. L. Silvestre, S. Wen, A. Ciulli, T. L. Blundell and C. Abell, *Angew. Chem., Int. Ed.*, 2009, **48**, 8452–8456.
67. H. L. Silvestre, T. L. Blundell, C. Abell and A. Ciulli, *Proc. Natl. Acad. Sci. U. S. A.*, 2013, **110**, 12984–12989.
68. P. Sledz, H. L. Silvestre, A. W. Hung, A. Ciulli, T. L. Blundell and C. Abell, *J. Am. Chem. Soc.*, 2010, **132**, 4544–4545.
69. N. K. Tonks, *Nat. Rev. Mol. Cell Biol.*, 2006, **7**, 833–846.
70. H. Bach, K. G. Papavinasasundaram, D. Wong, Z. Hmama and Y. Av-Gay, *Cell Host Microbe*, 2008, **3**, 316–322.
71. A. Koul, A. Choidas, M. Treder, A. K. Tyagi, K. Drlica, Y. Singh and A. Ullrich, *J. Bacteriol.*, 2000, **182**, 5425–5432.
72. R. Singh, V. Rao, H. Shakila, R. Gupta, A. Khera, N. Dhar, A. Singh, A. Koul, Y. Singh, M. Naseema, P. R. Narayanan, C. N. Paramasivan, V. D. Ramanathan and A. K. Tyagi, *Mol. Microbiol.*, 2003, **50**, 751–762.
73. D. Wong, J. D. Chao and Y. Av-Gay, *Trends Microbiol.*, 2013, **21**, 100–109.
74. A. F. Moretto, S. J. Kirincich, W. X. Xu, M. J. Smith, Z. K. Wan, D. P. Wilson, B. C. Follows, E. Binnun, D. Joseph-McCarthy, K. Foreman, D. V. Erbe, Y. L. Zhang, S. K. Tam, S. Y. Tam and J. Lee, *Bioorg. Med. Chem.*, 2006, **14**, 2162–2177.
75. W. J. Wood, A. W. Patterson, H. Tsuruoka, R. K. Jain and J. A. Ellman, *J. Am. Chem. Soc.*, 2005, **127**, 15521–15527.
76. M. B. Soellner, K. A. Rawls, C. Grundner, T. Alber and J. A. Ellman, *J. Am. Chem. Soc.*, 2007, **129**, 9613–9615.
77. K. A. Rawls, P. T. Lang, J. Takeuchi, S. Imamura, T. D. Baguley, C. Grundner, T. Alber and J. A. Ellman, *Bioorg. Med. Chem. Lett.*, 2009, **19**, 6851–6854.
78. M. F. Schmidt, M. R. Groves and J. Rademann, *ChemBioChem*, 2011, **12**, 2640–2646.
79. A. J. Barr, E. Ugochukwu, W. H. Lee, O. N. King, P. Filippakopoulos, I. Alfano, P. Savitsky, N. A. Burgess-Brown, S. Muller and S. Knapp, *Cell*, 2009, **136**, 352–363.
80. Y. Han, M. Belley, C. I. Bayly, J. Colucci, C. Dufresne, A. Giroux, C. K. Lau, Y. Leblanc, D. McKay, M. Therien, M. C. Wilson, K. Skorey, C. C. Chan, G. Scapin and B. P. Kennedy, *Bioorg. Med. Chem. Lett.*, 2008, **18**, 3200–3205.
81. G. Liu, Z. Xin, Z. Pei, P. J. Hajduk, C. Abad-Zapatero, C. W. Hutchins, H. Zhao, T. H. Lubben, S. J. Ballaron, D. L. Haasch, W. Kaszubska, C. M. Rondinone, J. M. Trevillyan and M. R. Jirousek, *J. Med. Chem.*, 2003, **46**, 4232–4235.
82. C. W. Murray, M. L. Verdonk and D. C. Rees, *Trends Pharmacol. Sci.*, 2012, **33**, 224–232.
83. Z. Yin, L. R. Whittell, Y. Wang, S. Jergic, M. Liu, E. J. Harry, N. E. Dixon, J. L. Beck, M. J. Kelso and A. J. Oakley, *J. Med. Chem.*, 2014, **57**, 2799–2806.

CHAPTER 9

Exploring Fragment Screening and Optimization Strategies Using Acetylcholine-Binding Protein

EWALD EDINK[a], GERDIEN E. DE KLOE[a], AND IWAN J. P. DE ESCH*[a]

[a]VU University Amsterdam, Division of Medicinal Chemistry, De Boelelaan 1083, 1081 HV, Amsterdam, Netherlands
*E-mail: i.de.esch@vu.nl

9.1 Introduction

9.1.1 Fragment-Based Approaches from an Academic Perspective

Fragment-based drug discovery (FBDD) has been hailed as an efficient approach to developing high-quality starting points for drug development.[1–4] From an academic perspective, fragment-based approaches are appreciated as a cost-effective alternative to high-throughput screening (HTS) that results in a much better understanding of ligand–protein interactions. As such, these approaches are also ideal for exploring very fundamental aspects of molecular recognition. Indeed, fragment-based approaches are useful in

RSC Drug Discovery Series No. 47
Fragment-Based Drug Discovery
Edited by Steven Howard and Chris Abell

Published by the Royal Society of Chemistry, www.rsc.org

studying detailed structure–activity relationships (SARs). For example, the well-known ligand efficiency (LE) metric evaluates the contribution of every atom to the binding affinity.[5] Such ligand-based considerations are combined with structural biology (crystal structures), molecular pharmacology and site-directed mutagenesis to evaluate the role of both the ligand and protein in order to accurately describe the actual ligand–protein-binding event. These approaches have been employed to study activity and affinity.[6–9] Currently, they are also being unleashed to unravel the underlying thermodynamic and kinetic aspects of ligand–protein interaction. Fragment-based approaches represent a rewarding academic research line that not only leads to new molecular tools and ligands with potential clinical applications but also results in exciting basic research in the field of molecular interactions.

9.1.2 Acetylcholine-Binding Protein as a Versatile Target to Explore Fragment-Based Approaches

The family of Cys-loop receptors includes therapeutically relevant ion channels such as the nicotinic acetylcholine receptors (nAChRs), $GABA_A$–, serotonin 5-HT_3–, and glycine receptors. These proteins are embedded in membranes and structural information is scarce. It is known that the ion channels are opened upon agonist binding at the extracellular ligand-binding domain (LBD) of these receptors. Structural understanding has been boosted by studying acetylcholine-binding proteins that have been derived from different snail and worm species such as *Lymnaea stagnalis* (Ls-AChBP),[10,11] *Aplysia californica* (Ac-AChBP)[12,13] and *Capitella teleta* (Ct-AChBP).[14,15] These water-soluble proteins have a similar protein structure in terms of size, architecture and pentameric assembly as the LBDs of membrane-bound Cys-loop receptors.[16] Ls-AChBP shares 24% sequence identity with the LBD of the human α7 nAChR and binds reference nicotinic receptor ligands such as acetylcholine (**1**) nicotine (**2**) and epibatidine (**3**).[10–12]

Currently, more than 40 X-ray structures of AChBPs from different species and with different co-crystallized ligands have been obtained. Together with site-directed mutagenesis studies on AChBPs and nAChRs subtypes, this has led to a good understanding of ligand recognition. For example, aromatic residues that align the binding pocket and are conserved in both AChBPs and nAChRs have been shown to be involved in binding nicotinic receptor ligands *via* cation–π interactions.[17–19] Besides providing a structural template for studying Cys-loop receptors, we have found AChBP to be an excellent protein to study various aspects of the fragment-based methodology. AChBPs are relatively easy to handle in biochemical screening assays and (especially when compared to membrane-bound proteins) in biophysical screening assays such as surface plasmon resonance (SPR) biosensor analysis and isothermal titration calorimetry (ITC). In addition, AChBPs are relatively easy to co-crystallize with ligands, enabling X-ray structural studies and structure-based considerations. For these reasons, AChBPs have been used in our labs for exploring methods to efficiently identify and optimize fragment hits.

9.2 Fragment Screening Techniques

9.2.1 Identification of Fragments for AChBP Using *in Silico* Screening Procedures

Provided that structural information of the protein target is available or can be readily generated (for example by generating reliable homology models), *in silico* screening (by means of molecular docking) is definitely the quickest to set up, although not necessarily the easiest to optimize. If anything, the virtual screening studies help to familiarize the scientists with the protein-binding site early in the project. At best, these studies lead to the prioritization of promising fragments that are validated as hits in subsequent biochemical or biophysical screening. Well-known complications in the virtual screening of fragments are the sampling of small chemical entities in large binding sites (especially considering the lower complexity of fragments)[20,21] and the actual scoring of the different binding modes of fragments.[22,23] Once the docking protocols have been set up, the throughput of *in silico* screening is enormous and it is tempting to not only evaluate the proprietary (fragment) libraries but also consider all *ca.* 300 000 commercially available fragments by means of virtual screening.[20] Admittedly, this violates the FBDD principle that only one or two thousand fragments need to be screened, but the aforementioned arguments regarding the limitations of the technology and the low costs might still warrant the use of bigger virtual libraries in the hope of overcoming the complications and uncertainties associated with *in silico* screening.[20] In the case of AChBP, Akdemir *et al.* were able to identify a diverse set of fragments that display similar or higher affinity for Ls-AChBP than the endogenous ligand acetylcholine (**1**) ($pK_i \geq 4.9$, determined by using a [^{3}H]epibatidine competitive binding assay, Figure 9.1) using a step-wise hierarchical *in silico* screening protocol against the nicotine-bound Ls-AChBP structure.[24] The procedure included a set of hierarchical filters, adapted to the specific characteristics of the Ls-AChBP-binding site, such as: (1) the presence of essential functional groups (cationic centers); (2) a pharmacophore filter that selects ligands with the proper size and shape and location of essential functional groups (cationic centers); and (3) post-processing of docking poses based on docking scores, automated analysis of essential Ls-AChBP interactions not taken into account in the docking scoring

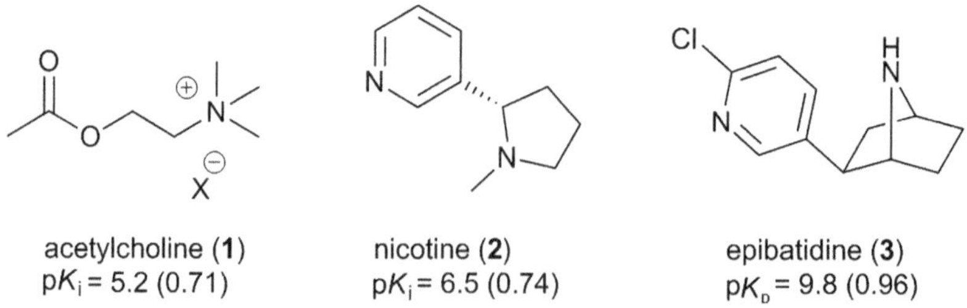

Figure 9.1 Chemical structures and binding affinities[12,52] of acetylcholine (**1**), nicotine (**2**) and epibatidine (**3**). LEs in kcal mol^{-1} Da^{-1} are depicted in brackets.

function (cation–π), and visual inspection of the top-ranked *in silico* hits (see Figure 9.2). This protocol was applied to our proprietary (containing 5315 compounds) and a commercially available compound screening collection (SPECS World Diversity Set containing 17 234 compounds). At this stage, both compound databases contained fragments as well as drug-like molecules. Of the 54 compounds that were predicted to bind to AChBP, 30 compounds were indeed validated as hits ($pK_i > 4.5$) against Ls-AChBP in an [^{3}H]epibatidine competitive binding assay. This represented a hit rate of 56%. Interestingly the majority of the hit compounds (71%) can be qualified as fragments (number of heavy atoms (HAs) ≤ 22, average HA = 18). Initially, we were surprised by the high number of fragment-like hits since docking scoring functions generally tend to exhibit a size dependence when ranking molecules.[25] However, a possible explanation for the high number of fragment-like *in silico* hits may be that the crystal structure used as structural template was the Ls-AChBP protein bound to the fragment-sized ligand nicotine (**2**) (this compound has only 12 HAs, see Figure 9.1) (PDB:1UWV).[10] Furthermore, the ligand induces a full closure of AChBP's flexible loop C (see Section 9.3.3) resulting in a relatively small binding-site cavity. As a result, docking of larger drug-like molecules is likely to result in steric clashes and/or energetically unfavorable binding poses compromising their docking scores. In general, it can be difficult to prioritize the right fragments on the basis of general scoring functions. To address these complications, the ranking of fragments using general docking scoring functions in combination with protein–ligand interaction fingerprint (IFP) scoring methods[22] can drastically improve the results. Very recently, de Graaf and co-workers applied this methodology on the G protein-coupled histamine H_1 receptor and identified structurally novel fragment-like compounds with unprecedentedly high hit rates (73%).[8] These results and studies by others on different target proteins demonstrate that virtual screening methods can be very successful in the identification of fragment hits and provide medicinal chemists with useful chemical starting points in the very early stages of a particular drug-discovery project.[20,26–28]

9.2.2 Screening for AChBP Fragment Hits Using Biochemical Screening Methods

Biochemical screening methods have been successfully used in drug discovery for many years and many highly sensitive technologies that can be adapted for identification of fragment-like ligands are available. When considering receptors or enzymes, various assay formats make use of considerable signal amplification that makes the readout more pronounced.[29] Also radioligand displacement assays represent very sensitive means to assess binding affinities. This approach obviously requires the availability of a labeled reference compound. Fragment library screening is rather straightforward in cases where the endogenous ligand or established reference compounds are fragment-like entities that can be converted into a radioligand (*e.g.*, examples can be found in the area of g-protein coupled receptors (GPCRs) such as histamine receptors and ligand-gated ion channels (LGICs)

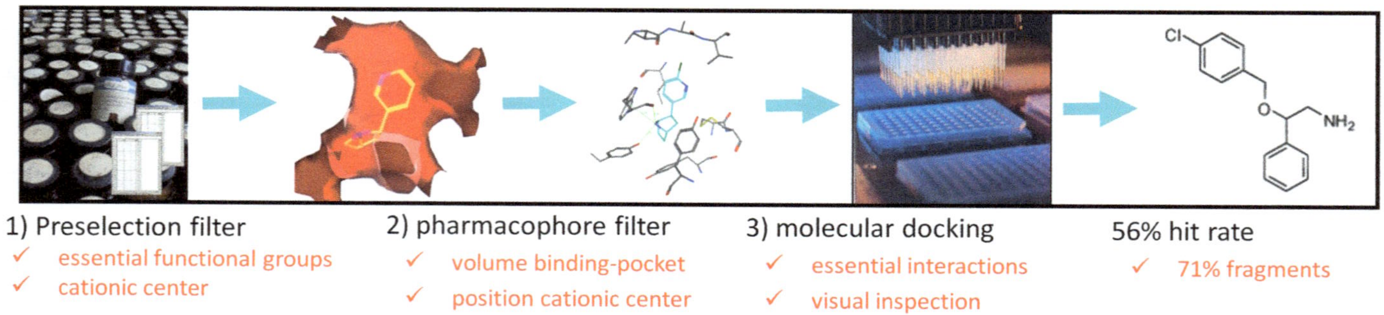

Figure 9.2 Schematic representation of the stepwise hierarchical *in silico* screening protocol against the nicotine (**2**) bound Ls-AChBP structure that resulted in a high hit rate of 56% from which the majority of hits were fragment-like compounds (71%, HA ≤ 22).

such as 5-HT_3 and nAChRs).[8,30] Considering the high LE of the endogenous ligands, library screening is typically performed at relatively low concentrations (10–40 μM). Whereas for some LGICs such as 5-HT_3 receptors, very sensitive biochemical assays were set up by monitoring second messengers such as Ca^{2+},[31] such a functional readout is not possible for AChBP (the role of the protein is to scavenge acetylcholine (**1**) but binding does not lead to a biochemical response in any cell). Instead, radioligand displacement assays can be performed for fragment library screening against AChBP.[32]

9.2.3 Screening for AChBP Fragment Hits Using Online Fluorescence Enhancement Assays

A fragment-screening assay for AChBP that utilizes a ligand that shows fluorescence enhancement upon binding to AChBP was recently developed by Kool and co-workers.[33] Following a SAR study around a benzylidene anabeseine core structure, that previously had been recognized as an AChBP fluorescence-enhancement ligand,[34] 4-diethylamino-2-hydroxybenzylidine anabeseine (DAHBA (**4**), see Figure 9.3) was identified as an optimal displacement ligand for an online bioaffinity assay system using HPLC equipment. The optimized online bioaffinity assay was used for screening the proprietary

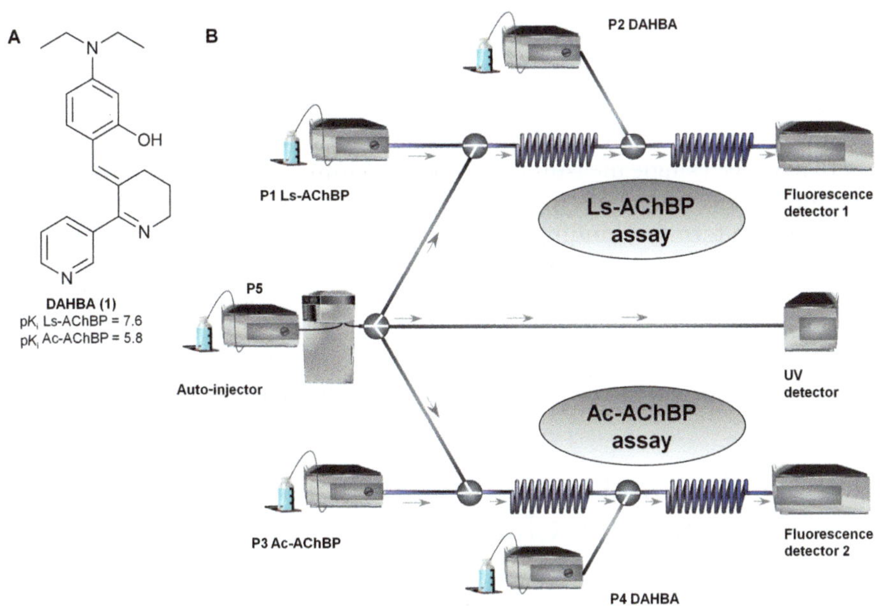

Figure 9.3 (A) Chemical structure and binding affinities of DAHBA (**4**) for Ls- and Ac-AChBP. (B) General schematic setup of the online bioaffinity analysis system that was used for screening the proprietary fragment library simultaneously on Ls- and Ac-AChBP. P1–P5 = HPLC pumps. Injection with split to the two online assays and the UV detector. Online addition of Ls- (P1), Ac-AChBP (P3), tracer ligand (P2 and P4), and test compounds (P5) *via* reaction coils to fluorescence detectors.

fragment library[35] containing 1010 fragments against Ls- and Ac-AChBP, simultaneously (Figure 9.3). Hit rates of 7.7% (78 hit compounds) and 1.1% (11 hit compounds) were observed for Ls- and Ac-AChBP, respectively. The lower hit rate for Ac-AChBP was expected as most known ligands have a lower affinity for Ac-AChBP compared to Ls-AChBP. In this case, the lower affinity and LE of ligands for Ac-AChBP is in-line with the druggability of these targets.[36–39] Of the 11 fragment hits identified for Ac-AChBP, 8 were also amongst the Ls-AChBP hits. Of the complete hit set, 23 diverse fragments that were considered novel chemotypes (dissimilar from known α7 nAChR ligands in the ChEMBL database,[40] Pipeline Pilot,[41] and ECFP4,[42] with Tanimoto coefficients ≤0.8) were selected for hit validation by an orthogonal assay. Binding affinities were determined by obtaining full dose–response curves using an [^{3}H]-epibatidine radioligand binding assay (RBA) and all 23 fragments displaced the radioligand with a pK_i > 4.5 and were confirmed as Ls-AChBP binders. It is interesting to note that the level of DAHBA displacement in the fluorescence-enhancement assay is proportional to the RBA-determined binding affinities of the fragment hits. By performing duplicate injections of a single concentration of each fragment hit (5×10^{-4} M, 40 μL) into the online assay system and comparing the average relative displacement of DAHBA to the signal that was derived by 100% displacement by nicotine, the binding affinity could be estimated (using the formula that describes sigmoidal dose–response curves; %displacement = $100\%/(1 + 10^{\log IC_{50} - \log[\text{ligand}]})$. These binding affinity estimations compared very well to the RBA-determined pK_i values ($R^2 = 0.80$). Altogether, these studies show that the online fluorescence-enhancement system provides an efficient screening assay that allows identification of fragment hits for AChBP with an estimated binding affinity as low as 10 mM. As a single measurement takes 1.5 min, screening a library of 1000 fragments with duplicate injections allowing for accurate ranking of fragment hits in terms of binding affinity can be performed in about 2 days. The robustness and ease of use was further demonstrated in the combichem-inspired fragment optimization approach described in Section 9.3.5. All in all, the online fluorescence-enhancement system represents an ideal assay that can easily be handled by synthetic chemists and placed within an organic chemistry laboratory to swiftly guide fragment hit optimization studies.

9.3 Orthogonal Screening Methods Used for Fragment Hit Characterization and Monitoring Optimization

9.3.1 Validation of SPR Biosensor-Based Screening of Fragments for AChBP

To investigate if SPR biosensor analysis is applicable in screening fragments that have affinity for AChBP, high-affinity α7 nAChR ligands were deconstructed to fragments of different sizes and subsequently analyzed using an SPR Biacore T100 system.[32] Furthermore, the deconstruction study would assist in benchmarking future fragment hits for AChBP with regards to LE, enabling better evaluation in the hit-selection phase as well as in the subsequent fragment-optimization stage. First, a fragment set was designed from high-affinity quinuclidine-containing α7 nAChR ligands.[43] Four series of compounds were synthesized, covering molecular weights from 100 to 400 Da (Figure 9.4). The obtained compounds were analyzed using an SPR-based biosensor interaction assay that was obtained by immobilizing Ls-AChBP and experimental conditions that were optimized by the labs of Prof. Danielson.[32] In order to validate the affinities that were obtained using the SPR biosensor system, the same set of compounds was also analyzed using a RBA ([^{3}H]-MLA) with Ls-AChBP. An excellent correlation (R^2 = 0.97) between the determined affinities of both assays was observed, validating the quality of the SPR biosensor-based assay for the detection of small fragments (126 Da) with binding affinities as low as 10 mM.[44]

By calculating the LEs[5] and group efficiencies (GEs)[45] from the experimentally determined binding affinities, the focused fragment set affords insights into the contributions that different parts of the original quinuclidine ligand make to its binding energy. It should be noted that in this particular study LEs and GEs were calculated using an adjusted formula in which binding energy (difference) is divided by the molecular weight (difference) in kDa. In all other studies described in this chapter, LEs are calculated by dividing the binding energy of a ligand by the number of HAs. Fragments that contain a

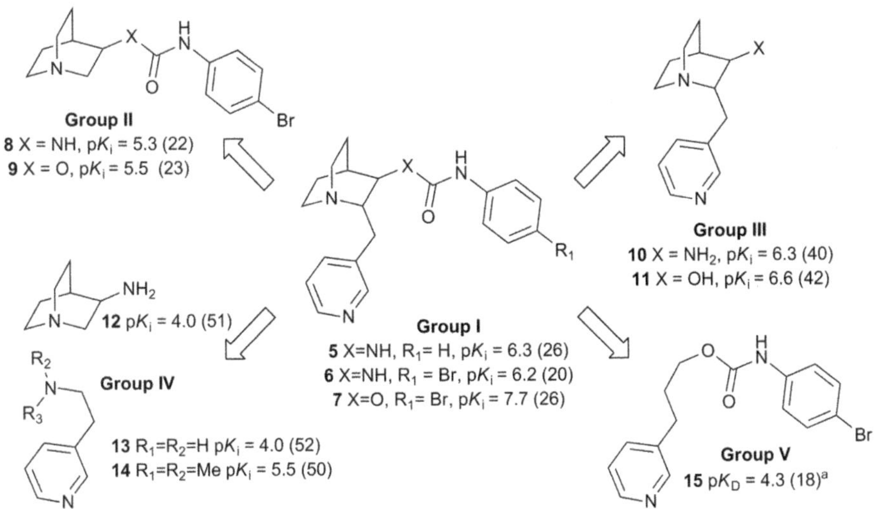

Figure 9.4 Generic structures of fragmentized α7 nAChR quinuclidine ligands (**5–15**) that were used for validation of the SPR biosensor AChBP interaction assay. Affinities for Ls-AChBP (pK_i) were determined using a [^{3}H]-epibatidine-displacement assay and LEs in kcal mol^{-1} kDa^{-1} are depicted in brackets. In this particular study, LEs were calculated using an adjusted formula from Hopkins *et al.*[5] LE = ΔG (in kcal mol^{-1})/M_w (in kDa). [a]Affinity for Ls-AChBP could not be determined using RBA and pK_D as determined by SPR biosensor analysis is reported here.

basic nitrogen atom display the highest LEs, whereas fragments lacking this functionality afford much lower LEs. Furthermore, it can be seen that decorating the quinuclidine 3-position with substituted phenyl moieties using polar linkers leads to compounds with high affinities for Ls-AChBP (as well as α7 nAChR) but results in significant reduction in LE. These results were not surprising as it is known from AChBP X-ray structures and site-directed mutagenesis studies on nAChRs that the basic nitrogen atom of nicotinic receptor ligands forms strong cation–π interactions with aromatic residues that align the binding site and form an aromatic cavity.[17–19]

These affinity differences between the studied ligands were translated into a LE hotspot map projected on the binding site of Ls-AChBP. The mapping of LE hotspots was performed by docking representative fragments into an ACHBP X-ray crystal structure (PDB: 1UW6). The molecular surface of the binding site was subsequently color-coded by projecting the LEs of the docked ligands and GEs of ligand moieties onto the protein atoms that are within a distance of 2.5 Å (Figure 9.5). The concept of LE hotspots implies that the fragment hit selection process that is often based on selecting hits

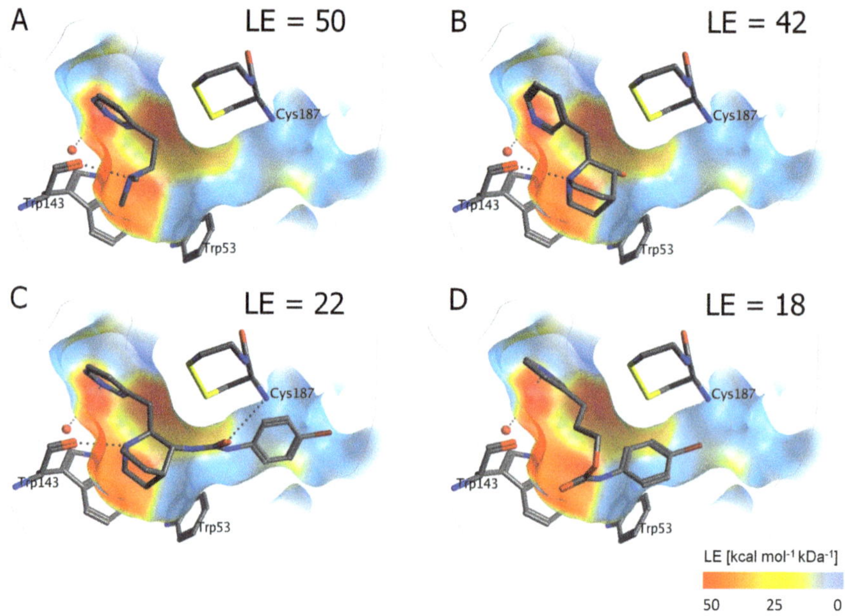

Figure 9.5 Binding modes that were obtained by molecular docking in Ls-AChBP (PDB:1UW6) of fragments **14** (A), **11** (B), **6** (C) and **15** (D) that were obtained by deconstruction of a high-affinity α7 nAChR quinuclidine ligand. The color coding of the surface was obtained by projecting the LEs of docked ligands and GEs of ligand moieties onto the molecular surface of the AChBP-binding site at 2.5 Å distances. In this particular study, LEs and GEs were calculated using adjusted formulae from Hopkins *et al.*[5] LE = ΔG (in kcal mol^{-1})/M_w (in kDa), and Verdonk and Rees.[45] GE = $\Delta\Delta G$ (in kcal mol^{-1})/ΔM_w (in kDa).

with the highest LEs may need reconsideration. For example, fragments that are expected to bind to LE hotspots are expected to display higher LEs, whereas fragments outside these hotspots display lower LEs. However, if a hit fragment is evidently binding to a region outside the LE hotspot (in the case of AChBPs this seems likely for hits that do not have a basic nitrogen atom), it might still be worth considering the entity for fragment growing towards the LE hotspot. This hypothesis also implies that LEs do not necessarily remain constant during fragment growing.[46]

In conclusion, the study described in this section shows that SPR biosensor analysis can be applied to detect AChBP fragments of low molecular weight and low binding affinity. In addition, it was shown that the deconstruction of a high-affinity ligand into smaller fragments can be used to map a protein-binding site with LE hotspots. In the next section, a study is summarized in which a LE hotspot map enabled the efficient optimization of a fragment hit.

9.3.2 Ligand Efficiency Hotspot Guided Optimization of an AChBP Fragment Hit

One of the fragment hits that was identified by the biochemical fragment screen against Ls-AChBP described in Section 9.2.2 is a thiazoline-substituted *N*-phenylpiperazine (**16**) that displays 10 μM affinity for Ls-AChBP (LE = 0.40) (Figure 9.6).

Since our proprietary fragment library originates from ligands and intermediates that have been synthesized in our historical drug-discovery programs, close analogs of identified fragment hits are available in the proprietary compound collection. As such, fragment hits can immediately be followed up by testing readily available structural analogs. As described by de Kloe and co-workers,[47] a similarity search in ChemBiofinder,[48] was used to identify close analogs (defined as Tanimoto co-efficient ≥0.85). One of these close analogs is fragment **17** in which the piperazine moiety is replaced by a piperidine moiety. Substituting one nitrogen atom for a carbon atom led to a 50-fold increase in binding affinity (Figure 9.6). This result illustrates that obtaining and testing close analogs of fragment hits (SAR-by-catalogue) can be a fruitful endeavor before embarking on fragment optimization by growing and/or linking. A subsequent close analog search using ROCS (Rapid Overlay of Chemical Structures) shape-based screening (based on three-dimensional shape and electronic properties)[49] against the in-house library of commercially available lead-like compounds of Solvay pharmaceuticals (acquired by Abbott in 2010) resulted in the identification of the tetrahydrothienopyridine **18**, known as the marketed antiplatelet drug Ticlopidine. Ticlopidine showed good binding affinity for Ls-AChBP (pK_i 6.6, LE = 0.53) and was selected for further optimization (Figure 9.6). As structural information on how Ticlopidine would bind to Ls-AChBP was missing, a binding model was generated using molecular docking experiments and the LE hotspot maps described in Section 9.3.1. Since Ticlopidine exhibits good LE for Ls-AChBP and contains a basic nitrogen atom, it was anticipated that the fragment would bind

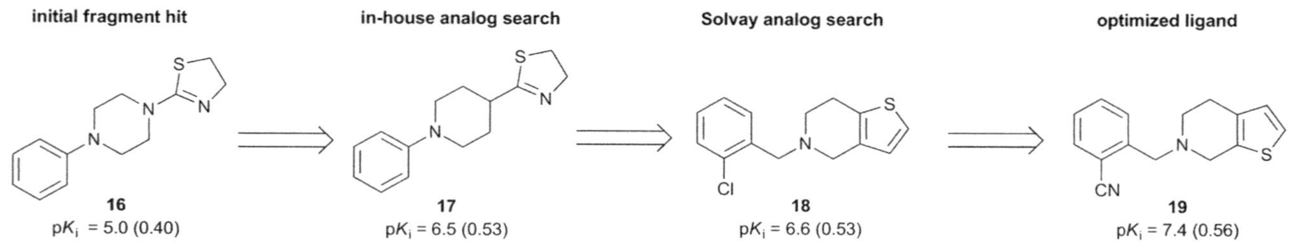

Figure 9.6 Optimization of fragment hit **16** for Ls-AChBP was established by a combination of analog searches and subsequent LE hotspot map guided optimization. Affinities for Ls-AChBP were determined using a [^{3}H]epibatidine-displacement assay, LEs in kcal mol^{-1} per HA are depicted in brackets.

at the hotspot region as depicted in Figure 9.5A. Based on this hypothetical binding mode, it was expected that incorporation of hydrogen-bond-accepting substituents on the 2-position of the benzyl moiety may lead to hydrogen-bond formation with a conserved water molecule. Several benzaldehydes were coupled to the tetrahydrothienopyridine scaffold by reductive amination. In accordance with the proposed binding mode, it was found that replacing the 2-chlorobenzyl of **18** for a 2-cyanobenzylmoiety resulted in a significant increase in binding affinity (pK_i 6.6 → 7.2). A subsequent investigation into optimization of the fragment's scaffold showed that moving the basic nitrogen atom one position within the tetrahydropyridine ring (**19**) afforded a modest additional increase in binding affinity for Ls-AChBP ($pK_i = 7.4$, LE = 0.56, Figure 9.6). With only 46 compounds synthesized, a 220-fold increase in binding affinity was established.

It is noted that in this fragment optimization exercise no X-ray crystal structures to guide fragment hit optimization were obtained, and the process could have been even more successful if structural data was acquired. Nevertheless it was shown that by using molecular docking in combination with binding mode prediction guided by LE hotspot maps, hit fragments such as **16** can be efficiently optimized.

9.3.3 SPR-Based Screening Reveals Ligand-Induced Conformational Changes in AChBP with Rapid and Slow Kinetics

Using a subset of the fragments that were obtained by deconstruction of the high-affinity α7 nAChR quinuclidine ligand (see Section 9.3.1 and Figure 9.4), an SPR biosensor-based assay was developed in close cooperation with our partners from Beactica.[44] The aim of setting up this biosensor-based assay was to characterize the interaction kinetics of fragments and ligands binding to Ls-AChBP. Unfortunately, the kinetics of many reference ligands are very fast and their binding affinities could not be dissected into the rate constants k_{on} and k_{off} due to mass transport limitations. For example, the binding of nicotine to Ls-AChBP in a buffer system without DMSO was completely governed by mass transport and values for k_{on} and k_{off} could not be determined. Interestingly, assay conditions proved to have a pronounced affect on the behavior of the sensitive biophysical assay. It was shown that when the buffer contains 5% DMSO, the binding kinetics are slower and the association and dissociation rate constants for nicotine (**2**) binding to AChBP could be estimated ($k_{on} = 12.5 \pm 6.21 \times 10^6\ M^{-1}\ s^{-1}$ and $k_{off} = 3.62 \pm 1.48\ s^{-1}$).

Furthermore, the SPR-based biosensor analysis allowed the detection of ligand-induced rapid and slow conformational changes of Ls-AChBP upon binding of fragment-sized ligands such as nicotine (**2**) and epibatidine (**3**) (Figure 9.1). By careful analysis of the obtained sensorgrams, Geitmann and co-workers revealed a ligand-induced slow conformational change with a rate constant in the 0.01 s^{-1} range.[44] The sensorgrams showed a dose-dependent slow signal reduction during the injection phase and the signal returned slowly back to baseline level with a rate that was slower than the apparent dissociation of the ligand. De-sensitization against this effect was observed when surfaces were exposed to a series of identical injections with short intervals. However, the surface regained sensitivity with longer waiting times between injections. Based on these observations and a carefully designed set of experiments that excluded methodological artifacts, it was concluded that the slow signal changes were caused by ligand-induced conformational changes that most likely are similar to conformational changes associated with agonist-induced de-sensitization of Cys-loop receptors. Although AChBP does not contain the membrane-spanning domain that constitutes the ion channel of mammalian Cys-loop receptors, the observed slow conformational changes in AChBP are likely to be relevant to conformational changes in Cys-loop receptors mediated by the binding of agonists. This is supported by experiments by Bouzat *et al.* showing that when AChBP is linked to an 5-HT_3 ion channel, a fully functional ion channel is obtained that opens upon binding of an agonist.[50] The detection of these slow conformational changes associated with ligand binding may be utilized in the design of new drugs for Cys-loop receptors as it would make a difference if a drug compound stabilizes an ion channel in the open or in the de-sensitized state.[51]

In addition to large signals resulting from slow conformational changes of AChBP, a more subtle effect arising from a form of fast ligand-induced SPR-based biosensor change was also observed. This effect was evident from large differences in the maximal signal weights for acetylcholine (**1**), nicotine (**2**) and epibatidine (**3**). Despite similar molecular weights, acetylcholine had a two times lower R_{max} value than nicotine, whereas usually SPR biosensor signal magnitudes are proportional to the molecular weight of the analyte. A similar discrepancy was observed for the ligands that were derived from the deconstruction of the quinuclidine-containing α7 nAChR ligands **5–7** (Figure 9.4). After normalization of the signals for the molecular weights, the determined R_{max} values for the different quinuclidine derivatives differed by almost 50%. Interestingly, the observed R_{max} values were correlated with the molecular volumes of the ligands ($r = 0.86$). In contrast to the slow changes described above, these changes did not show up as complexities in the sensorgrams and can therefore be concluded to occur fast, or faster, than the rates of the fragment or ligand binding to AChBP. Beactica scientists set up a modeling procedure that combined an induced-fit docking procedure that allowed flexibility of loop C that can be considered as a lid that encloses the binding site (see Figure 9.7). Individual binding mode hypotheses were generated for the investigated fragments and ligands. This modeling study supported a mechanism where the SPR-based biosensor signal was influenced by a graded structural response depending on the size of the ligand, as significant ligand-dependent conformational changes were observed. In addition, for the quinuclidine ligands with the largest substituents (*e.g.*, **5** and **6**), binding to AChBP was associated with side-chain rotations of the tyrosine residues that align the binding site. This study therefore exemplifies that optimization of fragment hits by growing can result in significant conformational changes of the receptor protein's binding site and that ligand-induced conformational changes may be detected using an SPR-based biosensor

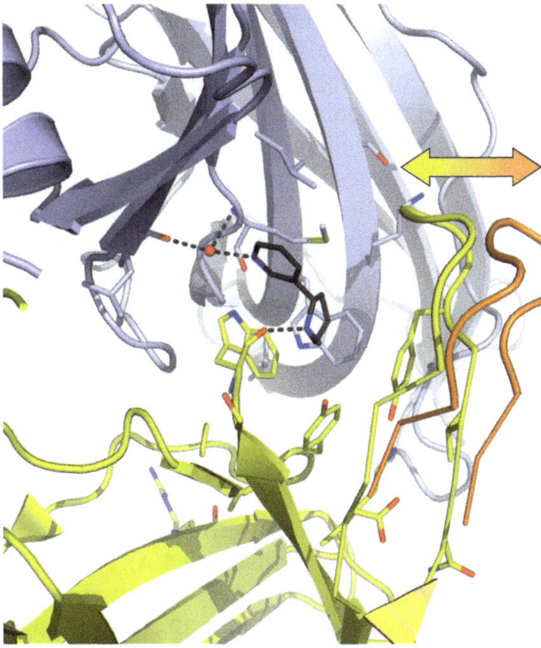

Figure 9.7 Superposition of the crystal structures of nicotine- (PDB:1UW6, yellow and blue ribbons) and methyllycaconitine-bound Ac-AChBP (PDB:2BYR, orange ribbon). For clarity, only nicotine (**2**) and not methyllycaconitine is displayed in sticks. This overlay exemplifies the ligand-dependent conformational changes in the flexible loop C (below the double-headed arrow) that could be detected using a biosensor-based assay for direct interaction studies of immobilized AChBP.

system. The SPR biosensor-based studies show the power of the SPR-based biosensor assay technology and indicate the wealth of molecular interaction data that can be inferred when these experiments are carefully performed.

9.3.4 Thermodynamic Analysis of Fragment Optimization: Growing into a Ligand-Inducible Subpocket of AChBP

Recently, we have monitored the optimization of a fragment (**20**) by growing with X-ray structures of ligand complexes and systematic thermodynamic analysis using both SPR and ITC.[52] Using AChBP crystal structures to guide fragment growing we were able to increase the binding affinity 150-fold with only a few compounds synthesized. Fragment **20** was identified by structural modification of an *in silico* hit (identified using virtual screening methods as described in Section 9.2.1) and subsequently co-crystallized with Ac-AChBP. In the 3.65 Å resolution crystal structure, the fragment was present in all five binding sites and the ability to use 5-fold restraints in the crystallographic refinement allowed confident building of the fragment into electron density despite the low resolution. A comparison with previously obtained co-crystal complexes revealed conformational changes of the target protein upon ligand binding, mainly with respect to the loop C which closes the binding site upon agonist binding.[12,53–56] However, a remarkable conformational change was observed for AChBP while binding to lobeline (**21**, see Figure 9.8), leading to the opening of a subpocket that enables the binding of the α-hydroxyphenethyl moiety of the ligand.[12] More specifically, this subpocket that we will refer to as the *lobeline pocket*, becomes accessible after a change in the rotameric state of Tyr91 (*g*- to *t*-conformation,[57] hereafter referred to as the *tyrosine-flip*). Using the structural information we predicted that further optimization of the hit fragment could be achieved by the opening of a ligand-inducible subpocket and growing the hit fragment into this binding region (see Figure 9.8).

Targeting the lobeline pocket by extending fragment **20** with an α-(*R*)-hydroxyphenethyl moiety yielded compound **22** and resulted in a ~50-fold increase in affinity (pK_i = 7.0 ± 0.1), although the LE dropped from 0.43 to 0.37 kcal mol^{-1} per HA (Figure 9.9A). A 3.59 Å resolution crystal structure of optimized fragment **22** bound to Ac-AChBP showed that fragment **20** was successfully grown into the lobeline pocket. The experimentally determined binding mode corresponded very well to the predicted binding mode that was obtained by molecular docking (rmsd of 1.1 Å). Nevertheless, the experimentally determined binding mode differed from the *in silico* predicted binding mode in that the hydroxyl group of **22** is not engaged in hydrogen bonding to two backbone carbonyl oxygen atoms, but instead is positioned in the hydrophobic area of the binding site that is formed by the side-chain of a tyrosine residue.

Since the hydroxyl group of the optimized fragment **22** is not involved in the formation of hydrogen bonds with the binding site, it was expected that its removal would lower the de-solvation penalty and thereby result in an increase in binding affinity. The norhydroxyl derivative (**23**) was therefore synthesized and screened for AChBP affinity. Indeed, **23** exhibited higher AChBP affinity (pK_i = 7.5 ± 0.1) than **22** and when compared with the starting fragment **20**, a 150-fold increase in binding affinity is observed (Figure 9.9A). To determine if **23** induces the tyrosine-flip and interacts with the lobeline pocket, an additional co-crystal structure was generated. The compound **23**–Ac-AChBP structure was solved with a resolution of 3.30 Å and shows an almost identical binding mode to **22** (root mean square deviation (rmsd) of 0.6 Å) validating that fragment **20** was also grown into the lobeline pocket by extending its nitrogen atom with a phenethyl moiety.

Having addressed the lobeline pocket successfully, the thermodynamic binding signature was explored, *i.e.*, does the binding affinity increase originate from favorable changes in enthalpy or entropy?[52] As such, the thermodynamic aspects of growing a fragment into the ligand-induced lobeline pocket of Ac-AChBP were measured by SPR biosensor analyses, using an assay developed by Geitmann *et al.*,[58,59] at five different temperatures (15, 20, 25, 30 and 35 °C) and also by using ITC (Figure 9.9B).

Thermodynamic binding analysis showed that extending fragment **20** with an α-(*R*)-hydroxyl-substituted phenethyl moiety (leading to **22**) results in a significant shift in the thermodynamic binding signature for Ac-AChBP

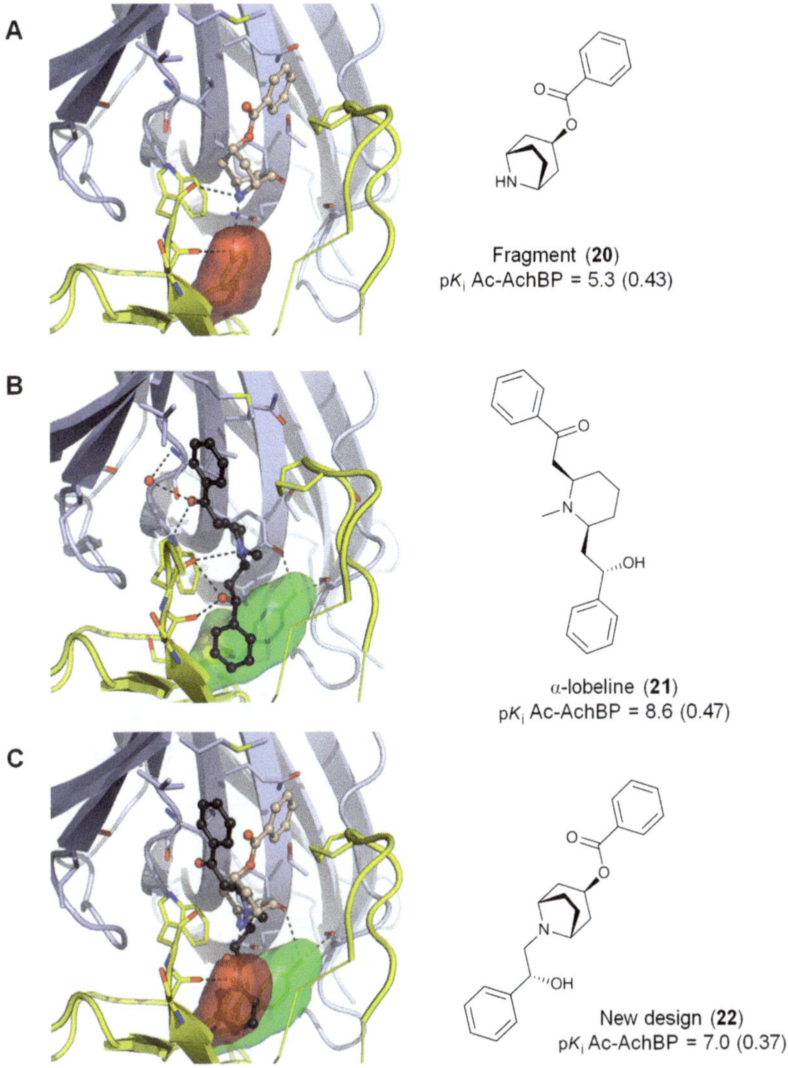

Figure 9.8 Fragment-optimization strategy. Cartoon ribbon representations of the crystal structures of fragment **20**-bound Ac-AChBP (A) and of lobeline (**21**) bound Ac-AChBP (PDB:2BYS) (B). Ac-AChBP affinities were determined using a [^{3}H]-epibatidine-displacement assay and the LE in kcal mol^{-1} per HA is depicted in brackets. (A) In the fragment **20**–Ac-AChBP complex, Tyr91 (red surface) is in the *g*-conformation rendering the lobeline pocket inaccessible. (B) However, Tyr91 adopts a *t*-conformation (green surface) in the lobeline–Ac-AChBP complex, leading to opening of the lobeline pocket. (C) The superposition of the fragment **20** and lobeline (**21**) molecules indicates that the fragment may be optimized by growing into the lobeline pocket.

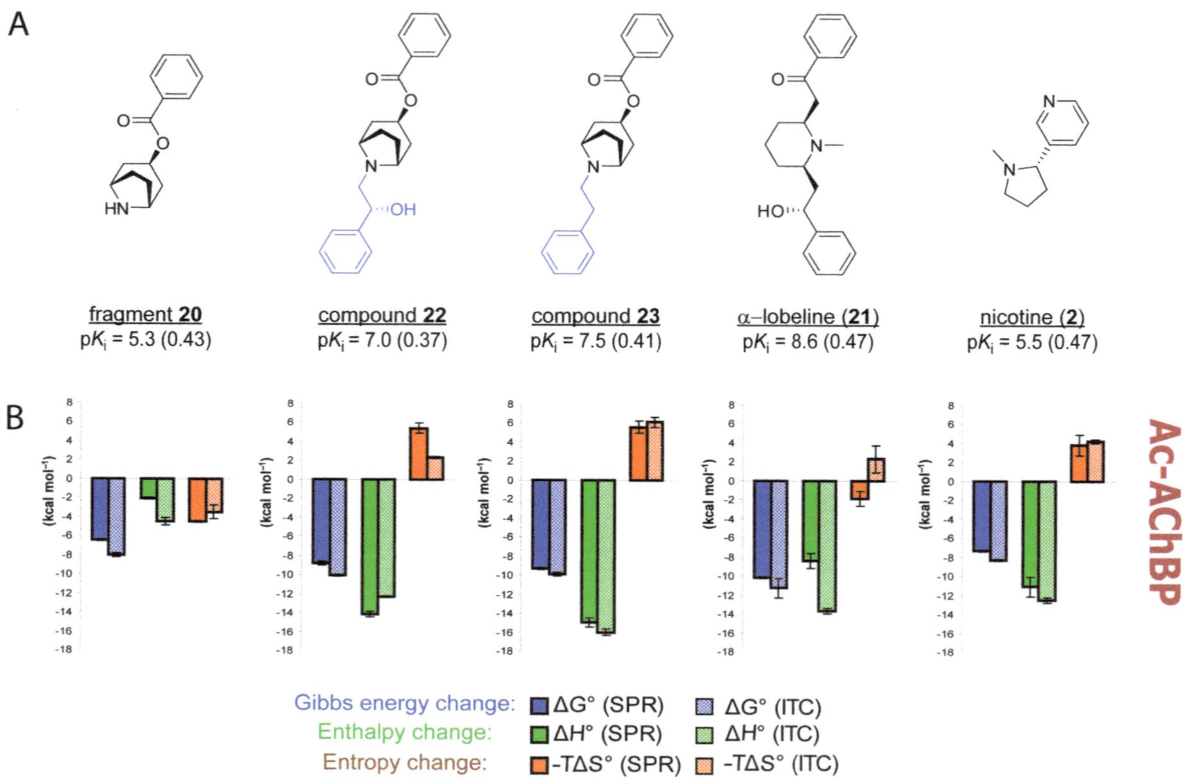

Figure 9.9 (A) Chemical structures and binding affinities for Ac-AChBP as determined by a [^{3}H]-epibatidine-displacement assay of the compounds that were evaluated using SPR biosensor analysis and ITC. LEs of each compound are depicted in brackets in kcal mol^{-1} per HA. (B) Thermodynamic profiles for binding to Ac-AChBP using SPR biosensor analysis (full bars ± standard error of the mean (SEM)) and ITC (hatched bars ± fitting errors), including the reference compounds lobeline and nicotine.

where a large increase in favorable enthalpic contribution (about −12 kcal mol^{-1}) to the binding is observed for the optimized fragment **22** (Figure 9.9B). An even more favorable enthalpic contribution was observed when the fragment was extended with a phenethyl moiety (leading to **23**). These favorable enthalpic contributions are compensated to a considerable extent by an unfavorable entropic contribution to the binding for both ligands. The observed favorable change in enthalpic contributions upon extending fragment **20** with a hydrophobic phenethyl moiety is in contrast with the classical view on the hydrophobic effect, in which increased burial of hydrophobic moieties in a hydrophobic pocket results in favorable changes in entropy. Studies on major mouse urinary protein suggest that enthalpy-driven hydrophobic association can result from poor solvation of the protein-binding site, prior to complex formation.[60–62] The apo-Ac-AChBP X-ray structure (PDB:2W8E)[55] shows that the residue Tyr91 is in the *g*-conformation and functions as a gatekeeper making the lobeline pocket inaccessible. It can therefore be speculated that prior to complex formation, the lobeline pocket is poorly solvated. A significant part of the favorable change in enthalpy may therefore result from strong van der Waals interactions between the phenethyl moiety of compound **23** and the protein-binding site, that are not compensated by the solvent.

As growing the fragment into the lobeline pocket resulted in a distinct change of the thermodynamic profile of binding, this study indicates that more than binding affinity alone, dissection of binding affinity into the separate enthalpic and entropic parameters, can provide valuable information with regards to the binding mode of a ligand. It is also apparent that our understanding of the underlying principles (such as structure–kinetics relationships) needs improving. Furthermore, using both SPR biosensor analysis and ITC, we obtained a thermodynamic dataset of ligands binding to Ac-AChBP[52] showing that these fundamentally different techniques result in very comparable thermodynamic profiles (Figure 9.9) in this particular case. A major advantage of using SPR biosensor-based analysis for thermodynamic analysis is that this method requires substantially lower amounts of protein compared to ITC. When using SPR biosensor-based screening assays for thermodynamic characterization, one should be aware that switching between temperatures, which is needed for thermodynamic analysis, significantly lengthens measurement times.

9.3.5 Rapid Optimization of an AChBP Fragment Hit Using Combinatorial Chemistry Approaches

One of the fragment hits (**24**) that was identified using the online fluorescence-enhancement bioaffinity system described in Section 9.2.3 showed a high LE value of 0.62 kcal mol^{-1} per HA for Ls-AChBP as well as for the α7 nAChR and was therefore considered for hit optimization (Figure 9.10). The fragment hit **24** contained an oxime functionality which inspired us to pursue a combinatorial chemistry approach based on the condensation of

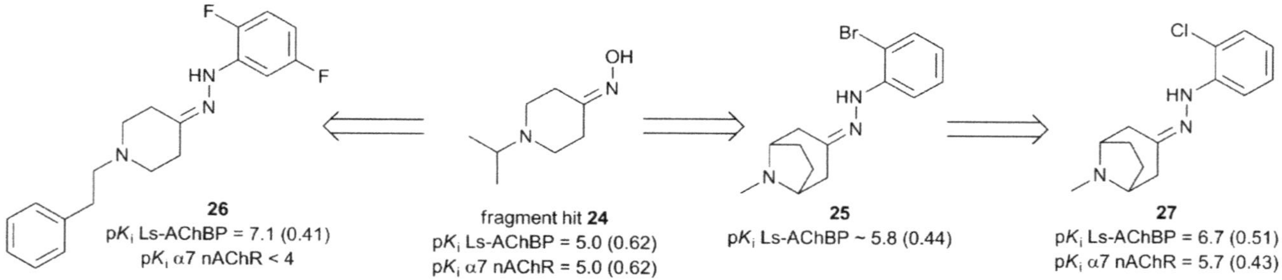

Figure 9.10 Chemical structures and pharmacological data for the initial fragment hit and compounds that were obtained by optimization using a combinatorial chemistry approach and the online bioaffinity fluorescence enhancement assay. Ls-AChBP and α7 nAChR affinities were determined using [^{3}H]-epibatidine- and [^{3}H]-methyllycaconitine-displacement assays, respectively. LEs in kcal mol^{-1} per HA are depicted in brackets.

ketones with hydroxylamines or hydrazines, enabling rapid SAR exploration. As such, two generations of focused chemical libraries were constructed in 96-well format. The first generation was based on three different piperidinones that were reacted with substituted hydroxylamines as well as hydrazines. The resulting products were screened for Ls- and Ac-AChBP affinity simultaneously in the robust online assay that is described in Section 9.2.3. The product with the highest estimated affinity originated from the combination of (2-bromophenyl)hydrazine and tropinone (compound **25**, Figure 9.10). Based on these results, the second-generation chemical library that was screened contained products derived from the condensation of mainly phenylhydrazines with different substitution patterns and a variety of piperidinones. The obtained condensation products that showed a high level of displacement of the fluorescent tracer ligand were subsequently re-synthesized on a larger scale. Subsequently their chemical structures were validated by NMR and LC-MS and their binding affinities determined accurately by obtaining full displacement curves. The two ligands that showed the highest affinity for Ls-AChBP were the phenethyl-substituted piperidine **26** (pK_i = 7.1) and the tropinone derivative **27** (pK_i = 6.7), see Figure 9.10. Thus, by synthesizing only two small combinatorial libraries in a 96-well format and subsequent screening in an online bioassay system, the binding affinity of hit fragment **24** could be increased 100-fold for Ls-AChBP. In addition, optimized fragment **27** showed good binding affinity for the homologous human α7 nAChR (pK_i = 5.7, LE = 0.43) illustrating that fragment screening and subsequent optimization against AChBP can deliver interesting fragment leads for the therapeutically relevant α7 nAChR. This study[63] shows that the use of a very robust screening assay in combination with combinatorial chemistry approaches can lead to rapid fragment hit exploration.

9.4 Conclusion and Perspectives

The approval of Zelboraf® (Vemurafenib) in 2011 is an important milestone showing that FBDD can enable an efficient hit-identification and optimization process (six years from the start of the program to FDA approval).[64,65] With an approved drug and multiple fragment-derived candidates in clinical studies, the methodology has proven to be of great value in the drug-discovery process.[1] Next to the promise of drug candidates, the fragment-based methodology is also of interest for identification of tool compounds and studying fundamental aspects of ligand–protein interactions.[8,32,44,66] These studies are enabled by the ever increasing number of screening techniques.[67–70]

Fragment-based studies that are described in this chapter include the optimization of *in silico* docking protocols for identification of fragment hits, the development of an online fluorescence-enhancement fragment screening assay and the biophysical characterization of conformational changes induced by fragment growing. These latter studies performed on AChBP exemplify that monitoring fragment-optimization processes with biophysical techniques such as SPR biosensor-based analysis can provide a wealth of information including kinetic and thermodynamic data and increases our knowledge of molecular interaction events. It is noted that determination of thermodynamic binding profiles, *i.e.*, dissecting binding energy into the separate enthalpic and entropic contributions, may be useful in fragment hit selection and guiding and monitoring (structure-based) fragment optimization.[71–74] On the other hand, in combination with structural data, the step-wise fragment-growing process provides an ideal dataset with which to improve our understanding of the physical forces that drive ligand–protein complexes. Therefore, both FBDD, and our understanding of the thermodynamics of ligand–protein complex formation have much to gain from the application of biophysical screening assays.

Due to its stability and water-soluble nature, we have found AChBP to be an excellent research tool to study various aspects of fragment-based approaches. The nAChR ligand-binding domain homolog provides relative easy access to X-ray co-crystal structures enabling structure-based optimization. Moreover, AChBP was found to be applicable in different biochemical and biophysical screening assays, allowing a comparison in terms of ease of setting up the assay, efficiency, and obtained information content. Combined with the availability of many LGIC reference compounds, the presented case allowed detailed characterization studies of molecular interactions. Especially, combining fragment-based methods with X-ray crystallography, *in silico* approaches, biophysical screening techniques and chemical genomics approaches provides a powerful tool for academic research groups to further investigate the molecular aspects of fragment and ligand recognition by protein drug targets.

References

1. M. Baker, *Nat. Rev. Drug Discovery*, 2013, **12**, 5.
2. G. Chessari and A. J. Woodhead, *Drug Discovery Today*, 2009, **14**, 668.
3. D. A. Erlanson, *Top. Curr. Chem.*, 2012, **317**, 1.
4. C. W. Murray and D. C. Rees, *Nat. Chem.*, 2009, **1**, 187.
5. A. L. Hopkins, C. R. Groom and A. Alex, *Drug Discovery Today*, 2004, **9**, 430.
6. Y. Amano, T. Yamaguchi and E. Tanabe, *Bioorg. Med. Chem.*, 2014, **22**, 2427.
7. A. Ciulli, G. Williams, A. G. Smith, T. L. Blundell and C. Abell, *J. Med. Chem.*, 2006, **49**, 4992.
8. C. de Graaf, A. J. Kooistra, H. F. Vischer, V. Katritch, M. Kuijer, M. Shiroishi, S. Iwata, T. Shimamura, R. C. Stevens, I. J. de Esch and R. Leurs, *J. Med. Chem.*, 2011, **54**, 8195.
9. D. Chen, J. C. Errey, L. H. Heitman, F. H. Marshall, A. P. Ijzerman and G. Siegal, *ACS Chem. Biol.*, 2012, **7**, 2064.
10. K. Brejc, W. J. van Dijk, R. V. Klaassen, M. Schuurmans, J. van Der Oost, A. B. Smit and T. K. Sixma, *Nature*, 2001, **411**, 269.

11. P. H. Celie, S. E. van Rossum-Fikkert, W. J. van Dijk, K. Brejc, A. B. Smit and T. K. Sixma, *Neuron*, 2004, **41**, 907.
12. S. B. Hansen, G. Sulzenbacher, T. Huxford, P. Marchot, P. Taylor and Y. Bourne, *EMBO J.*, 2005, **24**, 3635.
13. S. B. Hansen, T. T. Talley, Z. Radic and P. Taylor, *J. Biol. Chem.*, 2004, **279**, 24197.
14. B. Billen, R. Spurny, M. Brams, R. van Elk, S. Valera-Kummer, J. L. Yakel, T. Voets, D. Bertrand, A. B. Smit and C. Ulens, *Proc. Natl. Acad. Sci. U. S. A.*, 2012, **109**, 9173.
15. T. McCormack, R. M. Petrovich, K. A. Mercier, E. F. DeRose, M. J. Cuneo, J. Williams, K. L. Johnson, P. W. Lamb, R. E. London and J. L. Yakel, *Biochemistry*, 2010, **49**, 2279.
16. Y. Bourne, T. T. Talley, S. B. Hansen, P. Taylor and P. Marchot, *EMBO J.*, 2005, **24**, 1512.
17. D. A. Dougherty, *Chem. Rev.*, 2008, **108**, 1642.
18. S. Mecozzi, A. P. West Jr and D. A. Dougherty, *Proc. Natl. Acad. Sci. U. S. A.*, 1996, **93**, 10566.
19. X. Xiu, N. L. Puskar, J. A. Shanata, H. A. Lester and D. A. Dougherty, *Nature*, 2009, **458**, 534.
20. Y. Chen and B. K. Shoichet, *Nat. Chem. Biol.*, 2009, **5**, 358.
21. M. M. Hann, A. R. Leach and G. Harper, *J. Chem. Inf. Comput. Sci.*, 2001, **41**, 856.
22. G. Marcou and D. Rognan, *J. Chem. Inf. Model.*, 2007, **47**, 195.
23. D. Rognan, *Top. Curr. Chem.*, 2012, **317**, 201.
24. A. Akdemir, P. Rucktooa, A. Jongejan, R. Elk, S. Bertrand, T. K. Sixma, D. Bertrand, A. B. Smit, R. Leurs, C. de Graaf and I. J. de Esch, *Bioorg. Med. Chem.*, 2011, **19**, 6107.
25. Y. Pan, N. Huang, S. Cho and A. D. MacKerell Jr., *J. Chem. Inf. Comput. Sci.*, 2003, **43**, 267.
26. M. Sandor, R. Kiss and G. M. Keseru, *J. Chem. Inf. Model.*, 2010, **50**, 1165.
27. J. Oyarzabal, N. Zarich, M. I. Albarran, I. Palacios, M. Urbano-Cuadrado, G. Mateos, I. Reymundo, O. Rabal, A. Salgado, A. Corrionero, J. Fominaya, J. Pastor and J. R. Bischoff, *J. Med. Chem.*, 2010, **53**, 6618.
28. F. Sirci, E. P. Istyastono, H. F. Vischer, A. J. Kooistra, S. Nijmeijer, M. Kuijer, M. Wijtmans, R. Mannhold, R. Leurs, I. J. de Esch and C. de Graaf, *J. Chem. Inf. Model.*, 2012, **52**, 3308.
29. S. J. Pollack, K. S. Beyer, C. Lock, I. Muller, D. Sheppard, M. Lipkin, D. Hardick, P. Blurton, P. M. Leonard, P. A. Hubbard, D. Todd, C. M. Richardson, T. Ahrens, M. Baader, D. O. Hafenbradl, K. Hilyard and R. W. Burli, *J. Comput.-Aided Mol. Des.*, 2011, **25**, 677.
30. M. H. Verheij, C. de Graaf, G. E. de Kloe, S. Nijmeijer, H. F. Vischer, R. A. Smits, O. P. Zuiderveld, S. Hulscher, L. Silvestri, A. J. Thompson, J. E. van Muijlwijk-Koezen, S. C. Lummis, R. Leurs and I. J. de Esch, *Bioorg. Med. Chem. Lett.*, 2011, **21**, 5460.
31. A. J. Thompson, M. H. Verheij, R. Leurs, I. J. De Esch and S. C. Lummis, *Biotechniques*, 2010, **49**, 822.

32. G. E. de Kloe, K. Retra, M. Geitmann, P. Kallblad, T. Nahar, R. van Elk, A. B. Smit, J. E. van Muijlwijk-Koezen, R. Leurs, H. Irth, U. H. Danielson and I. J. de Esch, *J. Med. Chem.*, 2010, **53**, 7192.
33. J. Kool, G. E. de Kloe, B. Bruyneel, J. S. de Vlieger, K. Retra, M. Wijtmans, R. van Elk, A. B. Smit, R. Leurs, H. Lingeman, I. J. de Esch and H. Irth, *J. Med. Chem.*, 2010, **53**, 4720.
34. T. T. Talley, S. Yalda, K. Y. Ho, Y. Tor, F. S. Soti, W. R. Kem and P. Taylor, *Biochemistry*, 2006, **45**, 8894.
35. C. de Graaf, H. F. Vischer, G. E. de Kloe, A. J. Kooistra, S. Nijmeijer, M. Kuijer, M. H. Verheij, P. J. England, J. E. van Muijlwijk-Koezen, R. Leurs and I. J. de Esch, *Drug Discovery Today*, 2013, **18**, 323.
36. S. Barelier and I. Krimm, *Curr. Opin. Chem. Biol.*, 2011, **15**, 469.
37. I. J. Chen and R. E. Hubbard, *J. Comput.-Aided Mol. Des.*, 2009, **23**, 603.
38. P. J. Hajduk, J. R. Huth and S. W. Fesik, *J. Med. Chem.*, 2005, **48**, 2518.
39. T. H. Keller, A. Pichota and Z. Yin, *Curr. Opin. Chem. Biol.*, 2006, **10**, 357.
40. A. Gaulton, L. J. Bellis, A. P. Bento, J. Chambers, M. Davies, A. Hersey, Y. Light, S. McGlinchey, D. Michalovich, B. Al-Lazikani and J. P. Overington, *Nucleic Acids Res.*, 2012, **40**, D1100.
41. P. P. V. http://accelrys.com/products/pipeline-pilot.
42. D. Rogers and M. Hahn, *J. Chem. Inf. Model.*, 2010, **50**, 742.
43. A. Mazurov, J. Klucik, L. Miao, T. Y. Phillips, A. Seamans, J. D. Schmitt, T. A. Hauser, R. T. Johnson Jr and C. Miller, *Bioorg. Med. Chem. Lett.*, 2005, **15**, 2073.
44. M. Geitmann, K. Retra, G. E. de Kloe, E. Homan, A. B. Smit, I. J. de Esch and U. H. Danielson, *Biochemistry*, 2010, **49**, 8143.
45. M. L. Verdonk and D. C. Rees, *ChemMedChem*, 2008, **3**, 1179.
46. P. J. Hajduk, *J. Med. Chem.*, 2006, **49**, 6972.
47. G. E. de Kloe, PhD dissertation, VU University Amsterdam, The Netherlands, 2011, 183–208.
48. C. ChemBioFinder Ultra 12.0 from Cambridgesoft, Massachusetts, USA.
49. P. C. Hawkins, A. G. Skillman and A. Nicholls, *J. Med. Chem.*, 2007, **50**, 74.
50. C. Bouzat, F. Gumilar, G. Spitzmaul, H. L. Wang, D. Rayes, S. B. Hansen, P. Taylor and S. M. Sine, *Nature*, 2004, **430**, 896.
51. J. J. Buccafusco, J. W. Beach and A. V. Terry Jr., *J. Pharmacol. Exp. Ther.*, 2009, **328**, 364.
52. E. Edink, P. Rucktooa, K. Retra, A. Akdemir, T. Nahar, O. Zuiderveld, R. van Elk, E. Janssen, P. van Nierop, J. van Muijlwijk-Koezen, A. B. Smit, T. K. Sixma, R. Leurs and I. J. de Esch, *J. Am. Chem. Soc.*, 2011, **133**, 5363.
53. P. H. Celie, I. E. Kasheverov, D. Y. Mordvintsev, R. C. Hogg, P. van Nierop, R. van Elk, S. E. van Rossum-Fikkert, M. N. Zhmak, D. Bertrand, V. Tsetlin, T. K. Sixma and A. B. Smit, *Nat. Struct. Mol. Biol.*, 2005, **12**, 582.
54. P. H. Celie, R. V. Klaassen, S. E. van Rossum-Fikkert, R. van Elk, P. van Nierop, A. B. Smit and T. K. Sixma, *J. Biol. Chem.*, 2005, **280**, 26457.
55. C. Ulens, A. Akdemir, A. Jongejan, R. van Elk, S. Bertrand, A. Perrakis, R. Leurs, A. B. Smit, T. K. Sixma, D. Bertrand and I. J. de Esch, *J. Med. Chem.*, 2009, **52**, 2372.

56. S. Dutertre, C. Ulens, R. Buttner, A. Fish, R. van Elk, Y. Kendel, G. Hopping, P. F. Alewood, C. Schroeder, A. Nicke, A. B. Smit, T. K. Sixma and R. J. Lewis, *EMBO J.*, 2007, **26**, 3858.
57. S. C. Lovell, J. M. Word, J. S. Richardson and D. C. Richardson, *Proteins*, 2000, **40**, 389.
58. M. Geitmann, K. Retra, G. E. de Kloe, E. Homan, A. B. Smit, I. J. de Esch and U. H. Danielson, *Biochemistry*, 2010, **49**, 8143.
59. K. Retra, M. Geitmann, J. Kool, A. B. Smit, I. J. de Esch, U. H. Danielson and H. Irth, *Anal. Biochem.*, 2010, **407**, 58.
60. E. Barratt, R. J. Bingham, D. J. Warner, C. A. Laughton, S. E. Phillips and S. W. Homans, *J. Am. Chem. Soc.*, 2005, **127**, 11827.
61. R. J. Bingham, J. B. Findlay, S. Y. Hsieh, A. P. Kalverda, A. Kjellberg, C. Perazzolo, S. E. Phillips, K. Seshadri, C. H. Trinh, W. B. Turnbull, G. Bodenhausen and S. W. Homans, *J. Am. Chem. Soc.*, 2004, **126**, 1675.
62. S. D. Sharrow, M. V. Novotny and M. J. Stone, *Biochemistry*, 2003, **42**, 6302.
63. G. E. de Kloe, J. Kool, R. van Elk, J. van Muijlwijk-Koezen, A. B. Smit, H. Lingeman, H. Irth, W. M. A. Niessen and I. J. P. De Esch, *Med. Chem. Commun.*, 2011, **54**, 1693.
64. http://www.fda.gov/newsevents/newsroom/pressannouncements/ucm268241.htm.
65. http://practicalfragments.blogspot.nl/2011/08/first-fragment-based-drug-approved.html.
66. D. E. Scott, A. G. Coyne, S. A. Hudson and C. Abell, *Biochemistry*, 2012, **51**, 4990.
67. U. H. Danielson, *Curr. Top. Med. Chem.*, 2009, **9**, 1725.
68. M. D. Duong-Thi, E. Meiby, M. Bergstrom, T. Fex, R. Isaksson and S. Ohlson, *Anal. Biochem.*, 2011, **414**, 138.
69. S. Vanwetswinkel, R. J. Heetebrij, J. van Duynhoven, J. G. Hollander, D. V. Filippov, P. J. Hajduk and G. Siegal, *Chem. Biol.*, 2005, **12**, 207.
70. A. Vulpetti, U. Hommel, G. Landrum, R. Lewis and C. Dalvit, *J. Am. Chem. Soc.*, 2009, **131**, 12949.
71. F. E. Torres, M. I. Recht, J. E. Coyle, R. H. Bruce and G. Williams, *Curr. Opin. Struct. Biol.*, 2010, **20**, 598.
72. J. M. Lalonde, M. Le-Khac, D. M. Jones, J. R. Courter, J. Park, A. Schon, A. M. Princiotto, X. Wu, J. R. Mascola, E. Freire, J. Sodroski, N. Madani, W. A. Hendrickson and A. B. Smith 3rd, *ACS Med. Chem. Lett.*, 2013, **4**, 338.
73. A. D. Scott, C. Phillips, A. Alex, M. Flocco, A. Bent, A. Randall, R. O'Brien, L. Damian and L. H. Jones, *ChemMedChem*, 2009, **4**, 1985.
74. E. Edink, C. Jansen, R. Leurs and I. J. P. de Esch, *Drug Discovery Today: Technol.*, 2010, 7, e189–e201.

CHAPTER 10

Fragment-Based Approaches to Epigenetic Targets

CHUN-WA CHUNG*[a] AND PAUL BAMBOROUGH*[a]

[a]Computational and Structural Chemistry, Molecular Discovery Research, GlaxoSmithKline Medicines Research Centre, Gunnels Wood Road, Stevenage, Hertfordshire, SG1 2NY, UK
*E-mail: chun-wa.h.chung@gsk.com, paul.a.bamborough@gsk.com

10.1 Introduction to Epigenetics

Epigenetics refers to the heritable changes in gene expression, resulting in a change in phenotype, that do not involve alterations to the underlying DNA sequence. Epigenetic mechanisms include DNA methylation, production of non-coding RNA, and histone post-translational modifications (PTMs).[1] Together, these processes contribute to the epigenome, which governs when DNA is translated and adds context-dependency to gene regulation mechanisms.[2] In this chapter we will focus on one aspect of the epigenetic machinery, proteins involved in reading, writing and erasing the *histone code* of PTMs.[3]

10.1.1 Why Target Epigenetic Proteins?

Unlike the DNA-coded genome, the epigenetic state of cells is dynamic. It evolves through defined states during a cell's natural lifecycle of differentiation and development in a tightly regulated manner. However, there is

RSC Drug Discovery Series No. 47
Fragment-Based Drug Discovery
Edited by Steven Howard and Chris Abell

Published by the Royal Society of Chemistry, www.rsc.org

growing evidence that aberrant changes in our epigenome can occur, and that these are implicated in diseases such as cancer, inflammation and metabolic disorders.[4–6] Drug molecules that are able to target the essential epigenetic processes that maintain these dysfunctional states therefore offer the prospect of providing longer lasting, as well as more effective treatment, through the simultaneous control of multiple genes.

10.1.2 The Protein Components of the Histone Code

Genomic material within the nucleus is stored compactly in a form called *chromatin*. This consists of tightly wound DNA wrapped around a histone octamer, containing one tetramer of histones H3 and H4 (two copies each) and two histone H2A–H2B dimers.[7] The histone protein tails are exposed and subject to extensive dynamic covalent modification: these PTMs influence chromatin architecture and affect the accessibility of the DNA to transcription. Only a subset of the possible PTM combinations are found physiologically, because the enzymes (epigenetic writers and erasers) that add or remove these covalent marks do so in a sequence- and context-dependent manner. The deciphering of this PTM code uses the synergistic action of a number of reader domains (Figure 10.1).[8] Often, each reader is designed to bind a specific mark rather weakly. However, when brought together, either as domains within the same protein or as part of different proteins within the MDa protein–protein complexes associated with chromatin, the multivalent interaction of the reader proteins confers specificity and affinity for the specific pattern of modification recognised.[9] Here we have chosen to focus on two of the wide range of known histone modifications, acetylation and methylation, for which fragment-based reports of the design of inhibitors have been most prevalent.

10.2 Approaches to Epigenetic Chemical Probe Discovery

The difficulty of finding new drug targets that are both chemically tractable and biologically relevant has been held as one factor responsible for declining productivity within the pharmaceutical industry.[12] In the emerging area of epigenetics, where new protein families have only been revealed within the last decade, incomplete biological understanding, coupled with a lack of chemical tools make target selection especially challenging. This has prompted calls for development of well-characterised open-access chemical probes with which to interrogate the functions of potential targets.[13,14] These molecules need to have affinity *in vitro*, physico-chemical properties suitable for nuclear permeability, and enough selectivity to allow functional responses to be unambiguously attributed to specific targets. A comprehensive understanding of a compound's selectivity is rarely possible, so access to well-matched pairs of active molecule and inactive counterparts, ideally

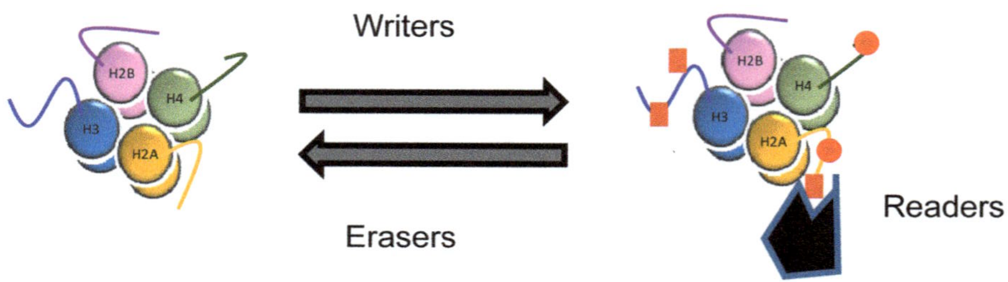

Modification	Writer	Eraser	Reader
Lysine acetylation (KAc)	Histone acetyl transferase (HAT)	Histone deacetylase (HDAC)	Bromodomain
Lysine methylation (KMe)	Histone methyl transferase (HMT)	Histone demethylase (HDM)	Chromodomain, chromo-barrel, tudor, tandem tudor, plant homeo domain, malignant brain tumour, WD40
Arginine methylation	Protein arginine methyl transferase (PRMT)	Histone demethylase (HDM)	ADD Tudor
Ser/Thr/Tyr phosphorylation	Kinase	Phosphatase	14-3-3 BRCT

Figure 10.1 Some common histone modifications, their modifying enzymes and some of their reader modules.[10,11]

from multiple diverse templates, is beneficial to avoid misinterpretation of phenotypes arising from unknown off-target effects.

To date, published probes have arisen from a variety of approaches including phenotypic, knowledge-based and fragment screening. Phenotypic screening has been especially productive for the discovery of compounds later found to act *via* histone deacetylase and acetyl-reader module inhibition.[15–17]

10.2.1 Fragment Screening

The lack of known binding partners, substrates or tool compounds for unexploited epigenetic targets makes the development of high-throughput *in vitro* assays highly challenging. Fragment screening offers one solution to this problem, with its use of direct binding methods and its promise of efficient coverage of chemical space with small compound libraries. Two more factors make these targets more tractable to fragment identification and optimisation. One is the ability to produce stable domains or truncates of many epigenetic enzymes and reader modules for biophysical assays and 3D structure determination. Another is the fact that although a large number of epigenetic modules exist, many of these domains belong to defined families that share motifs of common function and structure. Some fragments which bind to these conserved motifs may be common to multiple members of each domain target class, providing common chemical scaffolds which could be used as starting points for optimisation against several targets.

10.2.2 Fragment Library Composition

For domain families sharing common structural features, fragment libraries can be biased towards a family pharmacophore, which can increase the frequency of finding hits. However, for fragments binding in a conserved site, the introduction of selectivity will be a key consideration during optimisation. Diverse unbiased libraries can yield unexpected starting points, potentially away from conserved family sites, for example, allosteric starting points that may deliver intrinsic selectivity, but can prove more challenging to optimise, especially in the absence of structural data. Both targeted and diverse approaches have proved successful in different applications.[18]

10.3 Case Histories of Fragment-Based Drug Discovery

10.3.1 Histone Reader Modules

Readers of lysine acetylation and methylation are the most commonly investigated modules in this class. There are over 200 known methyl-reader domains, and chemical probes have been discovered for at least one case, L3MBTL1.[19,20] However, as we have been unable to find any examples of the use of fragment-based drug discovery (FBDD) for methyl-lysine readers, here we will concentrate on acetyl-lysine readers.

10.3.1.1 Acetyl-Lysine Readers: Bromodomains

Typically found within large chromatin-associated proteins, these small modules of around 110 amino acids in length are responsible for recognition of the acetyl-lysine (KAc) histone mark. At least 42 human genes encode one or more bromodomains. In the last few years it has become clear that small molecules can bind to bromodomains and block their recognition of acetylated histone tails, leading to profound biological effects.[21,22] The strongest phenotypes have been found with inhibitors of the bromodomain and extra terminal (BET) family of tandem-bromodomain-containing proteins (BRD2, BRD3, BRD4 and BRDT). Indeed, compounds were first identified using cell-based screens, and through chemoproteomics, biophysics and crystallography shown to act through bromodomain binding.[16,23–25] The mechanism of small-molecule interaction with bromodomain acetyl-lysine pockets was revealed at atomic resolution, and the potential to mimic the key features with fragment-sized compounds was immediately apparent.

10.3.1.1.1 Fragment Screening for Bromodomains. There are now several reports of fragment screening against bromodomains, and a smaller number of optimisation stories, summarised in Table 10.1. The first bromodomain

Table 10.1 Publications describing fragment screening against bromodomains.

Target	Primary screen	Library selection	References	Hits	Optimisation described
PCAF	2D ^{1}H–^{15}N HSQC NMR	Unknown	27	1,2	28 and 29
CREBBP	2D ^{1}H–^{15}N HSQC NMR	KAc-targeted (ligand-based)	30	3,4	Not published
BET (BRD2)	Fluorescence polarisation	KAc-targeted (ligand-based)	31	4–7	Not published
BET (BRD2)	Fluorescence polarisation	KAc-targeted (ligand-based)	32	8	Yes
BET (BRD4)	Not published	KAc-targeted (ligand-based)	33	9	34 and 35
BET (BRD4)	Not published	Not published	36	10	Yes
BET (BRD4)	AlphaScreen, then X-ray	KAc-targeted (ligand-based)	37	11	Yes
BET (BRD4)	X-ray	KAc-targeted (structure-based)	38	12–15	Yes
BET (BRD4)	X-ray	Not published	39	16	Not published
BAZ2B	AlphaScreen, then ligand NMR	Unbiased	40	17–19	Some

fragment screen was performed by the Zhou laboratory, using ^{15}N-NMR to first solve the solution structure of PCAF (P300/CBP-associated factor) and then to screen for small-molecule binders.[26,27] Several compounds were found which interacted with PCAF in or around the acetyl-peptide-binding site. Hits included the *N*1-aryl-propane-1,3-diamine (**1**), whose IC_{50} was estimated to be 5 µM. Follow up work aimed at increasing the activity of this hit led to a modest improvement in compound **2** (1.6 µM). Further analogues showed activity in an HIV-1 LTR-luciferase reporter assay,[28] and another group has since reported that members of this series inhibit HIV infection in a T-cell line.[29] These compounds do not contain an obvious acetyl-lysine mimic and their binding mode has been modelled with the help of the NMR data as no crystallographic complex has yet been successful. Subsequently, the Zhou group applied a similar approach to cAMP-response-element-binding-protein-binding protein (CREBBP). A small library of compounds selected to mimic the acetyl-lysine head-group were screened in pools, resulting in hits including **3** and **4**.[30] These fragments perturbed CREBBP residues in the BC and ZA loops, surrounding the acetyl-lysine site. Interestingly all but one did not bind to PCAF. This suggests that the fragments are able to differentiate between the acetyl-lysine sites of different bromodomains, which is encouraging given the extent of sequence conservation. The dissociation constant (K_D) of one fragment MS7972 (**3**) was estimated by tryptophan fluorescence to be around 20 µM (Figure 10.2).

Optimisation of these fragments against CREBBP has not been reported to our knowledge, although we have also reported **4** as an acetyl-lysine site binding fragment of the BET bromodomains.[31] In our example, a knowledge-based fragment library was selected using information about KAc-mimetic binding modes from our historical work on the BET bromodomain inhibitors. This was screened in a fluorescence polarisation assay, and 40 hits confirmed by crystallography. The availability of this number of structures allowed an in-depth analysis of binding modes to BRD2. It suggested, for example, that the indirect hydrogen bond to water in the KAc site is more important than the direct hydrogen bond to the conserved asparagine (Figure 10.3). It also showed the pattern of conservation of water molecules, highlighting three outside the acetyl-lysine recognition site which were present in all complexes. One fragment of interest was the indolizine **5**. This inspired the screening of acetaminophen (paracetamol, **6**) which binds both BRD2 and CREBBP. Two of the BET fragments, the dihydroquinazolinone **7**, and the phenyl dimethylisoxazole **8**, were found by others in parallel, as described below.

A group at the Structural Genomics Consortium had discovered that solvents including *N*-methyl pyrrolidine (NMP) bind in the KAc sites of many bromodomains. On the basis that other *N*-methylated rings might bind in a similar way, *N*-methyl dihydroquinazolinone **9** was tested against BRD4, and found active.[33] The story took an unexpected turn when the crystal structure of **9** was solved, revealing that the dihydroquinazolinone ring had been oxidised in the presence of ethylene glycol in the buffer, giving **9a**. In fact,

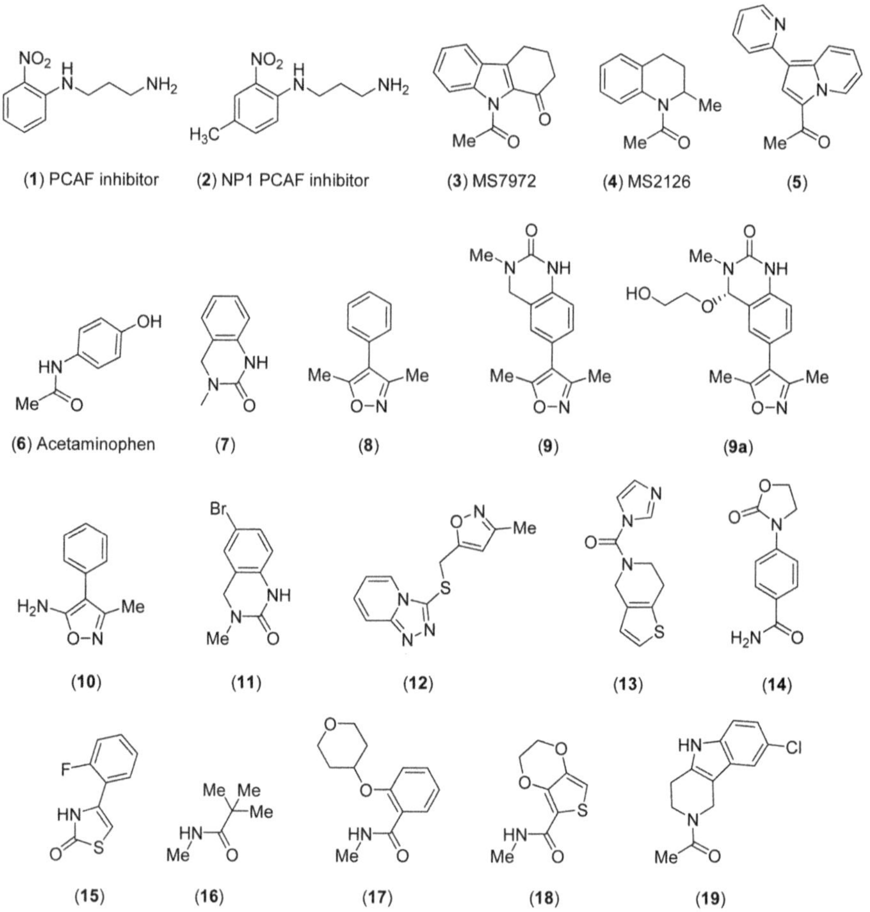

Figure 10.2 Bromodomain fragments.

it was the isoxazole of that molecule that bound in the KAc pocket.[33] Interestingly, our team at GSK had also discovered the dimethylisoxazole as an acetyl-lysine competitive inhibitor of the BET bromodomains by a different route, through our work with up-regulators of ApoA1 leading to I-BET151 (**20**, Figure 10.4).[25,42] We subsequently solved crystal structures of this series in BET bromodomains and chose this KAc-mimetic fragment (**8**) as the basis of a proof-of-concept optimisation study.[32] Recently, a third group reported the independent discovery of a closely related fragment **10** although the route that led to this finding was not explained.[36] The different approaches adopted by the groups to the optimisation of this fragment are discussed in Section 10.3.1.1.2.

Rational selection of acetyl-lysine mimetics featured heavily in the choice of fragment libraries in the above examples. The key pharmacophoric features of acetyl-lysine mimetics, revealed by X-ray crystallography, were condensed

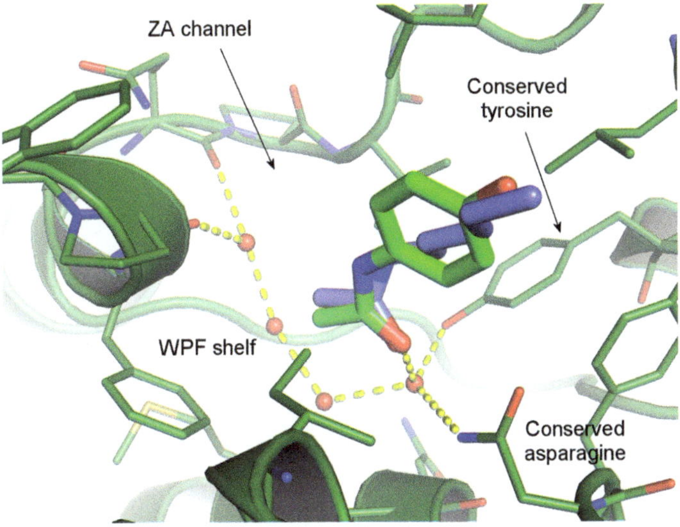

Figure 10.3 Acetyl-lysine site of BRD4, showing the binding mode of acetaminophen, superimposed on the acetyl-lysine side-chain of a histone peptide (blue).[41] The hydrogen-bonding pattern of the conserved water network is shown in yellow. The asparagine and tyrosine side-chains that participate in this network and that are conserved in the typical bromodomains are indicated.

into substructural definitions, and used as fragment database search terms. Medium- or high-throughput assays (NMR, or competition-binding methods utilising labelled peptides or small molecules) were then used to screen the targeted fragment sets, before X-ray crystallographic confirmation and binding mode determination. Zhao *et al.* also used a targeted fragment set, but unlike the ligand-based substructure design of the other groups they chose to filter a diverse fragment set by molecular docking to the BRD4 acetyl-lysine pocket. These were then tested directly by X-ray crystallography without any intervening screen.[38] This proved successful: from 41 fragments progressed into crystallography, 9 were identified in the site, including **12–15**. Direct X-ray crystallography was also used by Lolli and Battistutta to determine the binding modes of three very small fragments, also in BRD4, including *N*-methyltrimethylacetamide **16**.[39] Crystallographic screening against BRD4 has also been performed and reported in poster form by Hubbard *et al.*[43]

Compared to knowledge-based fragment libraries, unbiased fragment sets are relatively under-represented in the bromodomain literature so far. Ferguson *et al.* used such a library to screen for binders to the bromodomains of BAZ2B, BRD4 and CREBBP.[40] Hits from a high-concentration peptide-based AlphaScreen assay were progressed into three ligand-observed NMR screens. From 10 hits, 3 X-ray complex structures were solved in BAZ2B (**17–19**). They reported that the relative hit rates against the three bromodomains were consistent with their predicted ligandability.

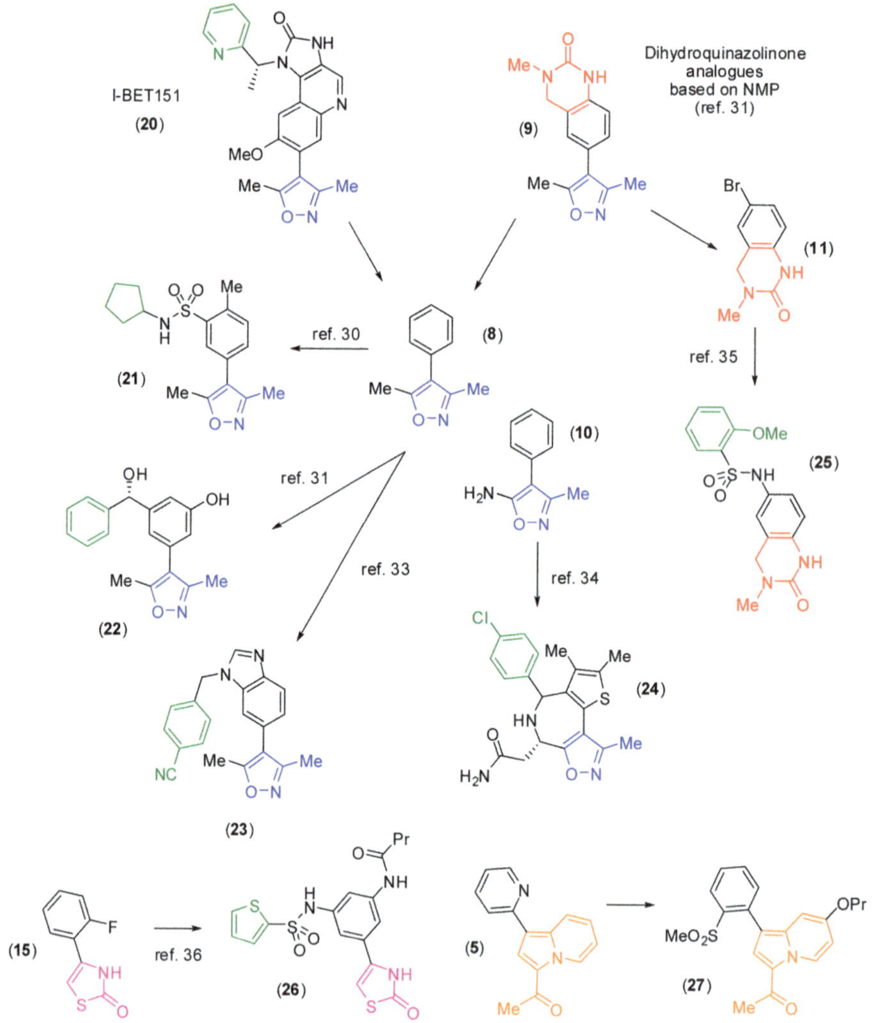

Figure 10.4 Optimisation of bromodomain KAc-site fragments. KAc-mimetic fragments are shown in blue (isoxazole), red (dihydroquinazolinone), magenta (thiazolidinone) or orange (indolizine). Groups introduced during optimisation to occupy the lipophilic WPF shelf are coloured green. All are inhibitors of the BET bromodomains except **27**, which is a BAZ2A/2B-selective molecule.

10.3.1.1.2 Fragment Optimisation for Bromodomains. To date, extensive optimisation case studies have only been reported for the BET bromodomains. The isoxazole fragments **8** and **10** are the most interesting, because they have been pursued by three different groups in different ways (Figure 10.4). One common theme is that starting from their respective fragments each group grew outwards to make additional interactions, targeting the lipophilic *WPF shelf* region.[16] The importance of this second site was not readily apparent, until the publication of crystal structures of compounds discovered and optimised through cellular assays, including I-BET and JQ1.[16,22] These span the KAc site and the WPF shelf, occupying the latter with their chlorophenyl rings. Furthermore, this site is not highly conserved between different bromodomains, where the WPF shelf and ZA channel regions (Figure 10.3) tend to be either more polar or relatively shallow, affording an opportunity to introduce selectivity for the BET subfamily over the other members even when starting from unselective AcK-site fragments. In this respect, the situation is reminiscent of the kinase family, where substituents can be used to tailor the selectivity of common ATP-site binding fragments.

From **8**, our team identified the optimised sulphonamide-substituted analogue **21** (Figure 10.4) through a pharmacophore search of commercial compounds.[32] Because of its preferred geometry, the sulphonamide is an efficient way to direct the substituent onto the shelf. Hewings and co-workers grew towards the same region, but chose to do so with different linkers, such as the benzyl alcohol **22** and the benzimidazole **23**.[33–35] Allowing for the different assays used, the compounds achieve similar potency and efficiency, but they are chemically distinct. The group at Constellation chose to follow a different strategy and incorporated the information from the isoxazole into the literature thienotriazolodiazepine BET bromodomain inhibitor (+)-JQ-1, to produce a hybrid isoxazole azepine **24**.[36]

Although the *N*-methyl dihydroquinazolinone turned out not to be the acetyl-lysine mimic group of **9**, this ring system is still a competent KAc mimic.[33] Medicinal chemists at Pfizer optimised the simple bromide analogue **11** to give the potent chemical probe PFI-1 (**25**).[37] They also achieved this by targeting the WPF shelf. However, because they approached this shelf from a different starting point and a different vector, they found different structure–activity relationships (SARs). For example, when the isoxazole **8** was optimised to the sulphonamide **21**, it was found that alkyl groups such as cyclopentyl were preferred. In contrast, from the dihydroquinazolinone **11**, higher potency was obtained with sulphonamides of the reverse directionality, in combination with aromatic substituents.

The novel 2-thiazolidinone KAc mimetic discovered by Zhao *et al.* was optimised against the BET bromodomains, exploring a range of substituents at several positions.[38] Like the Pfizer group, they found aryl sulphonamides to be optimal for targeting the WPF shelf. They also grew from a different substituent to explore another region of the site, the ZA channel, and in a final iteration combined this with the WPF shelf group, giving compound **26**.

10.3.2 Perspectives for Bromodomain FBDD

It might be thought that groups using knowledge-based fragment libraries and using similar optimisation strategies would arrive at similar elaborated molecules, but this has not been the case for the BET bromodomains. A comparison of the overlaid X-ray structures of these fragment-optimised compounds shows that the positions of the KAc mimetics and the WPF-shelf lipophilic rings are highly conserved (Figure 10.5). Yet, the way that they are linked is substantially different, chemically and three-dimensionally. It seems that (at least for the BET bromodomains) chemical novelty and selectivity as well as potency is quite achievable using FBDD.

The KAc sites of several members of the bromodomain family seem to be highly tractable to fragment screening. Even though fragment-based hit discovery seems to be feasible for many bromodomains, optimisation against these targets may yet prove more challenging than for the BETs. It is currently unclear whether high-affinity sites outside the conserved KAc pocket, such as the WPF shelf and ZA channel of the BET family, exist in other bromodomains. How fragment hits will be optimised for other bromodomains, and what additional interactions will be needed to boost their potency, remains to be seen. We have presented orally our FBDD hit-rates targeting a number of bromodomains beyond the BET family, including those considered relatively intractable where crystal structures reveal only shallow or polar pockets surrounding the KAc site. The rank-order of hit rates against different bromodomains that we obtained using ^{15}N-NMR screening of our diverse fragment set tracks with the hit rates from 10 μM compound concentration high-throughput screening using displacement assays. They also correlate with literature-predicted druggability scores based on X-ray structure site analysis.[44] Ferguson *et al.* have reported fragments that bind to BAZ2B, another less tractable bromodomain.[40] Encouragingly, the BAZ2A/B chemical probe GSK-2801 (**27**, Figure 10.4) originated from indolizine **5**, which was originally identified as a BET bromodomain inhibitor,[31] suggesting that not only are other bromodomains chemically tractable, but that common KAc fragments can be elaborated to give chemical probes with distinct selectivity profiles.

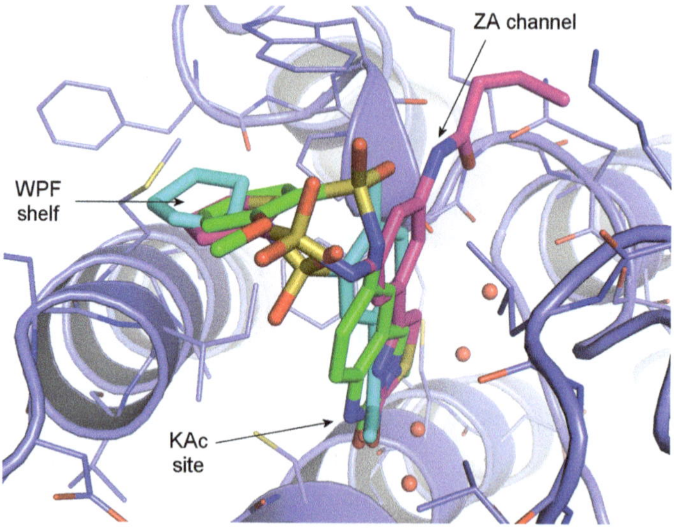

Figure 10.5 Three-dimensional superimposition of bound X-ray structures of BET inhibitors. Compounds are **21** (cyan), **25** (green) and **26** (magenta).

No reports have emerged of fragments that bind to other optimisable allosteric bromodomain sites beyond the KAc pocket. Perhaps this is not surprising, because screening efforts have typically used the isolated bromodomain helical bundle, in which the KAc sites are by far the deepest clefts. If additional druggable allosteric sites exist within bromodomain-containing proteins, these might only be revealed in the context of the rest of the poly-peptide chain.

10.4 Histone-Modifying Enzymes

The success of fragment screening for bromodomains rests on the detectable binding of fragment-sized isosteres of the natural recognition partner, the acetylated lysine side-chain, and the ability to optimise these productively by growing into surrounding pockets. A similar approach of mimicking post-translationally modified amino acid side-chains could be envisaged for histone-modifying enzymes. To date, there have been few reports where this has been attempted. However, a study on SET domains of protein methyl transferases reported that fragments of inhibitors that have been shown crystallographically to bind in the lysine channel of this class of enzyme may be very weak.[45] A few examples of other approaches for enzymes will be discussed below.

10.4.1 Lysine-Specific Demethylase 1

Two families of histone lysine demethylase (KDM) enzymes have been identified: the lysine-specific demethylases (LSD-1 and LSD-2), and the Jumonji (Jmj) proteins. Both have been implicated in cancer.[46] The first KDM to be discovered, LSD-1, is considered a target in human malignancies because its expression correlates with poor prognosis in a number of cancers including prostate, breast and bladder.[47] Fragment-sized inhibitors of LSD-1 include tranylcypromine (**28**, Figure 10.6), an antidepressant developed in the 1960s, which has activities against monoamine oxidases MAO-A and B.[48] These

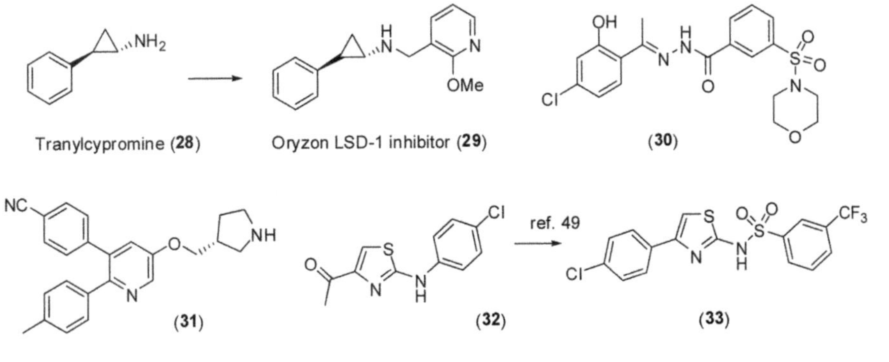

Figure 10.6 Some published inhibitors of LSD-1.

compounds act by irreversible covalent modification of the flavin adenine dinucleotide (FAD) co-factor in LSD-1 and MAO-A. The most potent reported analogues such as **29** have nM affinities *vs.* LSD-1 and >3000-fold selectivity *vs.* MAO-A/B.[49] Reversible competitive inhibitors such as **30** and **31** have been reported,[50–52] but the only inhibitors for which X-ray structures have been released to date are peptides with K_i 140 nM.[53]

10.4.1.1 *Fragment-Based Screening for LSD-1*

In the only report we have found of fragment screening against a lysine demethylase, aminothiazole **32** (IC_{50} = 249 μM) and a close analogue were identified using high-concentration biochemical screening, followed by confirmation in two orthogonal formats.[54] Through several iterations of SAR-driven optimisation the activity was improved to approximately 10 μM. Initial attempts to remove the perceived metabolic liability of the ketone group were unsuccessful, suggesting that this group was critical for activity. Modifications to the phenyl ring were mostly unproductive and highlighted steep SARs that made optimisation particularly challenging. It was found that replacement of the aniline with an aryl sulfonamide allowed the exchange of the ketone for a chlorophenyl ring. A further array resulted in **33** (IC_{50} 9.5 μM), which also showed selectivity over MAO-A. Rapid dilution experiments showed this compound to be reversible.

Unfortunately, despite the improvement in potency achieved, the compounds were inactive in cells. The authors commented on the difficulty of devising a strategy for hit optimisation due to the complexity of the SAR, compounded by the lack of structural data to provide clues as to the binding mode.[54]

10.4.1.2 *Perspective for LSD-1 FBDD*

In contrast to bromodomains, not all epigenetic proteins can be easily expressed in isolation or optimised to robust crystal systems suitable for small-molecule complex determination. In these instances, fragment screening may still be successful, but optimisation can be hindered by difficulties in interpreting emerging SARs or the lack of additional favourable interaction sites close to the initial fragment.

10.4.2 Histone-Modifying Metalloenzymes

In addition to LSD-1, few examples of fragment-based screening of epigenetic regulator enzymes can be found, which is surprising given the therapeutic interest of these as drug targets.[55] The classes that have received the most attention include the zinc-dependent histone deacetylases (HDACs: Class I, IIa, IIb, IV) and the iron-containing Jmjs. These enzymes share few common features apart from the presence of mechanistically important metal ions, which are frequently targeted by inhibitors.

10.4.2.1 Histone Deacetylases

The architecture of typical HDAC inhibitors can be summarised as a zinc-binding group (ZBG, typically a hydroxamate-, reverse hydroxamate- or benzamide-containing fragment) connected to a capping group *via* a linker (Figure 10.7). The hydroxamate ZBG has been used extensively as a starting point for metalloenzyme inhibition, for example the metalloprotease MMP-3.[56,57] This group also features in many of the most effective HDAC inhibitors to date (*e.g.*, Vorinostat, which itself originated from attempts to find polar fragments able to induce cell differentiation of MELCs into erythroid cells).[15,58]

10.4.2.1.1 FBDD for HDACs. In one study, a fragment-sized library built around literature ZBGs was screened against a panel of metalloenzymes. The hits against the HDAC1 and HDAC8 enzymes were found to mainly contain hydroxamates.[59] Elaboration of hydroxamate-containing fragments has led to potent HDAC inhibitors. In one recent example, a range of bicyclic aromatic hydroxamic acids was synthesised and screened in a catalytic assay. Elaboration of the benzothiophene **34** at the 5- and 6- positions led to potent analogues such as **35** with activity in a xenograft model.[60]

Despite the utility of the hydroxamic acid moiety, a desire to improve upon the selectivity and pharmacokinetic properties of this ZBG has stimulated research efforts aimed at finding alternative non-hydroxamate ZBGs. This is exemplified by the recent report of a novel Class II HDAC-selective ZBG, the trifluoromethyloxadiazole (TFMO) group.[61] This series (*e.g.* **36**), was discovered by traditional low (10 μM) concentration high-throughput screening. It seems that the relatively low potency of the TFMO fragment for zinc

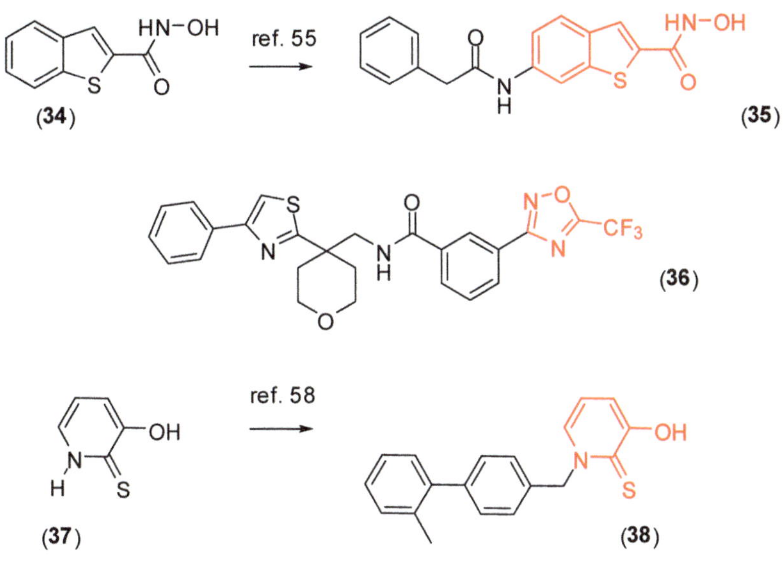

Figure 10.7 HDAC inhibitor fragments. Zinc-binding groups are shown in red.

is responsible for its greater selectivity. In a SILAC quantitative proteomics experiment comparing a TFMO-containing compound with its direct hydroxamate analogue, the TFMO predominantly bound to Class II HDACs whereas the hydroxamate bound to other metal-binding proteins.

The search for novel ZBGs for HDACs would appear to be an ideal opportunity for classical FBDD, but although this has been studied computationally, few experimental fragment screening studies have been published.[62] One example is that of Patil *et al.*, who tested two bidentate fragments against three HDAC enzymes, and reported the elaboration of 3-hydroxypyridin-2-thione (**37**) guided by molecular docking to 300 nM inhibitors of HDAC6, *e.g.*, **38**.[63]

10.4.2.2 *Jumonji Family Histone Lysine Demethylases*

Jmj enzymes are the largest class of KDMs.[64,65] These proteins regulate gene expression both by providing a vital 'scaffolding' role within transcriptional chromatin complexes, and by virtue of their enzymatic demethylase activity. The relative importance of these two roles and the therapeutic potential of Jmj proteins as drug targets have been much debated due to the lack of specific small-molecule inhibitors to demonstrate chemical tractability and pharmacological relevance.[66]

10.4.2.2.1 FBDD for Jmj Demethylases. The catalytic site of the Jmjs contains an Fe(II) ion that is doubly coordinated by the carboxylate and carbonyl of a 2-oxoglutarate co-factor (2-OG, **39**, Figure 10.8). Fragment-based work from the Structural Genomics Consortium has concentrated on 2-OG mimetics binding to the iron in a co-factor-competitive manner.[67] *N*-Oxalylglycine (NOG, **40**), a close analogue of 2-OG and a 24 μM inhibitor of JmjD2E, was optimised to single-figure μM IC_{50} *N*-oxalyl-D-tyrosinyl derivatives such as **41**

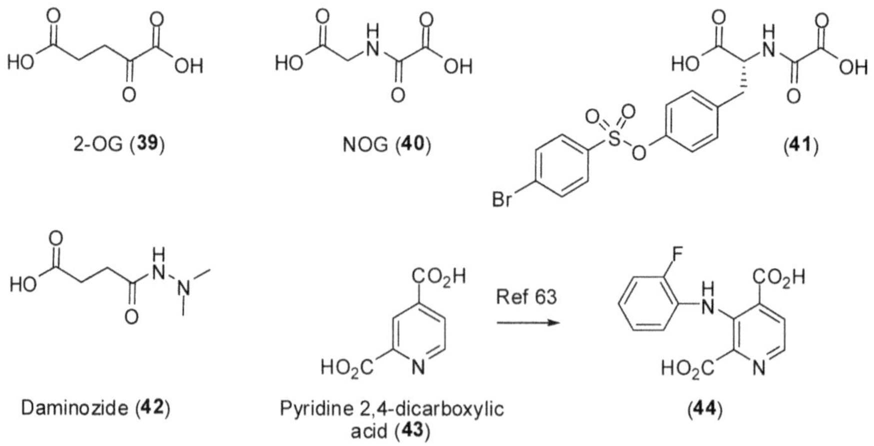

Figure 10.8 Fragments binding to Jmi family enzymes.

using structure-guided design.[68] More recently, the group has reported that another close 2-OG analogue, the plant growth retardant daminozide **42**, is a μM 2-OG competitive inhibitor of KDM2A/KDM7A with a novel iron-chelating binding mode.[69]

Pyridine 2,4-dicarboxylic acid (**43**) is another μM 2-OG competitive inhibitor of JmjD2E that was shown crystallographically to bind in the 2-OG site of the related enzyme JmjD2A.[67] Structure-guided optimisation of this fragment was attempted, but the JmjD2E activity could not be improved, although compounds such as **44** were selective over another 2-OG oxygenase PHD2.[70]

10.4.3 Perspectives for Histone-Modifying Enzymes

There may be many reasons for the lack of reported FBDD efforts for histone-modifying enzymes. Some have proven more difficult to produce, handle and assay than reader domains. The presence of domains outside the catalytic core can be vital to enzymatic activity and stability. Moreover, in some instances, their functional preferences can be dependent on additional partner proteins (*e.g.*, HDACs, LSD-1 and EZH2).[71–73] This can make establishing representative and robust assays and X-ray crystallographic systems more challenging. Where a catalytic metal exists, (HDACs, Jmjs) interaction with this group has been a key anchor for most inhibitors reported to date.[74] Yet, groups able to do this with high efficiency frequently have properties which are suboptimal in other respects, such as cellular permeability.[74,75] In addition, the discovery of metal-binding fragments that offer an appropriate balance of selectivity and opportunity for growth has not proven straightforward. Avoiding this metal-binding anchor point and obtaining fragment starting points that do not rely on metal chelation may offer greater potential if such compounds can be identified.[76,77]

10.5 Conclusions

Histone reader domain modules appear ideal candidates for FBDD. They tend to be stable as truncated domains and are amenable to biophysical screening and crystallography. For bromodomains, one remaining challenge is to understand the opportunities for bromodomains beyond the BET family to obtain affinity and selectivity by optimisation outside the conserved KAc recognition site. The extent to which this successful approach can be replicated in other reader module classes is, however, uncertain. Plant homeodomain (PHD) modules, for example, have relatively shallow binding sites as isolated domains, and attempts to find small-molecule starting points for these proteins have proven difficult. It may be more profitable to screen these modules in the context of physiological complexes or longer constructs as there is evidence that histone PTM recognition sites of these proteins may become more enclosed when additional domains and proteins are present.[71,78]

While we have found few reports of fragment-based approaches for histone-modifying enzymes it seems probable that these have also been attempted. The lack of published examples may suggest these enzymes have proven less tractable by these tactics, or that this field is still in its early days and findings have yet to be disclosed. Our experience suggests that both are likely to be factors. As our functional understanding of epigenetic targets along with our practical experience of handling them grows, it is likely that fragment-based methods will continue to play an important role in diversifying the range of chemical templates available. Therefore we expect it is only a matter of time before the array of chemical probes available for these exciting epigenetic modulators is expanded further by molecules originating from fragment approaches.

References

1. G. Egger, G. Liang, A. Aparicio and P. A. Jones, *Nature*, 2004, **429**, 457.
2. K. E. Gardner, C. D. Allis and B. D. Strahl, *J. Mol. Biol.*, 2011, **409**, 36.
3. B. D. Strahl and C. D. Allis, *Nature*, 2000, **403**, 41.
4. T. K. Kelly, D. D. De Carvalho and P. A. Jones, *Nat. Biotechnol.*, 2010, **28**, 1069.
5. E. Ballestar, *Clin. Rev. Allergy Immunol.*, 2010, **39**, 30.
6. R. G. Urdinguio, J. V. Sanchez-Mut and M. Esteller, *Lancet Neurol.*, 2009, **8**, 1056.
7. R. Margueron and D. Reinberg, *Nat. Rev. Genet.*, 2010, **11**, 285.
8. S. D. Taverna, H. Li, A. J. Ruthenburg, C. D. Allis and D. J. Patel, *Nat. Struct. Mol. Biol.*, 2007, **14**, 1025.
9. A. J. Ruthenburg, H. Li, D. J. Patel and C. D. Allis, *Nat. Rev. Mol. Cell Biol.*, 2007, **8**, 983.
10. C. A. Musselman, M. E. Lalonde, J. Cote and T. G. Kutateladze, *Nat. Struct. Mol. Biol.*, 2012, **19**, 1218.
11. A. J. Bannister and T. Kouzarides, *Cell Res.*, 2011, **21**, 381.
12. M. E. Bunnage, *Nat. Chem. Biol.*, 2011, **7**, 335.
13. S. V. Frye, *Nat. Chem. Biol.*, 2010, **6**, 159.
14. M. E. Bunnage, E. L. Chekler and L. H. Jones, *Nat. Chem. Biol.*, 2013, **9**, 195.
15. P. A. Marks and R. Breslow, *Nat. Biotechnol.*, 2007, **25**, 84.
16. C. Chung, H. Coste, J. H. White, O. Mirguet, J. Wilde, R. L. Gosmini, C. Delves, S. M. Magny, R. Woodward, S. A. Hughes, E. V. Boursier, H. Flynn, A. M. Bouillot, P. Bamborough, J. M. Brusq, F. J. Gellibert, E. J. Jones, A. M. Riou, P. Homes, S. L. Martin, I. J. Uings, J. Toum, C. A. Clement, A. B. Boullay, R. L. Grimley, F. M. Blandel, R. K. Prinjha, K. Lee, J. Kirilovsky and E. Nicodeme, *J. Med. Chem.*, 2011, **54**, 3827.
17. K. G. McLure, E. M. Gesner, L. Tsujikawa, O. A. Kharenko, S. Attwell, E. Campeau, S. Wasiak, A. Stein, A. White, E. Fontano, R. K. Suto, N. C. Wong, G. S. Wagner, H. C. Hansen and P. R. Young, *PLoS One*, 2013, **8**, e83190.
18. A. Schuffenhauer, S. Ruedisser, A. L. Marzinzik, W. Jahnke, M. Blommers, P. Selzer and E. Jacoby, *Curr. Top. Med. Chem.*, 2005, **5**, 751.
19. L. I. James, D. Barsyte-Lovejoy, N. Zhong, L. Krichevsky, V. K. Korboukh, J. M. Herold, C. J. MacNevin, J. L. Norris, C. A. Sagum, W. Tempel, E. Marcon,

H. Guo, C. Gao, X. P. Huang, S. Duan, A. Emili, J. F. Greenblatt, D. B. Kireev, J. Jin, W. P. Janzen, P. J. Brown, M. T. Bedford, C. H. Arrowsmith and S. V. Frye, *Nat. Chem. Biol.*, 2013, **9**, 184.
20. L. I. James, V. K. Korboukh, L. Krichevsky, B. M. Baughman, J. M. Herold, J. L. Norris, J. Jin, D. B. Kireev, W. P. Janzen, C. H. Arrowsmith and S. V. Frye, *J. Med. Chem.*, 2013, **56**, 7358.
21. E. Nicodeme, K. L. Jeffrey, U. Schaefer, S. Beinke, S. Dewell, C. W. Chung, R. Chandwani, I. Marazzi, P. Wilson, H. Coste, J. White, J. Kirilovsky, C. M. Rice, J. M. Lora, R. K. Prinjha, K. Lee and A. Tarakhovsky, *Nature*, 2010, **468**, 1119.
22. P. Filippakopoulos, J. Qi, S. Picaud, Y. Shen, W. B. Smith, O. Fedorov, E. M. Morse, T. Keates, T. T. Hickman, I. Felletar, M. Philpott, S. Munro, M. R. McKeown, Y. Wang, A. L. Christie, N. West, M. J. Cameron, B. Schwartz, T. D. Heightman, T. N. La, C. A. French, O. Wiest, A. L. Kung, S. Knapp and J. E. Bradner, *Nature*, 2010, **468**, 1067.
23. O. Mirguet, R. Gosmini, J. Toum, C. A. Clement, M. Barnathan, J. M. Brusq, J. E. Mordaunt, R. M. Grimes, M. Crowe, O. Pineau, M. Ajakane, A. Daugan, P. Jeffrey, L. Cutler, A. C. Haynes, N. N. Smithers, C. W. Chung, P. Bamborough, I. J. Uings, A. Lewis, J. Witherington, N. Parr, R. K. Prinjha and E. Nicodeme, *J. Med. Chem.*, 2013, **56**, 7501.
24. O. Mirguet, Y. Lamotte, C. W. Chung, P. Bamborough, D. Delannee, A. Bouillot, F. Gellibert, G. Krysa, A. Lewis, J. Witherington, P. Huet, Y. Dudit, L. Trottet and E. Nicodeme, *ChemMedChem*, 2014, **9**, 580.
25. J. Seal, Y. Lamotte, F. Donche, A. Bouillot, O. Mirguet, F. Gellibert, E. Nicodeme, G. Krysa, J. Kirilovsky, S. Beinke, S. McCleary, I. Rioja, P. Bamborough, C. W. Chung, L. Gordon, T. Lewis, A. L. Walker, L. Cutler, D. Lugo, D. M. Wilson, J. Witherington, K. Lee and R. K. Prinjha, *Bioorg. Med. Chem. Lett.*, 2012, **22**, 2968.
26. C. Dhalluin, J. E. Carlson, L. Zeng, C. He, A. K. Aggarwal and M. M. Zhou, *Nature*, 1999, **399**, 491.
27. L. Zeng, J. Li, M. Muller, S. Yan, S. Mujtaba, C. Pan, Z. Wang and M. M. Zhou, *J. Am. Chem. Soc.*, 2005, **127**, 2376.
28. C. Pan, M. Mezei, S. Mujtaba, M. Muller, L. Zeng, J. Li, Z. Wang and M. M. Zhou, *J. Med. Chem.*, 2007, **50**, 2285.
29. Q. Wang, R. Wang, B. Zhang, S. Zhang, Y. Zheng and Z. Wang, *MedChemComm*, 2013, **4**, 737.
30. Sachchidanand, L. Resnick-Silverman, S. Yan, S. Mutjaba, W. J. Liu, L. Zeng, J. J. Manfredi and M. M. Zhou, *Chem. Biol.*, 2006, **13**, 81.
31. C. Chung, A. W. Dean, J. M. Woolven and P. Bamborough, *J. Med. Chem.*, 2012, **55**, 576.
32. P. Bamborough, H. Diallo, J. D. Goodacre, L. Gordon, A. Lewis, J. T. Seal, D. M. Wilson, M. D. Woodrow and C. W. Chung, *J. Med. Chem.*, 2012, **55**, 587.
33. D. S. Hewings, M. Wang, M. Philpott, O. Fedorov, S. Uttarkar, P. Filippakopoulos, S. Picaud, C. Vuppusetty, B. Marsden, S. Knapp, S. Conway and T. D. Heightman, *J. Med. Chem.*, 2011, **54**, 6761.
34. D. S. Hewings, O. Fedorov, P. Filippakopoulos, S. Martin, S. Picaud, A. Tumber, C. Wells, M. M. Olcina, K. Freeman, A. Gill, A. J. Ritchie,

D. W. Sheppard, A. J. Russell, E. M. Hammond, S. Knapp, P. E. Brennan and S. J. Conway, *J. Med. Chem.*, 2013, **56**, 3217.
35. D. Hay, O. Fedorov, P. Filippakopoulos, S. Martin, M. Philpott, S. Picaud, D. S. Hewings, S. Uttakar, T. D. Heightman and S. J. Conway, *MedChemComm*, 2013, **4**, 140.
36. V. S. Gehling, M. C. Hewitt, R. G. Vaswani, Y. Leblanc, A. Côté, C. G. Nasveschuk, A. M. Taylor, J.-C. Harmange, J. E. Audia, E. Pardo, S. Joshi, P. Sandy, J. A. Mertz, R. J. Sims III, L. Bergeron, B. M. Bryant, S. Bellon, F. Poy, H. Jayaram, R. Sankaranarayanan, S. Yellapantula, N. B. Srinivasamurthy, S. Birudukota and B. K. Albrecht, *ACS Med. Chem. Lett.*, 2013, **4**, 835.
37. P. V. Fish, P. Filippakopoulos, G. Bish, P. E. Brennan, M. E. Bunnage, A. S. Cook, O. Federov, B. S. Gerstenberger, H. Jones, S. Knapp, B. Marsden, K. Nocka, D. R. Owen, M. Philpott, S. Picaud, M. J. Primiano, M. J. Ralph, N. Sciammetta and J. D. Trzupek, *J. Med. Chem.*, 2012, **55**, 9831.
38. L. Zhao, D. Cao, T. Chen, Y. Wang, Z. Miao, Y. Xu, W. Chen, X. Wang, Y. Li, Z. Du, B. Xiong, J. Li, C. Xu, N. Zhang, J. He and J. Shen, *J. Med. Chem.*, 2013, **56**, 3833.
39. G. Lolli and R. Battistutta, *Acta Crystallogr., Sect. D: Biol. Crystallogr.*, 2013, **69**, 2161.
40. F. M. Ferguson, O. Fedorov, A. Chaikuad, M. Philpott, J. Muniz, I. Felletar, D. F. von, T. D. Heightman, S. Knapp, C. Abell and A. J. Ciulli, *Med. Chem.*, 2013, **56**, 10183.
41. J. Morinière, S. Rousseaux, U. Steuerwald, M. Soler-López, S. Curtet, A. L. Vitte, J. Govin, J. Gaucher, K. Sadoul, D. J. Hart, J. Krijgsveld, S. Khochbin, C. W. Muller and C. Petosa, *Nature*, 2009, **461**, 664.
42. O. Mirguet, Y. Lamotte, F. Donche, J. Toum, F. Gellibert, A. Bouillot, R. Gosmini, V. L. Nguyen, D. Delannee, J. Seal, F. Blandel, A. B. Boullay, E. Boursier, S. Martin, J. M. Brusq, G. Krysa, A. Riou, R. Tellier, A. Costaz, P. Huet, Y. Dudit, L. Trottet, J. Kirilovsky and E. Nicodeme, *Bioorg. Med. Chem. Lett.*, 2012, **22**, 2963.
43. P. Hubbard, A. Ritchie, S. Hughes, M. Lipkin, W. Blackaby and A. MacLeod, http://www.biofocus.com/ugc-1/1/4/0/Acrystallographic-basedfragmentsc.pdf 2013.
44. L. R. Vidler, N. Brown, S. Knapp and S. Hoelder, *J. Med. Chem.*, 2012, **55**, 7346.
45. K. T. Nguyen, F. Li, G. Poda, D. Smil, M. Vedadi and M. Schapira, *J. Chem. Inf. Model.*, 2013, **53**, 681.
46. J. W. Hojfeldt, K. Agger and K. Helin, *Nat. Rev. Drug Discovery*, 2013, **12**, 917.
47. J. T. Lynch, W. J. Harris and T. C. Somervaille, *Expert Opin. Ther. Targets*, 2012, **16**, 1239.
48. M. G. Lee, C. Wynder, D. M. Schmidt, D. G. McCafferty and R. Shiekhattar, *Chem. Biol.*, 2006, **13**, 563.
49. N. Guibourt, M. A. Ortega and L. J. Castro-Palomino, Phenylcyclopropylamine derivatives and their medical use, WO2010084160, 2010.
50. C. Harrison, *Nat. Rev. Drug Discovery*, 2013, **12**, 899.

51. H. Vankayalapati, V. Sorna, S. L. Warner, D. J. Bearss, S. Sharma and B. Stephens, Substituted (*E*)-*N*′-(1-phenylethylidene)benzohydrazide analogs as histone demethylase inhibitors, WO2013025805, 2013.
52. D. Dhanak, Drugging the cancer epigenome (abstract), in *Proceedings of the 104th Annual Meeting of the American Association for Cancer Research*, AACR, Washington, DC, Philadelphia (PA), April 2013, pp. 6–10.
53. M. Tortorici, M. T. Borrello, M. Tardugno, L. R. Chiarelli, S. Pilotto, G. Ciossani, N. A. Vellore, S. G. Bailey, J. Cowan, M. O'Connell, S. J. Crabb, G. Packham, A. Mai, R. Baron, A. Ganesan and A. Mattevi, *ACS Chem. Biol.*, 2013, **8**, 1677.
54. J. R. Hitchin, J. Blagg, R. Burke, S. Burns, M. J. Cockerill, E. E. Fairweather, C. Hutton, A. M. Jordan, C. McAndrew, A. Mirza, D. Mould, G. J. Thomson, I. Waddell and D. J. Ogilvie, *Med. Chem. Commun.*, 2013, **4**, 1513.
55. K. Helin and D. Dhanak, *Nature*, 2013, **502**, 480.
56. P. J. Hajduk and J. Greer, *Nat. Rev. Drug Discovery*, 2007, **6**, 211.
57. P. J. Hajduk, S. B. Shuker, D. G. Nettesheim, R. Craig, D. J. Augeri, D. Betebenner, D. H. Albert, Y. Guo, R. P. Meadows, L. Xu, M. Michaelides, S. K. Davidsen and S. W. Fesik, *J. Med. Chem.*, 2002, **45**, 5628.
58. M. Tanaka, J. Levy, M. Terada, R. Breslow, R. A. Rifkind and P. A. Marks, *Proc. Natl. Acad. Sci. U. S. A.*, 1975, **72**, 1003.
59. S. Johnson, E. Barile, B. Farina, A. Purves, J. Wei, L. H. Chen, S. Shiryaev, Z. Zhang, I. Rodionova, A. Agrawal, S. M. Cohen, A. Osterman, A. Strongin and M. Pellecchia, *Chem. Biol. Drug Des.*, 2011, **78**, 211.
60. E. Marastoni, S. Bartoli, M. Berettoni, A. Cipollone, A. Ettorre, C. I. Fincham, S. Mauro, M. Paris, M. Porcelloni, M. Bigioni, M. Binaschi, F. Nardelli, M. Parlani, C. A. Maggi and D. Fattori, *Bioorg. Med. Chem. Lett.*, 2013, **23**, 4091.
61. M. Lobera, K. P. Madauss, D. T. Pohlhaus, Q. G. Wright, M. Trocha, D. R. Schmidt, E. Baloglu, R. P. Trump, M. S. Head, G. A. Hofmann, M. Murray-Thompson, B. Schwartz, S. Chakravorty, Z. Wu, P. K. Mander, L. Kruidenier, R. A. Reid, W. Burkhart, B. J. Turunen, J. X. Rong, C. Wagner, M. B. Moyer, C. Wells, X. Hong, J. T. Moore, J. D. Williams, D. Soler, S. Ghosh and M. A. Nolan, *Nat. Chem. Biol.*, 2013, **9**, 319.
62. K. Chen, L. Xu and O. Wiest, *J. Org. Chem.*, 2013, **78**, 5051.
63. V. Patil, Q. H. Sodji, J. R. Kornacki, M. Mrksich and A. K. Oyelere, *J. Med. Chem.*, 2013, **56**, 3492.
64. R. J. Klose, E. M. Kallin and Y. Zhang, *Nat. Rev. Genet.*, 2006, **7**, 715.
65. Y. Shi, *Nat. Rev. Genet.*, 2007, **8**, 829.
66. L. Kruidenier, C. W. Chung, Z. Cheng, J. Liddle, K. Che, G. Joberty, M. Bantscheff, C. Bountra, A. Bridges, H. Diallo, D. Eberhard, S. Hutchinson, E. Jones, R. Katso, M. Leveridge, P. K. Mander, J. Mosley, C. Ramirez-Molina, P. Rowland, C. J. Schofield, R. J. Sheppard, J. E. Smith, C. Swales, R. Tanner, P. Thomas, A. Tumber, G. Drewes, U. Oppermann, D. J. Patel, K. Lee and D. M. Wilson, *Nature*, 2012, **488**, 404.
67. N. R. Rose, S. S. Ng, J. Mecinovic, B. M. Lienard, S. H. Bello, Z. Sun, M. A. McDonough, U. Oppermann and C. J. Schofield, *J. Med. Chem.*, 2008, **51**, 7053.

68. N. R. Rose, E. C. Woon, G. L. Kingham, O. N. King, J. Mecinovic, I. J. Clifton, S. S. Ng, J. Talib-Hardy, U. Oppermann, M. A. McDonough and C. J. Schofield, *J. Med. Chem.*, 2010, **53**, 1810.
69. N. R. Rose, E. C. Woon, A. Tumber, L. J. Walport, R. Chowdhury, X. S. Li, O. N. King, C. Lejeune, S. S. Ng, T. Krojer, M. C. Chan, A. M. Rydzik, R. J. Hopkinson, K. H. Che, M. Daniel, C. Strain-Damerell, C. Gileadi, G. Kochan, I. K. Leung, J. Dunford, K. K. Yeoh, P. J. Ratcliffe, N. Burgess-Brown, D. F. von, S. Muller, B. Marsden, P. E. Brennan, M. A. McDonough, U. Oppermann, R. J. Klose, C. J. Schofield and A. Kawamura, *J. Med. Chem.*, 2012, **55**, 6639.
70. A. Thalhammer, J. Mecinovic, C. Loenarz, A. Tumber, N. R. Rose, T. D. Heightman and C. J. Schofield, *Org. Biomol. Chem.*, 2011, **9**, 127.
71. M. Bantscheff, C. Hopf, M. M. Savitski, A. Dittmann, P. Grandi, A. M. Michon, J. Schlegl, Y. Abraham, I. Becher, G. Bergamini, M. Boesche, M. Delling, B. Dumpelfeld, D. Eberhard, C. Huthmacher, T. Mathieson, D. Poeckel, V. Reader, K. Strunk, G. Sweetman, U. Kruse, G. Neubauer, N. G. Ramsden and G. Drewes, *Nat. Biotechnol.*, 2011, **29**, 255.
72. Y. J. Shi, C. Matson, F. Lan, S. Iwase, T. Baba and Y. Shi, *Mol. Cell*, 2005, **19**, 857.
73. A. Kuzmichev, T. Jenuwein, P. Tempst and D. Reinberg, *Mol. Cell*, 2004, **14**, 183.
74. P. Bertrand, *Eur. J. Med. Chem.*, 2010, **45**, 2095.
75. P. M. Lombardi, K. E. Cole, D. P. Dowling and D. W. Christianson, *Curr. Opin. Struct. Biol.*, 2011, **21**, 735.
76. L. Wang, J. Chang, D. Varghese, M. Dellinger, S. Kumar, A. M. Best, J. Ruiz, R. Bruick, S. Pena-Llopis, J. Xu, D. J. Babinski, D. E. Frantz, R. A. Brekken, A. M. Quinn, A. Simeonov, J. Easmon and E. D. Martinez, *Nat. Commun.*, 2013, **4**, 2035.
77. C. J. Vickers, C. A. Olsen, L. J. Leman and M. R. Ghadiri, *ACS Med. Chem. Lett.*, 2012, **3**, 505.
78. J. R. Horton, A. K. Upadhyay, H. H. Qi, X. Zhang, Y. Shi and X. Cheng, *Nat. Struct. Mol. Biol.*, 2010, **17**, 38.

Subject Index